IUTAM SYMPOSIUM ON COMPUTATIONAL METHODS FOR UNBOUNDED DOMAINS

FLUID MECHANICS AND ITS APPLICATIONS
Volume 49

Series Editor: R. MOREAU
MADYLAM
Ecole Nationale Supérieure d'Hydraulique de Grenoble
Boîte Postale 95
38402 Saint Martin d'Hères Cedex, France

Aims and Scope of the Series

The purpose of this series is to focus on subjects in which fluid mechanics plays a fundamental role.

As well as the more traditional applications of aeronautics, hydraulics, heat and mass transfer etc., books will be published dealing with topics which are currently in a state of rapid development, such as turbulence, suspensions and multiphase fluids, super and hypersonic flows and numerical modelling techniques.

It is a widely held view that it is the interdisciplinary subjects that will receive intense scientific attention, bringing them to the forefront of technological advancement. Fluids have the ability to transport matter and its properties as well as transmit force, therefore fluid mechanics is a subject that is particulary open to cross fertilisation with other sciences and disciplines of engineering. The subject of fluid mechanics will be highly relevant in domains such as chemical, metallurgical, biological and ecological engineering. This series is particularly open to such new multidisciplinary domains.

The median level of presentation is the first year graduate student. Some texts are monographs defining the current state of a field; others are accessible to final year undergraduates; but essentially the emphasis is on readability and clarity.

For a list of related mechanics titles, see final pages.

IUTAM Symposium on

Computational Methods for Unbounded Domains

Proceedings of the IUTAM Symposium
held in Boulder, Colorada, U.S.A.,
27–31 July 1997

Edited by

THOMAS L. GEERS

Department of Mechanical Engineering,
University of Colorado,
Boulder, U.S.A.

KLUWER ACADEMIC PUBLISHERS
DORDRECHT / BOSTON / LONDON

A C.I.P. Catalogue record for this book is available from the Library of Congress.

ISBN 978-90-481-5106-6

Published by Kluwer Academic Publishers,
P.O. Box 17, 3300 AA Dordrecht, The Netherlands.

Sold and distributed in the North, Central and South America
by Kluwer Academic Publishers,
101 Philip Drive, Norwell, MA 02061, U.S.A.

In all other countries, sold and distributed
by Kluwer Academic Publishers,
P.O. Box 322, 3300 AH Dordrecht, The Netherlands.

Printed on acid-free paper

Printed in the Netherlands.

Advisory Committee

T.L. Geers, Chairman, University of Colorado
[illegible], Carnegie Mellon University
[illegible]
P.M. Pinsky, Stanford University

Scientific Committee

T.L. Geers, Chairman, University of Colorado, USA
R.J. Astley, University of Canterbury, New Zealand
G.A. Athanassoulis, Nat'l Tech. Univ. of Athens, Greece
P. Bettess, University of Durham, UK
[illegible], USA
[illegible], Sweden
M.A. Ilgamov, [illegible], Russia
[illegible], Japan
[illegible]
M. Lenoir, Ecole Nat. Sup. de Tech. Avancées, France
[illegible]

TABLE OF CONTENTS

(Presenting Author in Italics)

PREFACE

During 27-31 July 1997, thirty-seven researchers in acoustics, aeronautics, elastodynamics, electromagnetics, hydrodynamics and mathematics participated in a Symposium on Computational Methods for Unbounded Domains. The symposium was sponsored by the International Union of Theoretical and Applied Mechanics and was held at the University of Colorado in the United States of America. The symposium was opened by Dr. Richard Byyny, Chancellor of the University's Boulder Campus, who concluded his remarks by reading a letter from Professor Bruno A. Boley, IUTAM Representative on the Scientific Committee.

Thirty-three papers were presented. About two-thirds of these focused on the classical wave equation of acoustics; however, three papers dealt with hydrodynamic surface waves, two with electromagnetic waves, three with elastodynamic waves, and four with waves in aerodynamics. Approximately two-thirds of the papers addressed steady-state problems, with the rest treating problems in the time domain. Extended abstracts of the papers appear in this volume, arranged in alphabetical order according to the last name of the presenting author.

A key unifying aspect of the symposium was the creation of four working groups that labored in parallel to formulate benchmark problems for evaluating computational boundaries. The working groups reviewed the papers presented each day, searching for benchmark candidates. Then they considered other possibilites and organized the ensemble into logical categories. At the end of the symposium, each group presented its benchmark candidates to the assembly of participants, which subsequently made a preliminary consolidation of the benchmarks. During the following months, the symposium chairman facilitated further consolidation and wrote the article appearing first in this volume.

I am grateful to the International Union of Theoretical and Applied Mechanics for including this symposium in its 1997 program. In addition, the generous financial support provided by the sponsoring organizations listed below is greatly appreciated.

Thomas L. Geers
Boulder, Colorado

Symposium Sponsors

Air Force Office of Scientific Resarch
International Association for Computational Mechanics
International Union of Theoretical and Applied Mechanics
National Aeronautics and Space Administration
National Science Foundation
Office of Naval Research
University of Colorado

SYMPOSIUM PARTICIPANTS: 1. Peter Monk, 2. Isaac Harari, 3. Assad Oberai, 4. Andrzej Safjan, 5. Toby Driscoll, 6. Clyde Scandrett, 7. Chongmin Song, 8. John Wolf, 9. Jean-Claude Nedelec, 10. Peter Pinsky, 11. Ehtesham Hayder, 12. Dan Givoli, 13. Murthy Guddati, 14. Ken Fyfe, 15. Leszek Demkowicz, 16. Hong-Kai Zhao, 17. Dave Burnett, 18. Klaus Gerdes, 19. Eiichi Watanabe, 20. Loukas Kallivokas, 21. Makis Athanassoulis, 22. Chris Damaren, 23. Hari Hariharan, 24. Semyon Tsynkov, 25. Bertil Gustafsson, 26. Sheryl Grace, 27. Bob Higdon, 28. Pete Bettess, 29. Jeremy Astley, 30. Greg Kriegsmann, 31. Jacobo Bielak, 32. Lothar Gaul, 33. Tom Hagstrom, 34. Ian Mathews, 35. Tom Geers, 36. Jean-Pierre Coyette

SYMPOSIUM PARTICIPANTS

Prof. R.J. Astley
Dept. of Mechanical Engineering
University of Canterbury
Private Bag 4800
Christchurch, NEW ZEALAND
j.astley@mech.canterbury.ac.nz

Prof. Gerassimos A. Athanassoulis
Dept. of Naval Architecture & Marine Eng'g
National Technical University of Athens
P.O. Box 64.033, 157 10
Zografos, Athens, GREECE
mathan@naval.ntua.gr

Prof. Peter Bettess
Shcool of Engineering
University of Durham
Science Laboratories, South Rd.
Durham City, DH1 3LE, UK
peter.bettess@durham.ac.uk

Prof. Jacobo Bielak
Dept. of Civil & Env. Engineering
Carnegie Mellon University
Pittsburgh, PA 15213-3890, USA
jbielak@cmu.edu

Dr. David S. Burnett
Rm 15F-329
General Dynamics / ATS
67 Whippany Road
Whippany, NJ 07981, USA
dsburnett@lucent.com

Dr. Jean-Pierre Coyette
LMS - Numerical Technologies
Interleuvenlaan, 70
B3001 Leuven, BELGIUM
jp@lmsnit.be

Dr. Christopher Damaren
Dept. of Mechanical Engineering
University of Canterbury
Private Bag 4800
Christchurch, NEW ZEALAND
c.damaren@mech.canterbury.ac.nz

Prof. Leszek Demkowicz
Texas Inst. for Comp.and Appl. Math.
Dept. of Aero. Eng'g and Eng' Mech.
University of Texas
Austin, TX 78712, USA
leszek@ticam.utexas.edu

Dr. Tobin A. Driscoll
Dept. of Applied Mathematics
University of Colorado
Boulder, CO 80309-0526, USA
tad@sturm.colorado.edu

Prof. Carlos A. Felippa
Dept. of Aerospace Eng'g Sciences
University of Colorado
Boulder, Colorado 80309-0428, USA
carlos@titan.colorado.edu

Prof. Ken R. Fyfe
Dept. of Mech. Eng'g
University of Alberta
Edmonton, Alberta, T6G ZG8 CANADA
ken.fyfe@ualberta.ca

Prof. Lothar Gaul
Institute A of Mechanics
University of Stuttgart
Pfaffenwaldring 9
70550 Stuttgart, GERMANY
l.gaul@mecha.uni-stuttgart.de

Prof. Thomas L. Geers
Dept. of Mechanical Engineering
University of Colorado
Boulder, CO 80309-0427, USA
geers@spot.colorado.edu

Dr. Klaus Gerdes
Semiinar for Applied Mathematics
ETH Zürich
Rämistr. 101
CH-8092 Zürich, SWITZERLAND
gerdes@sam.math.ethz.ch

Prof. Dan Givoli
Dept. of Aerospace Engineering
Technion - Israel Inst. of Technology
Haifa 32000, ISRAEL
givolid@aerodyne.technion.ac.il

Prof. Sheryl M. Grace
Dept. of Aerospace & Mechanical Eng'g
Boston University
Boston, MA 02101, USA
sgrace@bu.edu

Dr. Murthy N. Guddati
Schlumberger - Dowell
110 Schlumberger Drive
Sugarland, TX 77478, USA
mguddati@slb.com

Prof. Bertil Gustafsson
Dept. of Scientific Computing
Uppsala University, Box 120
S-75104 Uppsala SWEDEN
bertil@tdb.uu.se

Prof. Thomas Hagstrom
Dept. of Mathematics and Statistics
University of New Mexico
Albuquerque, NM 87131, USA
hagstrom@math.unm.edu

Prof. Isaac Harari
School of Engineering
Tel Aviv University
Tel Aviv, Tel Aviv 69978 ISRAEL
harari@eng.tau.ac.il

Prof. S.I. Hariharan
Dept. of Mathematical Sciences
University of Akron
Akron, OH 44325, USA
hari@math.uakron.edu

Dr. M. Ehtesham Hayder
Mail Stop 403, ICASE
NASA Langley Research Center
Hampton, VA 23681-0001, USA
hayder@icase.edu

Prof. Robert L. Higdon
Dept. of Mathematics
Oregon State University
Corvallis, OR 97331-4605, USA
higdon@math.orst.edu

Prof. Loukas F. Kallivokas
Dept. of Civil & Env. Engineering
Carnegie Mellon University
Pittsburgh, PA 15213-3890, USA
loukas+@cmu.edu

Prof. Gregory A. Kriegsmann
Dept. of Mathematics
New Jersey Institute of Technology
Newark, NJ 07105, USA
grkrie@ondas.njit.edu

Dr. Ian C. Mathews
Department of Aeronautics
Imperial College
London SW7 2BY, UK
i.mathews@ic.ac.uk

Prof. Peter Monk
Dept. of Mathematical Sciences
University of Delaware
Newark, DE 19716, USA
monk@math.udel.edu

Prof. Jean-Claude Nedelec
Math Appliquees
Ecole Polytechnique
91128 Palaiseau Cedex, FRANCE
nedelec@cmapx.polytechnique.fr

Dr. Assad A. Oberai
Div. of Mechanics & Computation
Stanford University
Stanford, CA 94305, USA
oberai@am-sun2.stanford.edu

Prof. Peter M. Pinsky
Div. of Mechanics & Computation
Stanford University
Stanford, CA 94305, USA
pinsky@cive.stanford.edu

Prof. Andrzej Safjan
Dept of Engineering Mechanics
University of Nebraska
Lincoln, Nebraska 68588-0347, USA
asafjan@unl.edu

Prof. Clyde Scandrett
Department of Mathematics
Naval Postgraduate School
Monterey, CA 93943, USA
cscand@nps.navy.mil

Dr. Chongmin Song
Dept. of Civil Engineering
Swiss Federal Institute of Technology
CH-1015 Lausanne, SWITZERLAND
cis@lchdec1.epfl.ch

Dr. John P. Wolf
Dept. of Civil Engineering
Swiss Federal Institute of Technology
CH-1015 Lausanne, SWITZERLAND
FAX: 41 21 693 22 64

Prof. Semyon V. Tsynkov
Dept. of Applied Mathematics
Tel Aviv University
Rawat Aviv, Tel Aviv 69978, ISRAEL
tsynkov@math.tau.ac.il

Prof. Hong-Kai Zhao
Dept. of Mathematics
Stanford University
Stanford, CA 94305-2125, USA
zhao@cartan.stanford.edu

Prof. Eiichi Watanabe
Dept. of Civil Engineering
Kyoto University
Sakyo-ku, Kyoto 606-01, JAPAN
watanabe@strsun1.kuciv.kyoto-u.ac.jp

BENCHMARK PROBLEMS

Thomas L. Geers

Center for Acoustics, Mechanics and Materials
Department of Mechanical Engineering
University of Colorado
Boulder, CO, USA

1. Introduction

The participants of the Symposium on Computational Methods for Unbounded Domains formed four working groups of nine members each to work in parallel on benchmark problems. Each group presented its benchmark candidates to the assembly of participants, which initiated a process of consolidation.

The assembly agreed to limit the benchmark problems to a moderate number. This presented a daunting challenge, given the variety of candidate problems existing across a broad spectrum of engineering and scientific disciplines. Also, it was decided to define each benchmark problem in the context of the field to which it is perhaps most pertinent, mentioning counterparts of interest in other fields. This position prevailed over employing a purely mathematical definition or defining every problem in the context of a single physical discipline. Finally, it was agreed to focus on assessing boundary performance and not be diverted by physical or mathematical features that do not directly impact such performance. While these decisions were not supported unanimously, they enjoyed the broadest support.

The benchmark problems presented here are offered as suggestions, not standards. They are designed to illuminate the salient characteristics of a computational boundary, but certainly not all characteristics of possible interest. Most of the benchmark problems pertain to the linear wave equations of acoustics, hydrodynamics, elastodynamics and electromagnetics. The last problem pertains to the Euler equations, and is therefore nonlinear. Inclusion of a benchmark problem for the Navier-Stokes equations was considered, but was not done because of the lack of interesting analytic or semi-analytic solutions for these equations. The reader interested in the Navier-Stokes equations is referred to the abstract by Tsynkov in this volume. Generally speaking, the references listed are either widely available books that discuss the problem or recent journal papers that address it; earlier references (perhaps even the seminal papers) are cited in those listed.

Each of the following problems consists of a finite system of interest surrounded by an infinite or semi-infinite domain, the latter represented by a finite computational domain D. The domain D is bounded by a computational boundary B, which represents to D the unbounded region outside of D. The inputs specified for most of the problems exhibit the following *common set* of temporal variations:

(a) A steady sinusoid: $v(t) = V \sin \omega t,\ 0.1 \le \omega a / c \le 10,$

(b) A windowed sinusoid: $v(t) = V H(t) H(T-t) \sin \omega t,\ 0.1 \le \omega a/c \le 10,\ cT/a = 10\pi c/\omega a,$ which produces five cycles in the window,

(c) A step-function: $v(t) = V H(t),$

where a is a characteristic length, c is a characteristic wave speed, and $H(\,)$ is the Heaviside step-function. However, Problem 4 has an input in addition to these, and Problem 6 has a completely different input specification.

T.L. Geers (ed.), IUTAM Symposium on Computational Methods for Unbounded Domains, 1–9.

Problem 1: Acoustic Radiation by a Piston in an Infinite, Rigid, Planar Baffle

This problem is pictured in Figure 1a for a circular piston and in Figure 1b for a rectangular piston. The input is a prescribed, uniform normal velocity $v(t)$ of the piston's surface defined by the common set of temporal variations. The output is calculated pressure $p(S,t)$ on the surface of the piston and the baffle and/or calculated pressure and velocity in the far field $(r >> a)$. Attributes of this problem are: scalar wave equation; simple geometry; exact solutions available or obtainable; linearity.

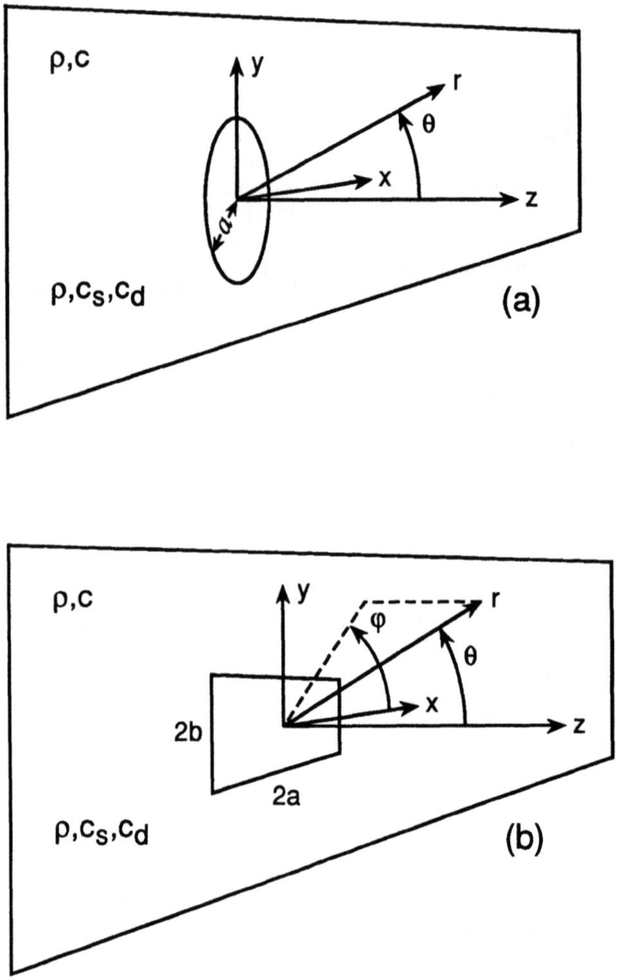

Figure 1. Circular and Rectangular Pistons Driving Half-Spaces

Remark A For steady-state sinusoids, the complex function $Z(\omega) = \int p(S,\omega)dS/V$ is the acoustic impedance of the fluid as seen by the piston. This impedance may be expressed as $Z(\omega) = R(\omega) + i\omega M(\omega)$. As ω approaches zero, $R(\omega)$ approaches zero and $M(\omega)$ approaches a constant value, which is the added mass for incompressible flow. As ω approaches infinity, $M(\omega)$ approaches zero and $R(\omega)$ approaches ρcA, where ρ and c are the fluid's mass density and speed of sound, and A is the surface area of the piston.

Remark B The aspect ratio b/a of the rectangular piston may be made large to produce the 2-D problem of a uniform line array.

References: 1.1 - 1.4

Problem 2: Acoustic Waves In and Around a Rigid Waveguide Opening into a Halfspace

This problem is depicted in Figure 2, where all surfaces are rigid. Either of two inputs may be specified: (1) a prescribed piston velocity $v(t)$, (2) an incident plane wave characterized by fluid velocity $v(t+z/c)$. When the piston is acting, the incident wave is absent; when the incident wave is acting, the piston is stationary. Each input is characterized by the common set of temporal variations. The output is calculated pressure on the surfaces of the piston, walls and plane and/or calculated pressure and velocity in the far field $(r >> a)$. Attributes of this problem are: scalar wave equation; simple geometry with an internal region; exact solutions available or obtainable; linearity.

Remark A For steady-state sinusoids, the acoustic impedance of the fluid as seen by the piston exhibits the asymptotic behavior described for the previous problem. The behavior exhibited between the asymptotes, however, is enriched by internal reflections.

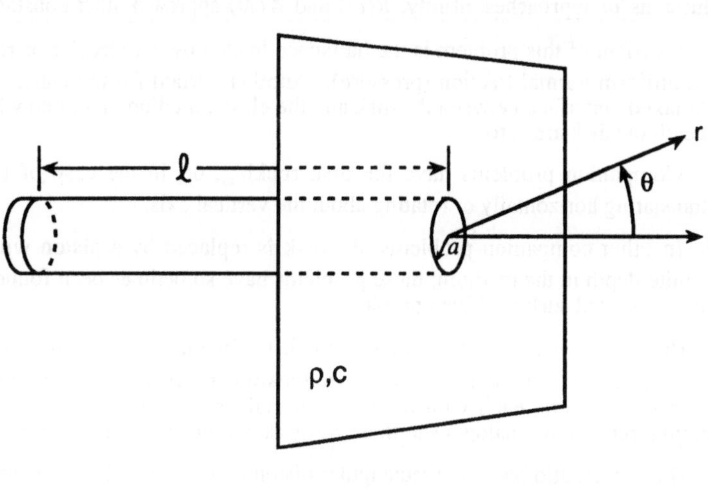

Figure 2. Circular Cylindrical Cavity with Rigid Walls Opening into an Infinite, Rigid Baffle and Terminated by a Piston

Remark B For $\ell/a << 1$, the behavior exhibited resembles that seen in Problem 1; however, for $\ell/a >> 1$, the behavior becomes that of a waveguide with a flanged end. The special case of $\ell/a \to \infty$ produces two coupled unbounded domains.

Remark C The associated problem in electromagnetics pertains to antenna radiation or radiation into a cavity.

References: 1.3, 2.1 - 2.6

Problem 3: Disk-Driven Hydrodynamic And Elastic Half-Spaces

This problem is portrayed for circular and rectangular disks by rotating Figures 1a and 1b such that the z axes point downward. The input is a prescribed, normal velocity v(t) of the rigid disk defined by the common set of temporal variations. In the case of the elastic half-space, the disk is welded to the medium. The output is calculated normal traction (pressure) on the surface of the disk and calculated displacement on the free surface of the half-space. Attributes of this problem are: surface waves; simple geometry; exact solutions available or obtainable; linearity.

Remark A For steady-state sinusoids, the complex function $Z(\omega) = \int p(S,\omega)dS/V$ is the hydrodynamic or elastodynamic impedance of the fluid as seen by the piston. The hydrodynamic impedance may be expressed as $Z(\omega) = R(\omega) + i\omega M(\omega)$. As ω approaches zero, $R(\omega)$ approaches zero and $M(\omega)$ approaches a constant value; as ω approaches infinity, $R(\omega)$ again approaches zero and $M(\omega)$ approaches another constant value. The relation for elastodynamic impedance is usually expressed as a dynamic stiffess $S(\omega) = i\omega Z(\omega) = K(\omega) + i\omega R(\omega)$. As ω approaches zero, $R(\omega)$ and $K(\omega)$ approach constant values; as ω approaches infinity, $R(\omega)$ and $K(\omega)$ approach other constant values.

Remark B A variant of this problem is the halfspace loaded over a circular or rectangular region with a uniform normal traction (pressure). Another variant for the elastic halfspace specifies a "relaxed" interface between the disk and the elastic medium, *i.e.*, one where shear tractions beneath the disk are zero.

Remark C Companion problems have the disk rocking, or, in the case of the elastic halfspace, translating horizontally or rotating about the vertical axis.

Remark D In other companion problems, the disk is replaced by a piston with its face located at a finite depth in the medium; these problems have sometimes been found easier to solve than the associated surface-driven problem.

Remark E The hydrodynamic problem may be modified by introducing a flat, rigid bottom at a depth comparable to a (and b); the elastodynamic problem may be modified by introducing uniform layers, each with distinct material properties. These modifications introduce internal reflections analogous to those discussed in the previous problem.

Remark F The aspect ratio b/a of the rectangular piston may be made large to produce a 2-D line excitation.

Remark G Important extensions pertain to viscoelastic and poroelastic media.

References: 3.1 - 3.14

Problem 4: Modal Acoustic Radiation by a Sphere

This problem is pictured in Figure 3. The input is a prescribed, axisymmetric normal velocity on the surface of the sphere given by $v(\theta,t) = v_n(t)P_n(\cos\theta)$, where $P_n(x)$ is the Legendre polynomial of degree n. The modal velocity $v_n(t)$ is defined by the common set of temporal variations, only with $0.1(n+1) \leq \omega a / c \leq 10(n+1)$. An additional modal-velocity specification is:

(d) The modal velocity for scattering of a (steady sinusoid, windowed sinusoid, step) plane wave by a rigid sphere: $v_n(t) = -[\mathit{incident-wave}\ v_n(t)]$.

The output is calculated modal pressure $p_n(t)$ on the surface of the sphere and/or calculated modal pressure and velocity fields in the far field (r >> a). Acoustic pressure and normal velocity are sometimes called conjugate quantities; their product is a work quantity, *viz.*, power. Attributes of this problem include: scalar wave equation; simple geometry admits uncoupled modes; exact solutions available or obtainable; linearity.

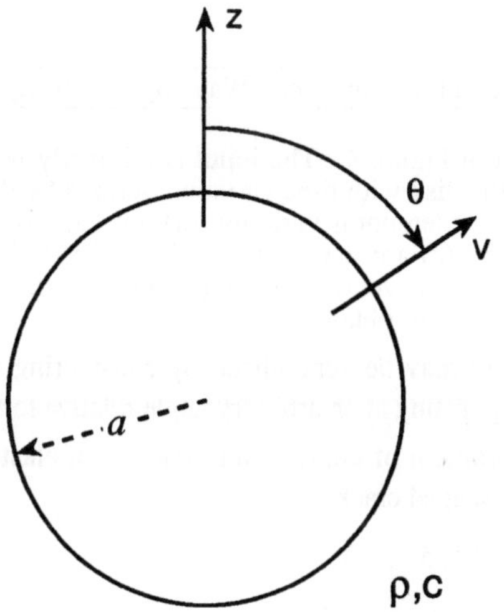

Figure 3. Spherical Radiator

Remark A For steady-state sinusoids, the complex function $z_n(\omega) = p_n(\omega)/v_n(\omega)$ is the modal specific acoustic impedance on that surface; this impedance may be expressed as $z_n(\omega) = r_n(\omega) + i\omega m_n(\omega)$. As ω approaches zero, $r_n(\omega)$ approaches zero and $m_n(\omega)$ approaches a constant value, which is the modal added mass for incompressible flow. As ω approaches infinity, $m_n(\omega)$ approaches zero and $r_n(\omega)$ approaches ρc. In recent years, the term "DtN map" has become popular as a generic descriptor of specific impedance, which is an engineering/physics term.

Remark B The corresponding problems in electromagnetics feature the surface electric and magnetic fields as conjugate quantities; those in elastodynamics feature the surface tractions and displacements. The modal specific impedances are matrices in these problems.

Remark C Modal solutions for an ensemble of the $v_n(t)$ defined in (d) may be superposed to obtain field solutions for scattering of plane waves by a rigid sphere, although care must be taken to secure convergence of the series. This is a fundamental problem in acoustics, electromagnetics and elastodynamics.

Remark D One might evaluate the performance of a computational boundary for a non-spherical radiating surface S' by utilizing the following artifice. Locate the spherical surface S inside S' and specify $v_n(t)$ on S as one from the common set of temporal variations. Then, by (converged) modal superposition, determine the pressure and normal-velocity fields and define the latter on S' as the input. Next, calculate the resulting acoustic fields with the computational procedure adopted for the surrounding domain D and the computational boundary B. Finally, compare the calculated acoustic fields with those determined previously.

Remark E This axisymmetric 3-D benchmark problem has a 2-D counterpart in which the sphere becomes a circle and $P_n(\cos\theta)$ becomes $\cos\theta$.

References: 1.2, 1.3, 2.2, 4.1 - 4.8

Problem 5: Scattering of an Electomagnetic Wave by a Perfectly Conducting Circular Disk

This problem is depicted in Figure 4. The input is a linearly polarized plane wave propagating along the axis of the disk with time variation defined by the common set of temporal variations. The outputs for temporal variation (a) are back-scattering and bistatic cross-sections; those for (b) and (c) are the calculated E- and H-fields in the far field (r >> a). Attributes of this problem are: vector wave equation; simple geometry featuring a sharp edge; exact solutions available or obtainable; linearity.

Remark A This problem may be generalized by considering as the input an elliptically polarized plane wave propagating at an arbitrary angle relative to the axis of the disk.

Remark B A related problem of considerable interest in elastodynamics is scattering of elastic waves by a penny-shaped crack.

References: 2.2, 4.1, 4.4, 4.5, 5.1, 5.2

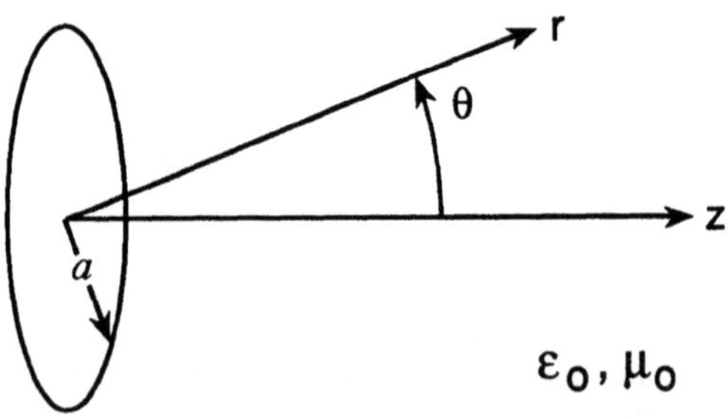

Figure 4. Disk in Free Space

Problem 6: Diverging Pulse Waves Generated in a Convecting Fluid

This problem is portrayed in Figure 5. The input is two Gaussian pulses: an acoustic pulse at position -25, 0 and a vorticity/entropy pulse at position +25, 0. The pulses are defined by

$$\rho(x,y,0) = exp\{-ln2[(x+25)^2 + y^2]/3^2\} + 0.1\, exp\{-ln2[(x-25)^2 + y^2]/4^2\}$$

$$p(x,y,0) = exp\{-ln2[(x+25)^2 + y^2]/3^2\}$$

$$u(x,y,0) = 0.05y\, exp\{-ln2[(x-25)^2 + y^2]/4^2\}$$

$$v(x,y,0) = -0.05(x-25)\, exp\{-ln2[(x-25)^2 + y^2]/4^2\}$$

The output is calculated density and velocity fields in the two-dimensional domain. Attributes of this problem are: Euler equations; simple geometry; exact solutions available or obtainable; nonlinearity.

Remark A This is one of the benchmark problems in Ref. 6.1.

Remark B Acoustic, entropy and vorticity waves must all be absorbed at the computational boundary.

Remark C Companion problems may be constructed on the basis of periodic sources, which may provide more rigorous testing of candidate boundaries.

References: 6.1 - 6.3

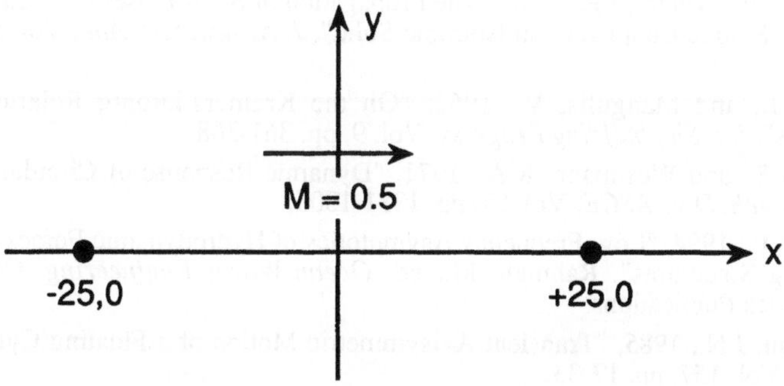

Figure 5. Pulse Points in a Convecting Fluid

References

1.1 Burnett, D.S., and Soroka, W.W., 1972, "Tables of Rectangular Piston Radiation Impedance Functions, with Application to Sound Transmission Loss through Deep Apertures", *J. Acoust. Soc. Am.*, Vol. 51, pp. 1618-1623.

1.2 Junger, M.C., and Feit, D., 1986, *Sound, Structures, and Their Interaction*, 2nd ed., MIT Press.

1.3 Morse, P.M., and Ingard, K.U., 1968, *Theoretical Acoustics*, McGraw-Hill.

1.4 Pierce, A.D., *Acoustics*, 1989, Acoustical Society of America.

2.1 Blaschak, J.G., Kriegsmann, G.A. and Taflove, A., 1989, "A Study of Wave Interactions with Flanged Waveguides and Cavities Using the On-Surface Radiation Condition", *Wave Motion*, Vol. 11, pp. 65-76.

2.2 Howard, A.Q., 1972, "On the Mathematical Theory of Electromagnetic Radiation from Flanged Waveguides", J. Math. Phys., Vol. 13, pp. 482-490.

2.3 Johnk, C.T.A., 1988, *Engineering Electromagnetic Fields and Waves*, Wiley.

2.4 Kinsler, et al., 1982, *Fundamentals of Acoustics*, 3rd ed., Wiley.

2.5 Morse, P.M., and Feshbach, H., 1953, *Methods of Theoretical Physics*, McGraw-Hill.

2.6 Shirai, H., and Felsen, L.B., 1987, "Rays, Modes and Beams for Plane Wave Coupling into a Large Open-Ended Circular Waveguide", *Wave Motion*, Vol. 9, pp. 461-482.

3.1 Buchner, 1993, "An Evaluation and Extension of the Shallow Draft Diffraction Theory", *Proc. 3rd Offshore and Polar Eng' Conf.*, Singapore, pp. 6-11.

3.2 Gaul, L. 1977, "Dynamic Interaction of a Foundation with a Viscoelastic Halfspace", *Ing.-Archiv*, Vol. 46, pp. 401-422.

3.3 Gazetas, G., 1983, "Analysis of Machine Foundation Vibration: State of the Art", *Soil Dyn. and Earthquake Eng'g*, Vol. 2, pp.2-42

3.4 Halpern, M.R., and Christiano, P., 1986, "Steady-State Harmonic Response of a Rigid Plate Bearing on a Liquid-Saturated Poroelastic Halfspace", *Earthquake Eng'g and Struc. Dyn.*, Vol. 14, pp. 439-454.

3.5 Halpern, M.R., and Christiano, P., 1986, "Response of a Poroelastic Half-Space to Steady-State Harmonic Surface Tractions", Int. J. Num. & Anal. Meth. in Geomech., Vol. 10, pp. 609-632.

3.6 Ilan, A., and Weight, J.P., 1990, "The Propagation of Short Pulses of Ultrasound from a Circular Source Coupled to an Isotropic Solid", *J. Acoust. Soc. Am.*, Vol. 88, pp. 1142-1151.

3.7 Kotik, J., and Mangulis, V., 1962, "On the Kramers-Dronig Relations for Ship Motions", *Int. Shipbuilding Progress*, Vol. 9, pp. 361-368.

3.8 Luco, J.E., and Westmann, R.A., 1971, "Dynamic Response of Circular Footings, *J. Eng'g Mech. Div., ASCE*, Vol. 97, pp. 1381-1305.

3.9 McIver, P., 1994, "Low-Frequency Asymptotics of Hydrodynamic Forces on Fixed and Floating Structures", Rahman, M., ed, *Ocean Waves Engineering*, Computational Mechanics Publications.

3.10 Newman, J.N., 1985, "Transient Axisymmetric Motion of a Floating Cylinder", *J. Fl. Mech.*, Vol. 157, pp. 17-33.

3.11 Newman, J.N., 1992 (7th printing), *Marine Hydrodynamics*, MIT Press.

3.12 Norwood, F.R., 1969, "Exact Transient Response of an Elastic Half Space Loaded over a Rectangular Region of its Surface", *J. Appl. Mech.*, Vol. 36, pp. 516-522.

3.13 Pak, R.Y.S., and Gobert, A.T., 1991, "Forced Vertical Vibration of Rigid Discs with Arbitrary Embedment", *J. Eng'g Mech. Div., ASCE*, Vol. 117, pp. 2527-2548.

3.14 Wolf, J.P., 1985, *Dynamic Soil-Structure Interaction*, Prentice-Hall.

4.1 Bowman, J.J., Senior, T.B.A. and Uslenghi, P.L.E., eds., 1970, *Electromagnetic and Acoustic Scattering by Simple Shapes*, North-Holland.

4.2 Geers, T.L., and Zhang, P.Z., 1988, "Doubly Asymptotic Approximations for Electromagnetic Scattering Problems", pp. 357-369, Tanaka, M., and Cruse, T.A., *Boundary Element Methods in Applied Mechanics*, Pergamon Press.

4.3 Geers, T.L., 1998, "Singly and Doubly Asymptotic Approximations", in this volume.

4.4 Ruck, G.T., *et al.*, 1970, *Radar Cross Section Handbook*, Plenum Press.

4.5 Stone, W.R., ed., 1990, *Radar Cross Sections of Complex Objects*, IEEE Press.

4.6 Stratton, J.A., 1941, *Electromagnetic Theory*, McGraw-Hill.

4.7 Van Bladel, J., 1964, *Electromagnetic Fields*, McGraw-Hill.

4.8 Pao, Y.-H., and Mow, C.-C., 1972, *Diffraction of Elastic Waves and Dynamic Stress Concentrations*, Rand Corp.

5.1 Budreck, D.E., and Achenbach, J.D., 1988, "Scattering from Three-Dimensional Planar Cracks by the Boundary Integral Equation Method, *J. Appl. Mech.*, Vol. 55, pp. 405-411.

5.2 Martynenko, S.V., 1987, "Scattering of a Longitudinal Wave in Normal Incidence on a Penny-Shaped Cavity in an Elastic Body", *Sov. Phys. - Acoustics*, Vol. 33, pp. 90-93.

6.1 Hardin, J.C., Ristorcelli, J.R., and Tam, C.K.W., eds, 1995, *ICASE/LaRC Workshop on Benchmark Problems in Computational Aeroacoustics*, NASA CP 3300.

6.2 Hu, F.Q., 1996, "On Absorbing Boundary Conditions for Linearized Euler Equations by a Perfectly Matched Layer", *J. Comp. Phys.*, Vol. 129, pp. 201-219.

6.3 Hagstrom, T., and Goodrich, J., 1988, "Experiments with Approximate Radiation Boundary Conditions for Computatioanl Aeroacoustics", to appear in *Appl. Num. Math.*

RECENT ADVANCES IN APPLYING WAVE-ENVELOPE ELEMENTS TO UNBOUNDED WAVE PROBLEMS

R. J. Astley

Department of Mechanical Engineering, University of Canterbury, Christchurch, New Zealand.

1. Introduction

The development of efficient numerical models for unbounded wave problems presents a major challenge for computation. The use of discrete node-based or grid-based schemes requires a suitable anechoic termination at the edge of a finite computational region. The more compact and manageable the inner region, the more demanding the numerical treatment that is required on its boundary. This difficulty is circumvented by a variety of Boundary Element (BE) techniques which represent the exterior solution as a surface distribution of source terms. These identically satisfy the field equations and radiate rather than absorb acoustical energy. BE schemes are intrinsically non-local in space and result in fully populated coefficient matrices, negating to some extent the economies implicit in using a surface rather than a volume discretisation. Moreover, when applied in the time domain, they generate differential or integral equations which are non-local in time *and* space. Various techniques have be used to reduce or ameliorate the computationally demanding character of such solutions (e.g. Geers 1978, Givoli and Cohen 1995).

The perception that the BE approach is the most effective numerical option for unbounded problems is challenged by an expanding repertoire of *domain-based* methods. These mostly involve finite element (FE) discretisations in an inner region coupled to local or non-local anechoic treatments at a finite boundary. They have higher algebraic dimensionality than comparable BE models but generate more sparsely connected systems of equations. Unbounded "infinite elements" have also been applied to wave problems (Bettess 1992). Here the unbounded outer region is modelled in its entirety by elements of infinite extent. The interpolation functions within such elements contain outwardly propagating wave-like factors. Early formulations of this type did not correctly model asymptotic behaviour (Bettess and Zienkeiwicz 1977) but subsequent "mapped" formulations (Zienkiewicz et al 1985) are asymptotically correct in the far field. Spheroidal coordinate systems have also been used in formulating infinite elements to facilitate a more compact inner discretisation (Burnett 1994). With the exception of the formulation of Olson and Bathe (1987) which is strictly valid only in the limits of high and low frequency, infinite elements have been formulated exclusively in the frequency domain.

Wave envelope elements (Astley 1983, Astley and Eversman 1988) and more recently mapped wave envelope elements (Bettess 1987, Astley et al 1994, Cremers and Fyfe 1995, Regan and Eaton 1995) employ interpolation functions similar to those used for infinite

T.L. Geers (ed.), IUTAM Symposium on Computational Methods for Unbounded Domains, 11–20.

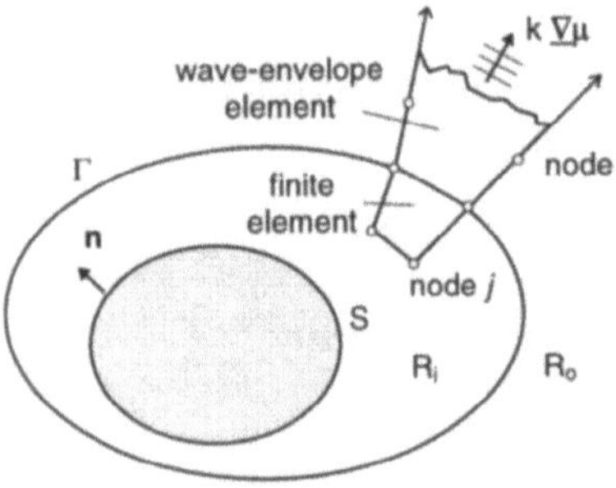

Figure 1(a). Problem geometry.

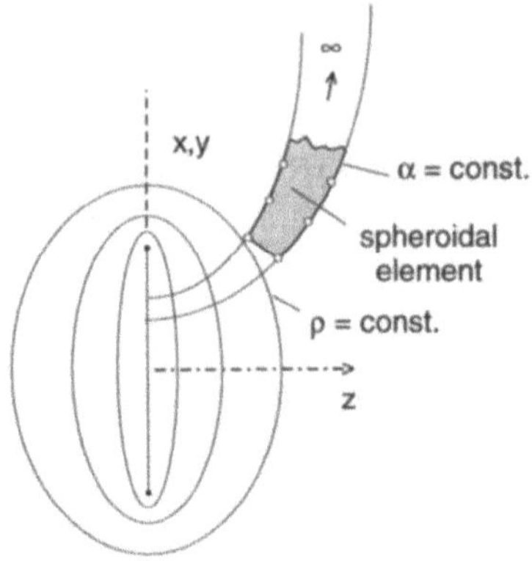

Figure 1(b). Topology of an oblate spheroidal element.

elements but use conjugated weighting functions. The trial and weighting functions used in this approach are consistent with those which have been shown to be admissible in an equivalent variational formulation (Demkowicz and Gerdes 1995). The wave envelope method has been developed primarily in the frequency domain but can readily be extended to the analysis of transient problems (Astley 1996).

In the remainder of this article, a wave-envelope element numerical scheme is applied to the solution of axisymmetric problems. The formulation is based on a Fourier transformation of a discrete model formulated in the frequency domain. The procedure yields a coupled system of ordinary differential equations in time which are local in space. Spheroidal element geometries are used. The accuracy of the scheme is demonstrated by a comparison of computed and analytic solutions for an axisymmetric test case. Time-harmonic and transient solution are presented. Initially an implicit time-integration scheme is used. The effect of element order and of the time increment are assessed. An indirect solution procedure is also presented which permits the sparsity of the transient equations to be utilised more effectively.

2. The exterior wave problem

A solution is sought for the exterior sound field generated by the motion of a closed surface S at rest at time $t=0$. The exterior region R is divided into a finite inner region R_i and an unbounded outer region R_o separated by an interface Γ (see figure 1(a)). The acoustic pressure $p(\mathbf{x},t)$ is initially zero at all points. For $t > 0$, $p(\mathbf{x},t)$ is governed by

$$\nabla^2 p = \frac{1}{c^2}\frac{\partial^2 p}{\partial t^2}, \qquad \text{in } R \qquad (1)$$

and $$\nabla p.\boldsymbol{n} = -\rho_0\, a_n, \qquad \text{on } S \qquad (2)$$

where $\boldsymbol{n}$ is a unit vector normal to S, ρ_0 and c are the mean fluid density and sound speed, and $a_n(\mathbf{x},t)$ is the normal acceleration of S. At large distances from S, the Sommerfield radiation condition holds, i.e.

$$r\{\partial p/\partial r + (1/c)\partial p/\partial t\} \to 0\,, \text{ as } r \to \infty\,, \tag{3}$$

where r is a spherical radial coordinate. This is satisfied by the transient initial value solution but must be imposed as a necessary condition for the time-harmonic problem. The transient acoustic pressure $p(\mathbf{x},t)$ and its complex Fourier transform, $\overline{p}\,(\mathbf{x},\omega)$ are related by the transform pair

$$\overline{p}(\mathbf{x},\omega) = \mathcal{F}\{p(\mathbf{x},t)\} = \int_{-\infty}^{+\infty} p(\mathbf{x},t)\,e^{-i\omega t}\,dt, \tag{4a}$$

and $$p(\mathbf{x},t) = \mathcal{F}^{-1}\{\overline{p}(x,\omega)\} = \frac{1}{2\pi}\int_{-\infty}^{+\infty} \overline{p}(\mathbf{x},\omega)\,e^{i\omega t}\,d\omega\quad . \tag{4b}$$

Equations (1)-(3) then transform to

$$\nabla^2\overline{p} + k^2\overline{p} = 0 \qquad \text{in } R\,, \tag{5}$$

$$\nabla\overline{p}\,.\boldsymbol{n} = -\ \rho_a\,\overline{a}_n \qquad \text{on } S\,, \tag{6}$$

and $$r^a\{\partial\,\overline{p}/\partial r + (ik)\overline{p}\,\} \to 0\,, \text{ as } r \to \infty\,, \tag{7}$$

where $k = \omega/c$ and $\overline{a}_n(\mathbf{x},\omega)$ is the complex Fourier transform of $a_n(\mathbf{x},t)$, Alternatively equations (5)-(8) define the steady harmonic problem where $\overline{p}\,(\mathbf{x},\omega)$ is the complex pressure amplitude.

3. The wave-envelope formulation.

The wave envelope (WE) formulation is based on a discrete representation of equations (5)-(7). A trial function is chosen of the form

$$\overline{p}(\mathbf{x},\omega) = \sum_{j=1}^{n} \overline{q}_j(\omega)\ \varphi_j(\mathbf{x},\omega)\,, \tag{8}$$

where $\varphi_j(\mathbf{x},\omega)$ $(j=1,..n)$ are known basis functions and $\overline{q}_j(\omega)$ $(j=1,..n)$ are unknown coefficients. Equation (9) is valid over the entire exterior region R. Application of a Galerkin procedure with weighting functions $W_j(\mathbf{x},\omega)$ $(j=1,2..n)$ then yields a system of linear equations;

$$[A(\omega)]\,\overline{\mathbf{q}} = \overline{\mathbf{f}} \tag{9}$$

where $\overline{\mathbf{q}}$ is the vector whose components are the trial coefficients $q_i(\omega)$ $(i=1,..n)$, $\boldsymbol{A}$ is an nxn coefficient matrix and $\overline{\mathbf{f}}$ an $nx1$ forcing vector. $\boldsymbol{A}$ and $\overline{\mathbf{f}}$ have components;

$$A_{ij} = \int_R \left\{ \nabla W_i\,.\,\nabla\varphi_j - k^2\,W_i\,\varphi_j \right\} dR\,, \quad \text{and} \qquad \bar{f}_i = \int_S \rho_a\,W_i\,\overline{a}_n\,dS\,. \tag{10a,b}$$

The weight and basis functions W_i and ϕ_j are chosen so that the above integrals are finite. They are defined as follows:

(a) In the inner region R_i , a regular FE model is used. i.e

$$W_j(\mathbf{x},\omega) = \varphi_j(\mathbf{x}) = N_j(\mathrm{x}), \tag{11}$$

where $N_j(\mathrm{x})$, is the global finite element shape function associated with node j (see figure 1(a)). The corresponding coefficients $\overline{q}_j(\omega)$ are nodal values of the pressure amplitude $\overline{p}\,(\mathbf{x},\omega)$.

(b) In the outer region, R_o, the basis functions $\varphi_l(\mathbf{x},\omega)$ corresponding to node l say (see figure 1(a)) are given by

$$\varphi_l(\mathbf{x},\omega) = P_l(\mathbf{x})\,e^{-ik\mu(\mathbf{x})} \tag{12}$$

where k is the scalar wave number ($=\omega/c$) and where the phase function $\mu(\mathbf{x})$ is chosen so that $\nabla\mu$ is aligned with the (estimated) direction of wave propagation (see figure 1(a)). In the case of an axisymmetric spheroidally formulated element, $\nabla\mu$ lies along a local radial direction in a prolate or oblate spheroidal coordinate system. The orientation of an oblate spheroidal element with respect to spheroidal coordinates ρ and α is indicated in figure 1(b). The interpolation function $P_l(\mathbf{x})$ then defines an "envelope" of waves of vector wavenumber $k\nabla\mu$. In the case of a spheroidal element the dependence of $P_l(\mathbf{x})$ on a spheroidal radius ρ is as an $m+1$ 'th order polynomial in $(1/\rho)$ where m is the number of nodes along the infinite side of the element. The phase function $\mu(\mathbf{x})$ is proportional to ρ. A mapping is used to define these functions (Astley 1997). The weighting functions within the element are then defined as

$$W_l(\mathbf{x},\omega) = D(\mathbf{x})\,[\varphi_l(\mathbf{x},\omega)]^* = D(\mathbf{x})\,P_l(\mathbf{x})\,e^{+ik\mu(\mathbf{x})}, \tag{13}$$

where [ ]* denotes a complex conjugate and $D(\mathbf{x})$ is a geometric factor which behaves as $(1/\rho^2)$ and takes the value of unity at the interface between the wave envelope region and the FE mesh. The presence of the conjugate term in the weighting functions greatly simplifies the integrals of equations (10) and permits a simple transformation to the time domain.

Upon substitution of expressions (11-13) into integral (10a) , it is not difficult to show that the coefficient matrix $A(\omega)$ can be written in terms of frequency-independent sub-matrices $\boldsymbol{K}$, $\boldsymbol{C}$, and $\boldsymbol{M}$. Equation (9) then becomes

$$[\mathbf{K} + i\omega\mathbf{C} - \omega^2\mathbf{M}]\,\bar{\mathbf{q}} = \bar{\mathbf{f}}, \tag{14}$$

which can be transformed directly into the time domain to give

$$\mathbf{Kq} + \mathbf{C}\dot{\mathbf{q}} + \mathbf{M}\ddot{\mathbf{q}} = \mathbf{f}(t), \tag{15}$$

where $q_i(t) = \mathcal{F}^{-1}\{\bar{q}_i(\omega)\}$, and $f_i(t) = \mathcal{F}^{-1}\{\bar{f}_i(\omega)\}$. The components $f_i(t)$ can moreover be assembled directly in the time domain, being given by

$$f_i = \int_S \rho\, N_i\, a_n(\mathbf{x},t)\, dS\,. \tag{16}$$

The components of the transient solution vector $\mathbf{q}$(t) within the inner region then correspond to instantaneous nodal values of acoustic pressure. In the WE region, they represent instantaneous nodal pressures at a delayed time equal to the time required for a disturbance to propagate from the edge of the inner mesh (Astley 1996).

4. Solution of the time harmonic equations.

A complex L-U solver is used to solve equation (14). A "skyline" storage algorithm takes advantage of the sparse structure within the coefficient matrix In calculating a frequency response over a range of frequencies, the stiffness, mass and damping matrices, $\mathbf{K}$, $\mathbf{C}$, and $\mathbf{M}$ are assembled once only. The coefficient matrix $[A(\omega)]$ must however be decomposed for each frequency of interest.

5. Direct solution of the transient equations.

In the transient case, a direct time-integration scheme is used to solve equation (15). In the absence of any clear strategy for lumping the mass and damping matrices - which would make an explicit scheme attractive - the implicit "average acceleration" (Newmark: $\beta = 1/4$, $\gamma = 1/2$) method is used. The algorithm at each time step is given by;

$$\mathbf{E}\mathbf{q}_{t+\delta t} = \mathbf{f}_{t+\delta t} + \mathbf{G}\mathbf{q}_t + \mathbf{H}\dot{\mathbf{q}}_t + \mathbf{J}\ddot{\mathbf{q}}_t \tag{17a}$$

$$\ddot{\mathbf{q}}_{t+\delta t} = (4/\delta t^2)(\mathbf{q}_{t+\delta t} - \mathbf{q}_t) - (4/\delta t)\dot{\mathbf{q}}_t - \ddot{\mathbf{q}}_t \tag{17b}$$

$$\dot{\mathbf{q}}_{t+\delta t} = \dot{\mathbf{q}}_t + (\delta t/2)(\ddot{\mathbf{q}}_t + \ddot{\mathbf{q}}_{t+\delta t}) \tag{17c}$$

where the subscripts "t" and "$t+\delta t$" denote values of $\mathbf{q}$ at successive time steps. The matrices **E**, **G**, **H** and **J** are given by

$$\mathbf{E}=\mathbf{K}+(2/\delta t)\mathbf{C}+(4/\delta t^2)\mathbf{M},\quad \mathbf{G}=(2/\delta t)\mathbf{C}+(4/\delta t^2)\mathbf{M},\quad \mathbf{H}=\mathbf{C}+(4/\delta t)\mathbf{M},\quad \text{and}\quad \mathbf{J}=\mathbf{M}. \tag{17d}$$

An L-U solver is used, the decomposition being performed once only at the first time step. This scheme is known to be unconditionally stable for symmetric problems but there is no assurance that this is the case in the current instance given that **K** and **C** are not symmetric. Stable solutions have been obtained however in all of the test solutions to date.

6. Iterative solution of the transient equations.

The calculation and storage of "fill terms", generated during the decomposition of **E**, constitute a significant computational overhead of the implicit scheme. This is avoided by splitting the diagonal and non-diagonal terms of **E** and solving equation (17a) iteratively. i.e by replacing equation (17a) by

$$\mathbf{q}^{n+1}_{t+\delta t} = \mathbf{D}^{-1}[\mathbf{f}_{t+\delta t} + \mathbf{G}\mathbf{q}_t + \mathbf{H}\dot{\mathbf{q}}_t + \mathbf{J}\ddot{\mathbf{q}}_t - \mathbf{L}\mathbf{q}^{n}_{t+\delta t} - \mathbf{U}\mathbf{q}^{n+1}_{t+\delta t}] \tag{18}$$

where $\mathbf{q}^{n}_{t+\delta t}$ denotes the value of the solution vector at time $t+\delta t$ after n iterations have been performed. Here **D**, **L**, and **U** are diagonal, lower and upper matrices such that **E** = **D** + **L** + **U**. Convergence is achieved when

$$\left|\mathbf{q}^{n+1}_{t+\delta t} - \mathbf{q}^{n}_{t+\delta t}\right| / \left|\mathbf{q}^{n}_{t+\delta t}\right| \le \varepsilon \tag{19}$$

where ε is a prescribed value. The speed of convergence will be shown to be rapid provided that the time step is sufficiently small.

7. Results

Solutions are presented for the sound field generated by a cylindrical piston in a plane baffle. Axisymmetric elements are used. Computed solutions are presented for the two FE/WE meshes shown in figure 2. The WE elements are based on an oblate spheroidal formulation. Mesh (b) is clearly more compact and of a more pronounced spheroidal character than mesh (a). Both models possess axial symmetry about the line *ABC*. In the steady case the piston

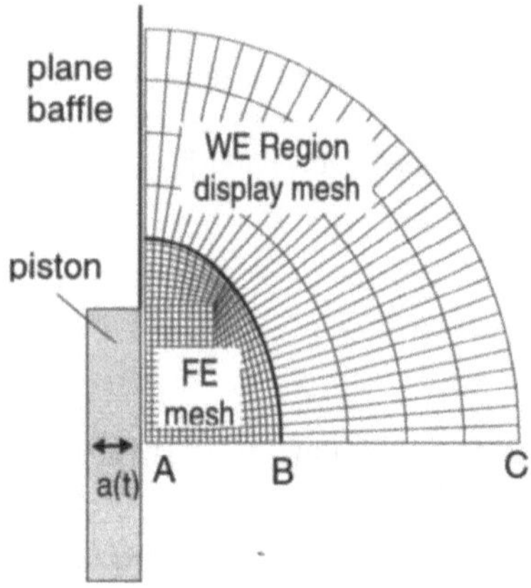

Figure 2(a). Test case mesh (a).

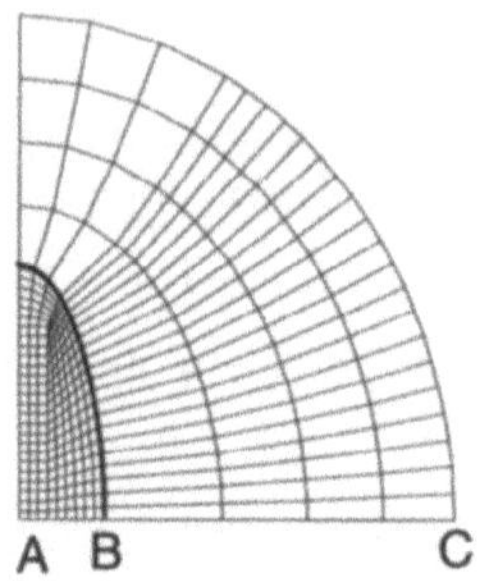

Figure 2(b). Test case mesh (b)

undergoes time harmonic excitation, $a(t) = a_0 e^{i\omega t}$. In the unsteady case, it experiences a single pulse given by : $a(t) = 0,\ t < 0,\ a(t) = a_0 \sin\omega t,\ 0 < t < \pi/\omega,$ and $a(t) = 0,\ t > \pi/\omega.$

7.1 Steady solutions, the effect of element order.

The effect of element order on the accuracy of the computed solution is demonstrated in figures 3(a) and 3(b) which show a comparison of computed and analytic frequency responses at the points *A*, *B* and *C* for each model. These results are obtained by solving equation (14). The non-dimensional acoustic pressure amplitude $|\bar{p}\,\omega/\rho_0 c a_0|$ is plotted against non-dimensional wavenumber *kD* where *D* is the diameter of the piston. Results are presented for WE elements of radial orders 2 and 3.

In the case of mesh (a), elements of radial order 3 give a good approximation for the entire frequency range ($0 < kD < 15$) over which the inner FE mesh can reasonably be expected to provide adequate resolution ($kD = 15.0$ corresponds to 8 nodes per wavelength). Elements of order 2 give acceptable results only at lower frequencies, say kD < 5.0. In the case of mesh (b) the accuracy of both solutions deteriorates. Clearly however in both models the quality of the solution increases dramatically with element order. This suggests a practical limitation on the effectiveness of the spheroidal formulation as a means of decreasing the overall dimensionality of the discrete problem i.e if an increase in the radial order of the WE region is required to maintain a prescribed level of accuracy as the inner FE mesh is compressed - as seems to be the case - this will offset any computational advantage derived from the reduction in FE node numbers. The same effect is apparent in other test cases for which solutions have been computed and affects both prolate and oblate spheroidal elements.

7.2 Transient solutions, the effect of step size.

Exact and computed transient solutions for the impulsively excited piston are shown in figures 4(a)-(c). Pressure histories are plotted against dimensionless time $T\,(=tc/D)$ at points *A*, *B* and *C* . The piston acceleration takes the form of a single half-wave pulse of duration *0.5* units of dimensionless time. The more compact of the two meshes - mesh (b) - is used. Computed

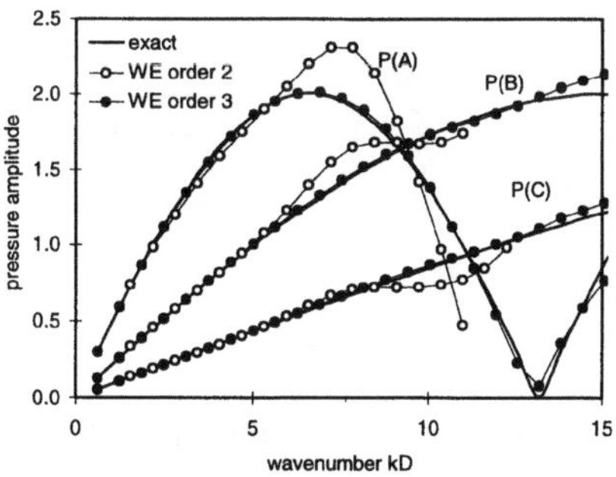

Figure 3(a). Steady response, mesh (a).

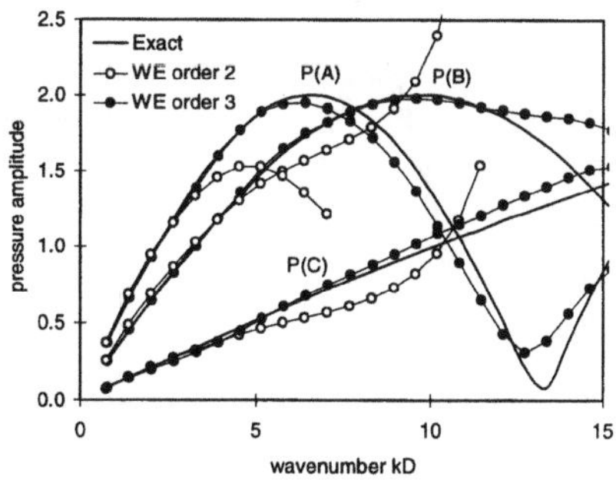

Figure 3(b). Steady response, mesh(b).

results are obtained in both instances by using WE elements of order 3. Results are shown for time steps: $\delta T=.025$, $\delta T=.05$ and $\delta T=.10$. The smallest of these ($\delta T=.025$) is equal to the time which elapses as a disturbance propagates between adjacent nodes on the face of the piston. All solutions are stable but those obtained by using the larger time step ($\delta T=.10$) are marginal in terms of accuracy. This is to be anticipated given that this value corresponds to approximately one fifth of the duration of the pulse. The results of a numerical exercise such as this do not constitute proof that the average acceleration scheme is unconditionally stable when applied to such problems but do lend weight to this supposition.

7.3 The performance of the iterative solver.

The performance of the iterative scheme is illustrated in figures 5(a)-(c). The data presented in these figures are obtained by applying the iterative scheme of equation (18) to the solution of the transient test problem. The pressure histories for mesh (b) when re-computed in this way are indistinguishable from those shown in figure 4 to four significant provided that $\varepsilon < 10^{-3}$. The number of iterations required for convergence at each time step of such a procedure are shown in figure 5 for various conditions. In all cases, the iterative solver runs for 100 time steps and an iteration count is performed at each step. The results are presented in bar-chart format as the percentage of solution steps for which convergence was achieved after a given number of iterations. The effect on convergence of varying the parameter ε is shown in figure (a). The effects of varying the step size and model size are shown in figures (b) and (c).

The results of figure (5a) are obtained by fixing the time step at $\delta T=.025$ and varying ε in the range 10^{-3} - 10^{-6}. As ε decreases the convergence criterion becomes more stringent and more iterations are required. Given that the results obtained for $\varepsilon = 10^{-3}$ are indistinguishable for all practical purposes from those obtained from the implicit scheme, this data indicates that seven iterations per time step is a realistic estimate of the computational effort which is required when the iterative solver is used in place of the direct solution scheme.

The effect of step size is illustrated in figure 5(b). Here the convergence parameter ε is fixed - at 10^{-3} - and the step size is varied from 0.025 to 0.10. The use of a time step smaller than $\delta T=.025$ does not further reduce the number of iterations. As the time step increases, however, the convergence of the scheme deteriorates. This is to be anticipated since the number of surrounding nodes which participate in the solution will increase with the step size

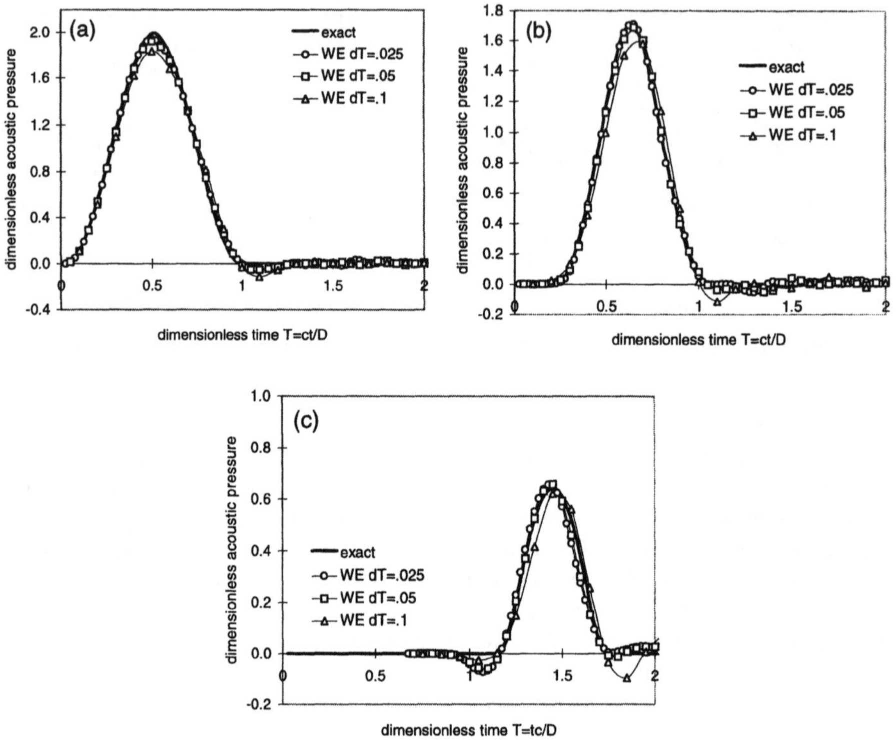

Figure 4. Pressure histories at A, B, and C
(figs (a), (b) and (c)).

once this exceeds the time taken for a disturbance to traverse one element (in this case $\delta T=.025$). This is confirmation that the iterative scheme performs most efficiently - as one would expect - when the time step is comparable to the grid spacing divided by the wave speed.

The effect of model size on the convergence of the scheme is indicated in figure (5c) The convergence of the scheme for three different models of varying size is presented for comparison. The three models have 276, 603 and 948 degrees of freedom respectively (the first two are models (b) and (a) of figure 2. The third is a larger model which is similar to model (a) but with a more extensive inner mesh. All three have the same grid spacing in the region adjacent to the piston. The convergence parameter ε is set to 10^{-3} and the dimensionless step size to *0.025*. There is no indication that convergence deteriorates with increasing model size.

8. Conclusions.

The analysis and results presented in this article support the following conclusions:

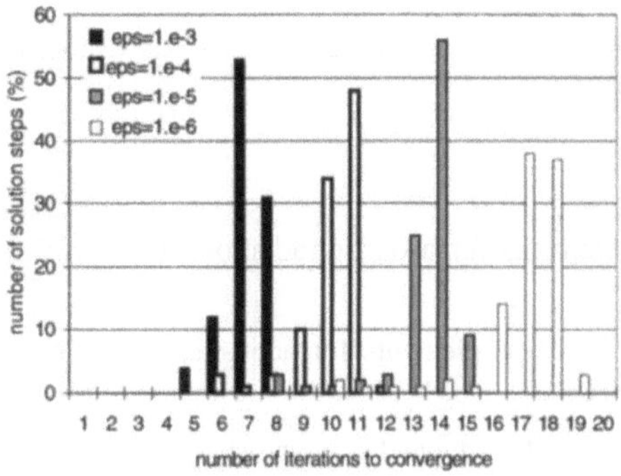

Figure 5(a). Convergence of the iterative scheme, the effect of the parameter ε.

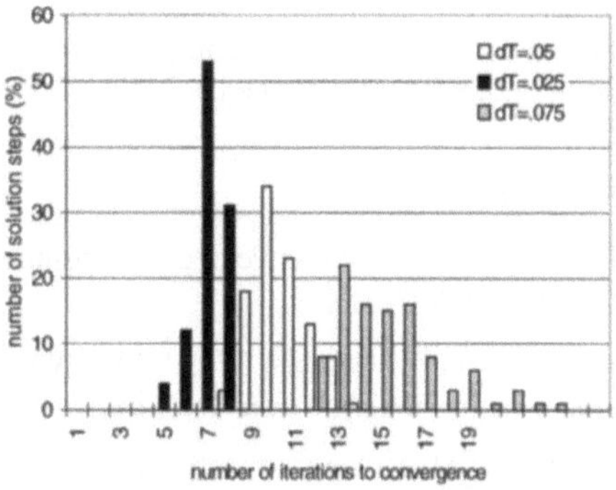

Figure 5(b). Convergence of the iterative scheme, the effect of step size.

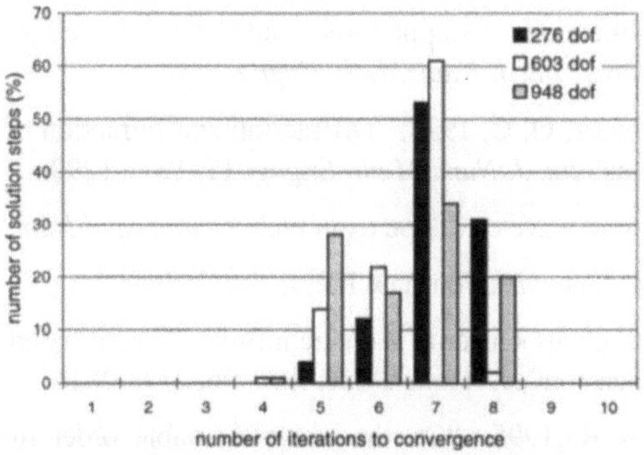

Figure 5(c). Convergence of the iterative scheme, the effect of model size.

1) The potential reduction in the size of the inner mesh which is made possible by the use of spheroidal rather than spherical elements may be offset by a necessary increases in the radial order - and hence dimensionality - of the exterior elements.
2) An implicit scheme has been shown to give accurate and (apparently) stable transient solutions for axisymmetric test problems.
3) Selection of step size in the above scheme is governed by considerations of accuracy in resolving the temporal variation rather than by time-scales associated with the mesh.
4) An iterative scheme which does not require a matrix inversion has been tested for the solution of axisymmetric problems and has been shown to converge rapidly provided that the time step is sufficiently small.

Further studies are suggested by the current work. These include;

- Parametric studies to clarify conclusion 1) above.
- Rigorous analysis of the stability of the time-integration scheme for non-symmetric matrices.

- An assessment of relative computational efficiency of the implicit and iterative schemes using optimally programmed codes.

9. References

Astley, R J 1983 "Wave envelope and infinite elements for acoustic radiation" *Int. J. Num. Meth. Fluids* **3**, 507-526.

Astley, R J and Eversman, W , 1988 "Wave envelope elements for acoustical radiation in inhomogeneous media", *Comp. and Struct.* **30,** 801-810 .

Astley, R. J , Macaulay G. J. and Coyette, J. P, 1994 "Mapped wave envelope elements for acoustic radiation and scattering", *J. Sound Vib.* **170,** 97-118.

Astley, R. J. 1996 "Transient wave envelope elements for wave problems", *J. Sound Vib.* **192**, 245-261

Astley, R J, 1997, in publication, " Mapped Spheroidal Wave-Envelope Elements for Unbounded Wave Problems" *Int. J. Num. Meth. Engng.*

Bettess, P and Zienkiewicz, O, C, 1977, "Diffraction and refraction of surface waves using finite and infinite elements" *Int. J. Num. Meth. Engng.* **11**, 1271-1290.

Bettess, P., 1987, "A simple wave envelope example". *Comm.Appl.Num. Meth.* **3**, 77-80.

Bettess, P, 1992, *Infinite Elements*. Penshaw Press, Sunderland.

Burnett, D S, 1994, "A three dimensional acoustic infinite element based on a prolate spheroidal multipole expansion", *J. Acoust. Soc. Am.* **96**, 2798-2816 .

Cremers, L and Fyfe, K R ,1995 "On the use of variable order infinite wave envelope elements for acoustic radiation and scattering *J. Acoust. Soc. Am.* **97**, 2028-2040

Demkowicz L and Gerdes K, 1995, "Solution of 3-D Laplace and Helmholtz equations in exterior domains using hp infinite element methods" *TICAM Report 95-04*, University of Texas at Austin.

Geers, T L, 1978, "Doubly asymptotic approximations for transient motions of submerged structures", *J. Acoust. Soc. Am.* **64,** 1500-1508.

Givoli, D and Cohen, D, 1995 "Nonreflecting boundary conditions based on Kirchhoff-type formulae" *J. Comp. Physics* **117,** 102-113.

Olson, L G and Bathe, K J, 1985, "An Infinite Element for the Analysis of Fluid-Structure Interaction", *Eng. Comput.* **2**, 319-329.

Regan, B A and Eaton, J A , 1995, "Application of an Efficient Iterative 3D Finite element Scheme to the Fan Noise radiation Problem", *CEAS/AIAA-95-012, 16'th AIAA Aeroacoustics conference,Munich, Germany.*

Zienkiewicz, O. C.,Bando, K, Bettess, P , Emson C and Chiam, T.C. 1985 "Mapped infinite elements for exterior wave problems", *Int. J. Num. Meth. Engng.* **21**, 1229-1252.

WATER-WAVE GREEN'S FUNCTION FOR A 3D UNEVEN-BOTTOM PROBLEM WITH DIFFERENT DEPTHS AT $x \to +\infty$ AND $x \to -\infty$

Gerassimos A. Athanassoulis and Konstadinos A. Belibassakis

Ship and Marine Hydrodynamics Laboratory
Department of Naval Architecture and Marine Engineering
National Technical University of Athens

1. Introduction

The interaction of floating and/or immersed bodies with the uneven bottom topography in the coastal environment is a mathematically difficult problem that, recently, has also become practically important, as the dimensions of large floating obstacles increase. That means that the characteristic lengths: horizontal dimension of the body L, wave length λ, bottom variation length λ_b, and average depth h may be all comparable and, thus, the ratio h/λ may be neither large enough (> 0.5) nor small enough (< 0.07) so that the deep or the shallow water models to be applicable. Furthermore, all existing shallow water models, developed for coastal environment applications, are based on the assumption of small or mild bottom slope. See, e.g., the relevant surveys by [Mei, 1983] and [Massel, 1989, 1993]. Thus, it seems timely to investigate water wave problems in a marine environment such that λ, λ_b and h are all of the same order of magnitude, being also non-constant because of a bottom topography variation. Although the non-linearity of water waves is of great importance, especially in the case of shallow water, it seems unavoidable to start studying a linearised version of the problem.

Consider the marine environment illustrated in Fig. 1a. This environment consists of a water layer D_{3D} bounded above by the free surface $\partial D_{F,3D}$ and below by a rigid bottom $\partial D_{\Pi,3D}$. It is assumed that the bottom slope exhibits an arbitrary one dimensional variation in a subdomain of finite length, i.e. the bathymetry is characterised by parallel, straight bottom contours lying between two regions of constant but different depth. The liquid is assumed homogeneous, inviscid and incompressible, and the liquid flow is excited by a pulsating monochromatic point source of angular frequency ω, located at a point in the variable bathymetry region. (This is not a restriction. The constant depth is included as a special case).

Our main concern in formulating and studying this problem is to investigate the far-field wave pattern which, in this case, is azimuthally strongly anisotropic. In this sense, this work is a prerequisite to the study of radiation and diffraction problems by floating or immersed bodies (scatterers) lying in a water layer with variable bathymetry characterised by a non-constant depth at infinity.

Theoretical aspects of the problem of small-amplitude water waves traveling in a region of varying depth, mainly uniqueness results, have been presented, under various geometric assumptions concerning the bottom topography, by [Vainberg and Mazja, 1973], [Fitz-Gerald and Grimshaw, 1979] and other authors. See also the survey by [Simon and Ursell, 1984].

T.L. Geers (ed.), IUTAM Symposium on Computational Methods for Unbounded Domains, 21–32.

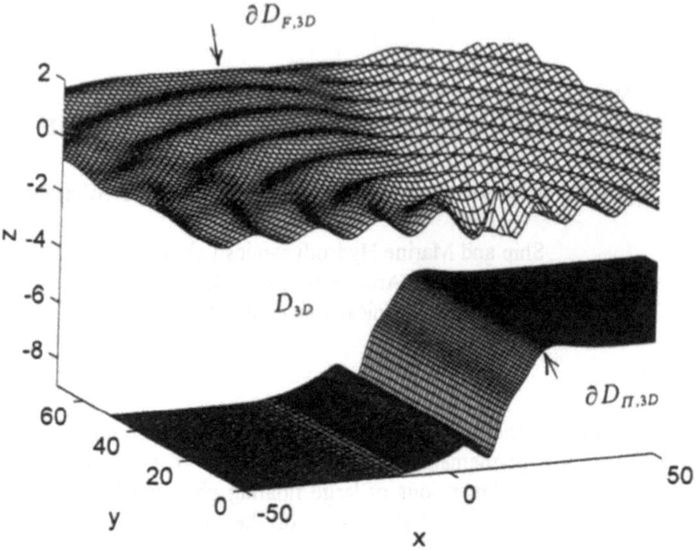

Figure 1(a). Geometrical configuration

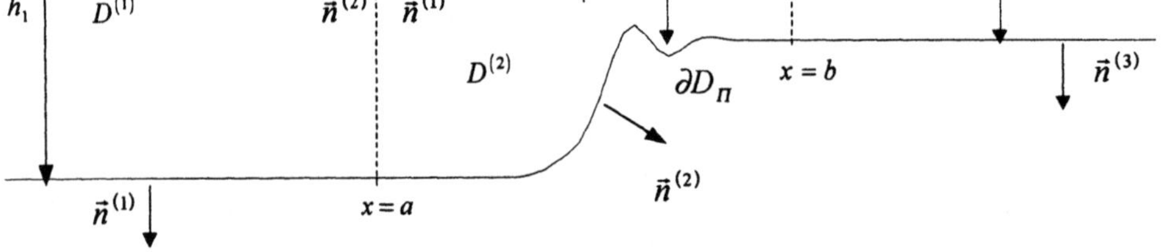

Figure 1(b). Domain decomposition and basic notation

2. Differential formulation of the problem

Before proceeding to the formulation of the problem we shall introduce some geometrical notation. A Cartesian coordinate system is introduced, with its origin at some point on the mean water level (in the variable bathymetry region), the z-axis pointing upwards and the y-axis being parallel to the bottom contours. See Fig. 1b. The liquid domain D_{3D} will be represented by $D_{3D} = D \times R$, where D is the (two-dimensional) intersection of D_{3D} by a vertical plane perpendicular to the bottom contours, and $R = (-\infty, +\infty)$ is a copy of the real line. To make the notation clear we explicitly write the analytical definitions of both the three-dimensional liquid domain D_{3D} and the two-dimensional vertical intersection D:

$$D_{3D} = \{(x,y,z):\ (x,y) \in R^2, -h(x) < z < 0\}, \qquad D = \{(x,z):\ x \in R, -h(x) < z < 0\}.$$

The function $h(x)$ appearing in the above definitions of D_{3D} and D, represents the local depth, measured from the mean water level. The following technical facts, concerning the depth function $h(x)$, are given here for later use:

$$h(x) \in C^2(R), \quad \operatorname{supp} h'(x) \subseteq (a,b), \tag{2.1a}$$

$$h(x) = h(a) = h_1 \quad \text{for all} \quad x \le a, \qquad h(x) = h(b) = h_3 \quad \text{for all} \quad x \ge b. \tag{2.1b}$$

The liquid domain D_{3D} is decomposed in three subdomains $D_{3D}^{(i)} = D^{(i)} \times R$, $i = 1,2,3$, defined as follows: $D_{3D}^{(1)}$ is the constant-depth subdomain characterised by $x < a$, $D_{3D}^{(3)}$ is the constant-depth subdomain characterised by $x > b$, and $D_{3D}^{(2)}$ is the variable bathymetry subdomain lying between $D_{3D}^{(1)}$ and $D_{3D}^{(3)}$. The analytic definitions of the vertical intersections $D^{(i)}$, $i = 1,2,3$, are as follows:

$$D^{(i)} = \left\{(x,z): \begin{Bmatrix} x < a \ (\text{for } i = 1) \\ x > b \ (\text{for } i = 3) \end{Bmatrix}, -h_i < z < 0\right\}, \qquad D^{(2)} = \{(x,z): a < x < b, -h(x) < z < 0\}.$$

Without loss of generality, we assume that $h_1 > h_3$, i.e. $D^{(1)}$ is the deep-water region and $D^{(3)}$ is the shallow-water region. The decomposition is also applied to the boundaries $\partial D_{F,3D} = \partial D_F \times R$ and $\partial D_{\Pi,3D} = \partial D_\Pi \times R$. The lines ∂D_F and ∂D_Π are decomposed in three pieces each, for example, $\partial D_F = \partial D_F^{(1)} \cup \{(a,\,0)\} \cup \partial D_F^{(2)} \cup \{(b,\,0)\} \cup \partial D_F^{(3)}$, where $\partial D_F^{(i)}$ belongs to the boundary of $D^{(i)}$, and similarly for ∂D_Π.

Finally, we define the vertical interfaces $\partial D_{I,3D}^{(12)} = \partial D_I^{(12)} \times R$ and $\partial D_{I,3D}^{(23)} = \partial D_I^{(23)} \times R$, which are the common vertical boundaries of subdomains $D_{3D}^{(1)}$ and $D_{3D}^{(2)}$, and $D_{3D}^{(2)}$ and $D_{3D}^{(3)}$, respectively. Clearly, $\partial D_I^{(12)}$ and $\partial D_I^{(23)}$ are vertical segments (between the bottom and the mean water level) at $x = a$ and $x = b$, respectively.

After introducing the geometrical notation we need, we are proceeding to the mathematical formulation of the wave problem we described at the beginning of this section. Assuming that the surface elevation and the velocity field are small, the linearised water wave equations can be used. See, e.g., [Stoker, 1957] or [Wehausen and Laitone, 1960]. Then, the velocity field is time harmonic with angular frequency ω, the same as the pulsating source, and can be represented by a velocity potential of the form

$$\mathscr{P}(x,y,z;t) = \operatorname{Re}\{\Phi(x,y,z;\mu)\exp(-i\omega t)\},$$

where μ is an appropriate frequency parameter; (see below, equation (2.2b)). The wave problem, formulated in terms of the frequency-dependent potential $\Phi = \Phi(x,y,z;\mu)$, reads as follows:

$$\nabla^2 \Phi = -\delta(\vec{x} - \vec{x}_0), \quad \vec{x} \in D_{3D}, \tag{2.2a}$$

$$\frac{\partial \Phi}{\partial z} - \mu \Phi = 0, \quad \vec{x} \in \partial D_{F,3D}, \qquad \frac{\partial \Phi}{\partial n} = 0, \quad \vec{x} \in \partial D_{\Pi,3D}, \tag{2.2b,c}$$

$$\Phi \sim \text{outgoing waves as } r = \sqrt{(x - x_0)^2 + (y - y_0)^2} \to \infty, \tag{2.2d}$$

where $\mu = \omega^2/g$, g is the acceleration due to gravity, and $\vec{x}_0 = (x_0, 0, z_0)$ is the location of the point source. The boundary value problem (2.2a,b,c,d) will be referred to as the *original wave problem* $\mathcal{P}_{3D}(D_{3D}, \mu, \vec{x}_0)$.

The characteristic length scales involved (directly or indirectly) in the above problem are: the two far-field water depths h_i, $i = 1,3$, the corresponding far-field wave lengths $\lambda_i = 2\pi/k^{(i)}$, $i = 1,3$, obtained as the positive roots of the equations (dispersion relations) $\mu h_i = k^{(i)} h_i \tanh(k^{(i)} h_i)$, $i = 1,3$, the bottom variation length $\lambda_b = b - a$, as well as an average amplitude of bottom corrugations. From these characteristic lengths we can form various non-dimensional characteristic numbers, the most important of which are: The *shallowness ratios* h_i/λ_i, $i = 1,3$, the *mean* and the *maximum bottom slopes* s_{mean}, s_{max}, and the *shoaling ratio* h_3/h_1. In the present work all these non-dimensional numbers are considered to be of the same order of magnitude. That is, no asymptotic assumptions are made for them.

Since the geometry of the liquid domain is of cylindrical character, i.e. $D_{3D} = D \times R$, it is possible to reduce the dimensionality of the problem by taking the Fourier transform with respect to y. The transformed potential will be denoted by $\varphi(x,z;\xi,\mu) = \varphi(x,z;\xi)$, the μ-dependence being suppressed. The following pair of equations clarifies which are our choices as regards the (inessential) 2π-factors and exponent signs in the Fourier transform:

$$\varphi(x,z;\xi) = \frac{1}{2\pi} \int_{-\infty}^{+\infty} \Phi(x,y,z) e^{-iy\xi} \, dy, \qquad \Phi(x,y,z) = \int_{+\infty}^{-\infty} \varphi(x,z;\xi) e^{+iy\xi} \, d\xi. \tag{2.3a,b}$$

Applying the Fourier transform to the problem $\mathcal{P}_{3D}(D_{3D}, \mu, \vec{x}_0)$, we obtain the following two-dimensional wave problem:

$$\nabla^2 \varphi - \xi^2 \varphi = -\frac{1}{2\pi} \delta(x - x_0) \delta(z - z_0), \quad \text{in } D, \tag{2.4a}$$

$$\frac{\partial \varphi}{\partial z} - \mu \varphi = 0 \quad \text{on } \partial D_F, \qquad \frac{\partial \varphi}{\partial n} = 0 \quad \text{on } \partial D_{\Pi}, \tag{2.4b,c}$$

$$\varphi \sim \text{outgoing waves as } |x| \to \infty. \tag{2.4d}$$

The boundary value problem (2.4a,b,c,d) will be referred to as the *transformed wave problem* $\mathcal{P}_{\xi}(D, \mu, \vec{x}_0)$.

3. Reformulation as a transmission problem in a bounded domain

The transformed wave problem will be reformulated as a transmission problem in the bounded subdomain $D^{(2)}$ with the aid of the following representation of the potential $\varphi(x,z;\xi)$ in the half-strips $D^{(1)}$, $D^{(3)}$:

Left half-strip $D^{(1)}$ $(x \le a)$:

$$\varphi^{(1)}(x,z;\xi) = C_0^{(1)}(\xi)\exp\left(-xK_0^{(1)}(\xi)\right)Z_0^{(1)}(z) + \sum_{n=1}^{\infty} C_n^{(1)}(\xi)\, Z_n^{(1)}(z)\exp\left((x-a)K_n^{(1)}(\xi)\right), \quad (3.1a)$$

Right half-strip $D^{(3)}$ $(x \ge b)$:

$$\varphi^{(3)}(x,z;\xi) = C_0^{(3)}(\xi)\exp\left(+xK_0^{(3)}(\xi)\right)Z_0^{(3)}(z) + \sum_{n=1}^{\infty} C_n^{(3)}(\xi)\, Z_n^{(3)}(z)\exp\left(-(x-b)K_n^{(3)}(\xi)\right), \quad (3.1b)$$

where

$$K_0^{(i)}(\xi) \equiv K_0^{(i)} = \sqrt{\xi^2 - \left(k_0^{(i)}\right)^2}\, sign\left(k_0^{(i)} - |\xi|\right), \quad K_n^{(i)}(\xi) \equiv K_n^{(i)} = \sqrt{\xi^2 + \left(k_n^{(i)}\right)^2}, \quad \begin{cases} i = 1,3, \\ n = 1,2,\ldots \end{cases} \quad (3.1c)$$

$k_n^{(i)}$, $n = 0,1,2,\ldots$, $i = 1,3$, are defined below, and $C_n^{(i)}(\xi) \equiv C_n^{(i)}$, $i = 1,3$, $n = 0,1,2,\ldots$, are undefined coefficients which will be determined by means of the appropriate matching conditions at the interfaces $\partial D_I^{(12)}$ and $\partial D_I^{(23)}$.

In the expansions (3.1a,b) the sets of numbers $\left\{i\,k_0^{(i)}, k_n^{(i)},\ n = 1,2,\ldots\right\}$, $i = 1,3$, and the sets of vertical functions $\left\{Z_n^{(i)}(z),\ n = 0,1,2,\ldots\right\}$, $i = 1,3$, are, respectively, the eigenvalues and the corresponding eigenfunctions of the following regular Sturm-Liouville problems

$$\frac{d^2 Z^{(i)}(z)}{dz^2} + \left(k^{(i)}\right)^2 Z^{(i)}(z) = 0, \quad -h_i < z < 0 \quad (3.2a)$$

$$\frac{dZ^{(i)}(0)}{dz} - \mu Z^{(i)}(0) = 0, \qquad \frac{dZ^{(i)}(-h_i)}{dz} = 0, \quad (3.2b,c)$$

obtained by separation of variables in the half-strips $D^{(i)}$, $i = 1,3$. The eigenvalues $\left\{i\,k_0^{(i)}, k_n^{(i)}\right\}$ are obtained as the roots of the dispersion relation

$$\mu h_i = -k^{(i)} h_i \tan\left(k^{(i)} h_i\right), \quad i = 1,3, \quad (3.3a)$$

and the eigenfunctions $\left\{Z_n^{(i)}(z),\ n = 0,1,2,\ldots\right\}$ are given by

$$Z_0^{(i)}(z) = \frac{\cosh\left(k_0^{(i)}(z+h_i)\right)}{\cosh\left(k_0^{(i)} h_i\right)}, \quad Z_n^{(i)}(z) = \frac{\cos\left(k_n^{(i)}(z+h_i)\right)}{\cos\left(k_n^{(i)} h_i\right)}, \quad n = 1,2,\ldots, \quad i = 1,3. \quad (3.3b)$$

The correctness (completeness) of the expansions (3.1a,b) follows by the standard theory of regular eigenvalue problems. See, e.g., [Coddington and Levinson, 1955, Section 7.4]. We only recall here that the series expansion in terms of $Z_n^{(i)}(z)$, $n = 0,1,2,\ldots$, of any $C^2(-h_i, 0)$-function satisfying the boundary conditions (3.2b,c) is uniformly convergent in $[-h_i, 0]$, while for a C^2-function not satisfying the boundary condition (3.2b or c) the convergence is uniform only in $[-h_i + \varepsilon, -\varepsilon]$ and the limit of the series at $z = 0$ or $-h_i$ may not coincide with the end value(s) of the function.

Exploiting the expansions (3.1a,b) the problem $\mathscr{P}_\xi(D, \mu, \vec{x}_0)$ can be reformulated as a boundary value (transmission) problem in the bounded subdomain $D^{(2)}$ as follows:

Find the potential $\varphi^{(2)}(x,z;\xi)$ defined in $D^{(2)}$ and the coefficients $C_n^{(i)}(\xi)$, $i=1,3$, $n=0,1,2,\ldots$, satisfying the equations:

$$\nabla^2\varphi^{(2)}-\xi^2\varphi^{(2)}=-\frac{1}{2\pi}\delta(x-x_0)\delta(z-z_0),\quad \vec{x}\in D^{(2)} \tag{3.4a}$$

$$\frac{\partial\varphi^{(2)}}{\partial n^{(2)}}-\mu\varphi^{(2)}=0,\quad \vec{x}\in\partial D_F^{(2)},\qquad \frac{\partial\varphi^{(2)}}{\partial n^{(2)}}=0,\quad \vec{x}\in\partial D_\Pi^{(2)}, \tag{3.4b,c}$$

$$\varphi^{(2)}=\varphi^{(1)},\quad \frac{\partial\varphi^{(2)}}{\partial n^{(2)}}=-\frac{\partial\varphi^{(1)}}{\partial n^{(1)}},\quad \vec{x}\in\partial D_I^{(12)}, \tag{3.4d,e}$$

$$\varphi^{(2)}=\varphi^{(3)},\quad \frac{\partial\varphi^{(2)}}{\partial n^{(2)}}=-\frac{\partial\varphi^{(3)}}{\partial n^{(3)}},\quad \vec{x}\in\partial D_I^{(23)}, \tag{3.4f,g}$$

where $\vec{x}=(x,z)$ is the generic field point, $\vec{x}_0=(x_0,z_0)$ is the location of the point source, and $\vec{n}^{(i)}$ is the outward unit normal of the boundary of $D^{(i)}$, $i=1,2,3$; see Fig. 1b. The problem (3.4a-g) will be referred to as *the transmission problem* $\mathcal{P}_\xi\left(D^{(2)},C_n^{(1)},C_n^{(3)},\mu,\vec{x}_0\right)$.

4. Variational formulation and the coupled-mode system of equations

The transmission problem admits a variational formulation in accordance with Bai's construction [Bai and Yeung, 1974], [Mei, 1978], [Aranha *et al.*, 1979], initially called "hybrid element method". Consider the functional $\mathcal{F}=\mathcal{F}\left(\varphi^{(2)},\{C_n^{(1)}\}_{n=0,1,\ldots},\{C_n^{(3)}\}_{n=0,1,\ldots}\right)$ defined by

$$\mathcal{F}=\frac{1}{2}\int_{D^{(2)}}\left[\left(\nabla\varphi^{(2)}\right)^2+\xi^2\left(\varphi^{(2)}\right)^2-\frac{1}{\pi}\delta(x-x_0)\delta(z-z_0)\varphi^{(2)}\right]dV-\frac{1}{2}\mu\int_{\partial D_F^{(2)}}\left(\varphi^{(2)}\right)^2 dS+$$

$$+\int_{\partial D_I^{(12)}}\left(\varphi^{(2)}-\frac{1}{2}\varphi^{(1)}\right)\frac{\partial\varphi^{(1)}}{\partial n^{(1)}}dS+\int_{\partial D_I^{(23)}}\left(\varphi^{(2)}-\frac{1}{2}\varphi^{(3)}\right)\frac{\partial\varphi^{(3)}}{\partial n^{(3)}}dS. \tag{4.1}$$

Remarks: (i) $\varphi^{(1)},\varphi^{(3)}$ are considered only on the interfaces $\partial D_I^{(12)}$, $\partial D_I^{(23)}$, respectively. They should be thought of in terms of their expansions (3.1a,b) which permit us to keep as arguments in the functional $\mathcal{F}$ the coefficients $C_n^{(i)}$, $i=1,3$, instead of the functions $\varphi^{(i)}$, $i=1,3$, themselves. (ii) The appearance of the δ-function in the functional is just a matter of conciseness. It can be avoided if we introduce a decomposition of $\varphi^{(2)}$ of the form $\varphi^{(2)}=\varphi_{\text{reg}}^{(2)}+G(\vec{x};\vec{x}_0)$, where $G(\vec{x};\vec{x}_0)$ is the free-space Green function of the operator $\Delta-\xi^2$. In this case the functional $\mathcal{F}$ is defined on $\varphi_{\text{reg}}^{(2)}$ instead of $\varphi^{(2)}$ and, thus, no singular terms appear, at the expense of some additional (known) boundary terms, coming from the contribution of $G(\vec{x};\vec{x}_0)$. This approach, which is also beneficial from the numerical point of view, will be presented elsewhere.

Theorem: The functions $\varphi^{(2)}(x,z;\xi)$ in $D^{(2)}$, and $\varphi^{(i)}\left(x,z;\{C_n^{(i)}(\xi)\}_{n=0,1,\ldots},\xi\right)$ in $D^{(i)}$, $i=1,3$, form a solution of the transmission problem (3.4) iff they render the functional $\mathcal{F}$ stationary, i.e. iff they satisfy the variational equation

$$\delta \mathscr{F}\left(\varphi^{(2)},\ \{C_n^{(1)}\}_{n=0,1,\dots},\ \{C_n^{(3)}\}_{n=0,1,\dots}\right)=0. \tag{4.2}$$

The proof is straightforward, although lengthy. It goes along the lines of Bai's original proof, with some modifications in order to take into account the different far-field depths.

The above variational principle will now be used to obtain an alternative, semi-discrete (Kantorovich) formulation in terms of *local modes*, i.e. in terms of an x-dependent local vertical L^2-basis $\{Z_n(z;x),\ n=0,1,2,\dots\}$. This family of basis functions is obtained by formulating and solving a local vertical Sturm-Liouville problem of the form (3.2) in the interval $[-h(x),0]$. The form of functions $Z_n(z;x)$, $n=0,1,2,\dots$, is similar to (3.3b), with h_i and $k_n^{(i)}$, $n=0,1,2,\dots$ being replaced by the local depth $h(x)$ and the local eigenvalues $k_n(x)$, $n=0,1,2,\dots$. In this connection, a first choice of a semi-discrete expansion of the $\varphi^{(2)}(x,z;\xi)$ in the variable-bathymetry domain $D^{(2)}$ – that has been applied to similar problems in hydroacoustics, e.g., [Fawcett, 1992], and water waves, e.g., [Massel, 1993, 1996], [Porter and Chamberlain, 1997] – is the following:

$$\varphi^{(2)}(x,z;\xi)=\varphi_0(x;\xi)\cdot Z_0(z;x)+\sum_{n=1}^{\infty}\varphi_n(x;\xi)\cdot Z_n(z;x). \tag{4.3}$$

However, this expansion is incompatible with the bottom boundary condition (3.4c) whenever $dh(x)/dx\neq 0$. To remedy this inconsistency we introduce an additional vertical mode, denoted by $\varphi_{-1}^{(2)}(x,z;\xi)$ and called *sloping-bottom mode*, so that the difference $\varphi^{(2)}(x,z;\xi)-\varphi_{-1}^{(2)}(x,z;\xi)$ satisfies both boundary conditions (3.4b,c). Then, the series expansion of the difference $\varphi^{(2)}-\varphi_{-1}^{(2)}$ converges (at any fixed x) not only in L^2 sense, but also uniformly in the closed interval $[-h(x),0]$. Accordingly, the representation (4.3) is replaced by the representation

$$\varphi^{(2)}(x,z;\xi)=\varphi_{-1}(x;\xi)\cdot Z_{-1}(z;x)+\varphi_0(x;\xi)\cdot Z_0(z;x)+\sum_{n=1}^{\infty}\varphi_n(x;\xi)\cdot Z_n(z;x), \tag{4.4}$$

where $Z_{-1}(z;x)$ is an appropriate vertical mode satisfying the conditions $Z_{-1}(0;x)=0$, $dZ_{-1}(0;x)/dz=0$ and $Z_{-1}(-h(x);x)=0$, $dZ_{-1}(-h(x);x)/dz=1$.

Remarks: (i) The above discussion concerning the convergence of series expansion (4.4) is based on the assumption that $\varphi^{(2)}(x,z;\xi)\in C^2(D^{(2)})\cap C^1(\bar{D}^{(2)})$, for each $\xi\in R$. (ii) Each term in the expansion (4.4) satisfies the free surface condition (3.4b) individually. Thus, representation (4.4) renders the free surface condition an *essential condition* in relation to the variational formulation, which proves to be very beneficial from the numerical point of view. (iii) Not only the first term of the expansion (4.4), which has been introduced for the first time in this work, is given a special name (sloping bottom mode), but also the other terms of the expansion (4.4) were given names expressing their physical behaviour. Thus, $\varphi_0(x;\xi)Z_0(z;x)$ is called the *propagation mode*, and $\varphi_n(x;\xi)Z_n(z;x)$, $n=1,2,\dots$, are called the *evanescent modes*. Furthermore, $Z_n(z;x)$ are called the (local) *vertical modes*, and $\varphi_n(x;\xi)$ are called the *modal amplitude functions*. (iv) A suitable choice for $Z_{-1}(z;x)$ is

$$Z_{-1}(z;x)=h(x)\left[(z/h(x))^3+(z/h(x))^2\right],\qquad a<x<b. \tag{4.5}$$

(v) The term $\varphi_{-1}(x;\xi)\cdot Z_{-1}(z;x)$ is not needed when the bottom is flat. Thus, in connection with conditions (2.1), we can impose the following end conditions on the amplitude of the sloping bottom mode: $\varphi_{-1}(a;\xi)=\varphi_{-1}(b;\xi)=0$. For more details concerning the sloping bottom mode see [Athanassoulis and Belibassakis, 1997].

Having constructed a complete local-mode (semi-discrete) representation of the function $\varphi^{(2)}(x,z;\xi)$, we proceed by using it in order to change arguments of the functional $\mathcal{F}$ and obtain an alternative formulation of our problem. Thus, after substituting (4.4) in (4.1), making the necessary algebraic manipulation and using the standard arguments of the Calculus of Variations with respect to the new arguments (degrees of freedom of the system), which are now: $\varphi_n(x;\xi)$, $a<x<b$, $n=-1,0,1,2,\ldots$, $\varphi_n(a;\xi)$, $\varphi_n(b;\xi)$, $n=0,1,2,\ldots$, and $C_n^{(i)}(\xi)$, $i=1,3$, $n=0,1,2,\ldots$, we obtain the following result (a prime denotes differentiation with respect to x):

<u>*Theorem*</u>: The variational equation (4.2) and, thus, the transmission problem (3.4), are equivalent to the following system of second-order ordinary differential equations (ODEs)

$$\sum_{n=-1}^{\infty} a_{mn}(x)\varphi_n''(x;\xi)+b_{mn}(x)\varphi_n'(x;\xi)+\left(c_{mn}(x)-\xi^2 a_{mn}(x)\right)\varphi_n(x;\xi)=-\frac{1}{2\pi}\delta(x-x_0)Z_m(z_0), \tag{4.6a}$$

in $a<x<b$, $m=-1,0,1,\ldots$, supplemented by the boundary conditions

$$\varphi_{-1}(a;\xi)=\varphi_{-1}'(a;\xi)=0, \qquad \varphi_{-1}(b;\xi)=\varphi_{-1}'(b;\xi)=0, \tag{4.6b}$$

$$\varphi_0'(a;\xi)+K_0^{(1)}(\xi)\varphi_0(a;\xi)=0, \quad \varphi_n'(a;\xi)-K_n^{(1)}(\xi)\varphi_n(a;\xi)=0, \quad n=1,2,3,\ldots \tag{4.6c}$$

$$\varphi_0'(b;\xi)-K_0^{(3)}(\xi)\varphi_0(b;\xi)=0, \quad \varphi_n'(b;\xi)+K_n^{(3)}(\xi)\varphi_n(b;\xi)=0, \quad n=1,2,3,\ldots \tag{4.6d}$$

The coefficients $C_n^{(i)}(\xi)$ are then obtained by the equations:

$$C_0^{(1)}(\xi)=\varphi_0(a;\xi)\exp\left(a K_0^{(1)}(\xi)\right), \qquad C_n^{(1)}(\xi)=\varphi_n(a;\xi), \qquad n=1,2,3,\ldots \tag{4.7a}$$

$$C_0^{(3)}(\xi)=\varphi_0(b;\xi)\exp\left(-b K_0^{(3)}(\xi)\right), \qquad C_n^{(3)}(\xi)=\varphi_n(b;\xi), \qquad n=1,2,3,\ldots \tag{4.7b}$$

The x-dependent coefficients $a_{mn}(x)$, $b_{mn}(x)$ and $c_{mn}(x)$ are given in Table 1. The symbols $\|\cdot\|$ and $<\cdot,\cdot>$, appearing in Table 1, stand for the norm and the inner product of the space $L^2(-h(x),0)$, the integration being always performed with respect to the vertical variable $z\in[-h(x),0]$. Furthermore, the coefficients $K_n^{(i)}(\xi)$, $n=0,1,2,\ldots$, $i=1,3$, which realise the interaction between the near-field and the far-field dynamics through the boundary conditions (4.6c,d), are given by equations (3.1c).

<u>Remarks</u>: (i) The end conditions (4.6b) for the sloping bottom mode $\varphi_{-1}(x;\xi)$ have been introduced above as essential conditions. (ii) The parametric ξ-dependence of the coefficients of the system (4.6a) is restricted only to the coefficient of $\varphi_n(x;\xi)$ and is quadratic. The ξ-dependence of the boundary coefficients $K_n^{(i)}(\xi)$ is of the form $\pm\sqrt{\xi^2\pm(k_n^{(i)})^2}$ and thus $K_n^{(i)}(\xi)$ can be a complex number. (iii) System (4.6a) can be replaced by two homogeneous systems in $a<x<x_0$ and $x_0<x<b$, and the matching condition (or source condition):

$$\sum_{n=-1}^{\infty} a_{mn}(x_0)\left[\varphi_n'(x_0^+;\xi)-\varphi_n'(x_0^-;\xi)\right]=-\frac{1}{2\pi}Z_m(z_0), \qquad m=-1,0,1,2,\ldots \tag{4.8}$$

(iv) Under the smoothness assumptions (2.1), all coefficients $a_{mn}(x)$, $b_{mn}(x)$, $c_{mn}(x)$ of the system (4.6a) are continous functions of x and can be calculated by means of the solution of the local (vertical) eigenvalue problem. (v) Discontinuities of $h''(x)$, $h'(x)$ or $h(x)$ can also be treated by introducing appropriate domain decomposition with matching boundaries at the points of discontinuities.

TABLE 1: Coefficients of system (4.6a)

	$m=-1$ $n=-1,0,1,2,\ldots$	$m=0,1,2,\ldots$ $n=-1$	$m=0,1,2,\ldots$ $n=0,1,2,\ldots$
$a_{mn}(x)$	$\langle Z_{-1},Z_n\rangle$	$\langle Z_m,Z_{-1}\rangle$	$\delta_{nm}\Vert Z_m\Vert^2$
$b_{mn}(x)$	$2\langle Z_{-1},\partial Z_n/\partial x\rangle$	$2\langle Z_m,\partial Z_{-1}/\partial x\rangle$	$2\langle Z_m,\partial Z_n/\partial x\rangle+$ $+h'(x)Z_m(-h)Z_n(-h)$
$c_{mn}(x)$	$\langle Z_{-1},\Delta Z_n\rangle$	$\langle Z_m,\Delta Z_{-1}\rangle+$ $+(1+h'(x))Z_m(-h)$	$\langle Z_m,\Delta Z_n\rangle+$ $+h'(x)Z_m(-h)(\partial Z_n(-h)/\partial x)$
NOTES:	$Z_n=Z_n(z;x)$ is parametrically dependent on x through $h(x)$.		
	$\Delta\,\bullet=\frac{\partial^2\bullet}{\partial z^2}+\frac{\partial^2\bullet}{\partial x^2}$, $\langle f,g\rangle=\int_{-h(x)}^{0}f(z)g(z)dz$, $\Vert f\Vert^2=\langle f,f\rangle$		

The theoretical investigation of the problem (4.6), i.e. the appropriate function space setting, uniqueness and existence results, asymptotic results concerning the ξ-dependence and other properties of the solution, seems to be very interesting and useful. However, to the best of authors' knowledge, no attempts to study these questions have been undertaken up to now.

As regards its numerical solution, the infinite system of ODEs (4.6) is solved by truncation, retaining only the first N_m evanescent modes, the total number of equations being reduced to N_m+2. In order to obtain first numerical results, a second-order finite-difference scheme has been applied for the approximation of x-derivatives, and the system of ODEs has been reduced to an algebraic system. The matrix of this system is banded 5-diagonal and its size is $N\cdot(N_m+2)$, where N is the total number of equidistant points used to disretize the interval $a\le x\le b$. An important aspect of the numerical solution is the specific way $\varphi(x,z;\xi)$ is dependent on ξ. Note first that $\varphi(x,z;\xi)$ is an even function of ξ, since ξ enters the system (4.6) as a function of ξ^2. Thus, the ξ-dependence should be studied only for $\xi\ge 0$. Extensive numerical experiments have shown that, for $x\ne x_0$, the transformed wave potential $\varphi(x,z;\xi)$ exhibits a fast (exponential) decay as $\xi\to\infty$. However, in the neighborhood of the singularity $(x=x_0)$, the transformed potential decays much slower. Numerical evidence indicates that $\varphi(x=x_0,z;\xi)\approx c(z)/\xi$, as $\xi\to\infty$. Numerical results obtained by means of the above procedure are presented in the next section.

5. Numerical solution of the 3D problem by Fourier inversion

Being able to solve numerically the boundary value problem (4.6) for any value of the Fourier parameter ξ, we can reconstruct the three-dimensional wave potential $\Phi(x,y,z)$ by applying the inverse Fourier transform (2.3b). Taking into account the rapid decay of $\varphi(x,z;\xi)$, $x\ne x_0$, for large values of ξ, the Fourier inversion is effectively performed by FFT in a bounded interval of the form $[0, \Xi]$. For all applications examined in this work $\Xi \leq 4k_0^{(3)}$. Only in the neighbourhood of the singularity $(x = x_0)$, where the ξ-decay of $\varphi(x_0, y; \xi)$ is slow, the Fourier inversion requires special treatment. In this case, the interval $[0, \infty)$ is subdivided into the subintervals $[0, \Xi]$ and $[\Xi, \infty)$, and the inversion is performed by applying FFT to the restriction of $\varphi(x_0, y; \xi)$ in $[0, \Xi]$, and by analytic/asymptotic integration in $[\Xi, \infty)$ (exploiting the estimate $\varphi \approx c/\xi$ and using the exponential integral).

Numerical results are now presented for the environment

$$\begin{cases} h(x) = h_1 = 6 \text{ m}, & x < a = 0, \\ h(x) = \dfrac{h_1 + h_3}{2} - \dfrac{h_1 - h_3}{2} \tanh\left(3\pi\left(\dfrac{x}{b} - \dfrac{1}{2}\right)\right), & a < x < b, \\ h(x) = h_3 = 2 \text{ m}, & x > b = 20 \text{ m}, \end{cases} \qquad (\mathfrak{F})$$

which models a smooth but steep underwater shoaling, with $s_{max} = 0.94$ and $s_{mean} = 0.2$. The pulsating source is located at $(x_0, y_0, z_0) = (10\text{m}, 0, -1\text{m})$ and its angular frequency is $\omega = 2$ rad/sec. In this case, the ratios $h_1/\lambda_1 = 0.4$ and $h_3/\lambda_3 = 0.17$, well outside the deep or shallow water cases.

In Fig. 2 the real and the imaginary part of the modal amplitude functions $\varphi_n(x; \xi)$ for $n = -1, 0, 12$, are presented as functions of the variable x, at a small value of the Fourier parameter ξ, (where they are more significant). The number N_m of evanescent modes required for the numerical convergence is, in this case, about 10. (In general, N_m increases with s_{mean} and s_{max} as far as h_{min}/λ is not higher than 0.5.)

The horizontal structure of the wave potential (and also of the free-surface elevation η, since $\eta = i\omega\Phi/g$), is shown in Fig. 3. From this figure we observe that a shadow zone is formed along the bottom irregularity $(0 < x < 20\text{m})$. The highly desirable asymptotic determination of the (azimuthally anisotropic) far-field wave pattern, is currently studied by the present authors.

Finally, in Fig. 4, the vertical structure of the wave potential in the xz-plane passing through the source is visualised with the aid of equipotential contours. From this figure we are able to observe the correct representation of the singularity, which is located at $z_0 = -1$m, and the satisfaction of the bottom boundary condition $\partial\varphi/\partial n = 0$, which is equivalent to the orthogonality between the equipotential contours and the bottom.

6. Conclusions

The water wave field generated by a pulsating point source which is located in a region with variable bottom bathymetry has been studied. The main features of the problem are: (i) The far-field water depths for $x \to +\infty$ and $x \to -\infty$ are different. (ii) No asymptotic simplification based on deep or shallow water assumptions are made. The main results of this work are: (a) The problem is reformulated as a boundary value problem for a system of second-order ODEs in a bounded interval $(a \leq x \leq b)$. (b) The far-field wave pattern of the initial problem presents a strong azimuthal anisotropy, its main features being a shadow zone centered in area of variable bathymetry. As $x \to \pm\infty$, in sectors not including the variable bathymetry region, the asymptotic behaviour of the far field approaches the standard one, corresponding to the constant depth h_1 or h_3, respectively.

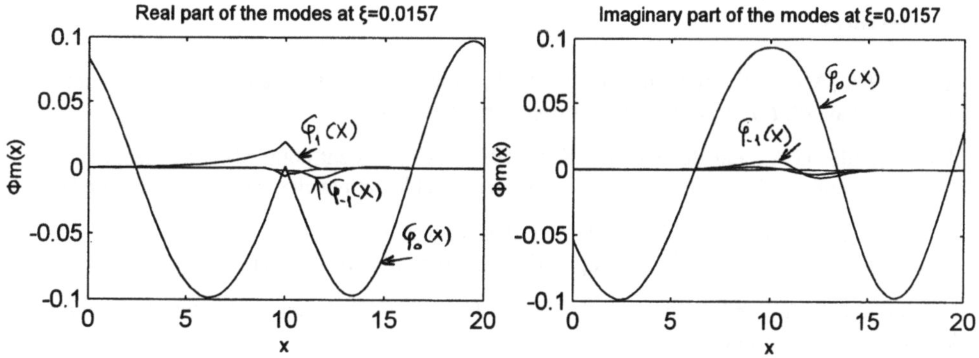

Figure 2. Modal amplitude functions $\varphi_n(x;\xi)$, for $n=-1,0,1,2$, versus x, at $\xi = 0.0157$. Source location $x_0 = 10\text{m}$.

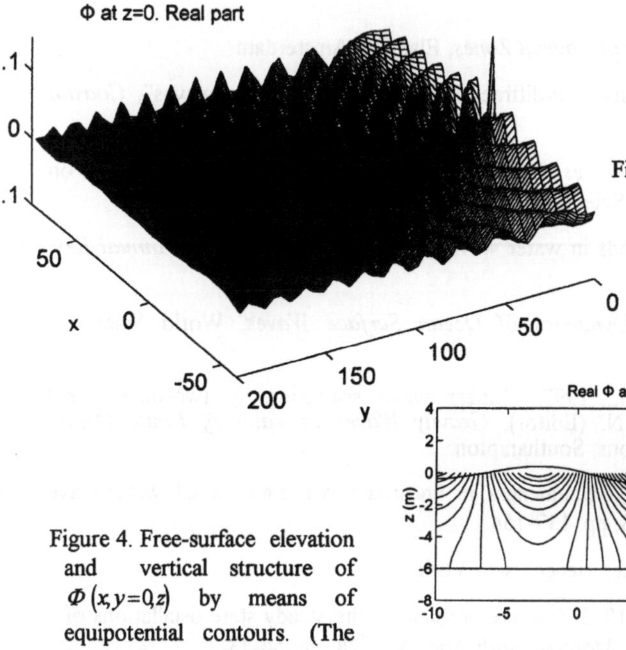

Figure 3. Real part of $\Phi(x,y,z=0)$ obtained by numerical inversion of $\varphi(x,z;\xi)$. Source location $(x_0, y_0, z_0) = (10\text{m}, 0, -1\text{m})$.

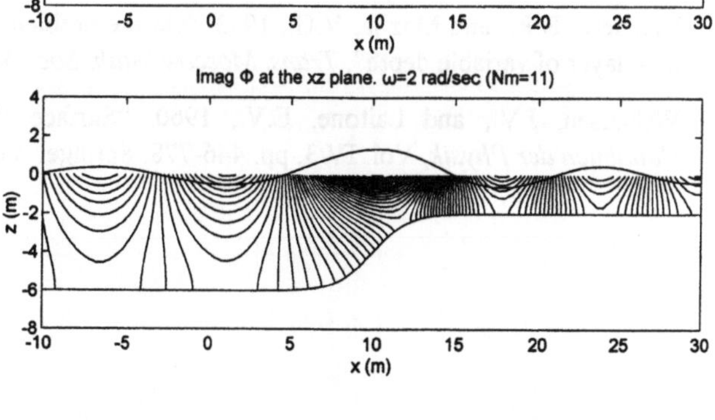

Figure 4. Free-surface elevation and vertical structure of $\Phi(x, y=0, z)$ by means of equipotential contours. (The free-surface elevation is exaggerated for clarity).

7. References

Aranha, J.A., Mei, C.C., and Yue, D.K.P., 1979, "Some properties of a hybrid element method for water waves", *Int. J. Num. Methods Eng.*, pp. 1627-1641.

Athanassoulis, G.A., and Belibassakis, K.A., 1997, "A consistent coupled-mode theory for the propagation of small-amplitude water waves over variable bathymetry regions", (submitted).

Bai, K.J., and Yeung, R., 1974, "Numerical solutions of free-surface and flow problems", *Proc. 10th Symp. Naval Hydrodyn.*, pp. 609-641, Office of Naval Research.

Coddington, E.A., and Levinson, N., 1955, *Theory of Ordinary Differential Equations*, McGraw-Hill, New York.

Fawcett, J.A., 1992, "A derivation of the differential equations of coupled-mode propagation", *J. Acoust. Soc. Am.*, Vol. 92(1), pp. 290-295.

Fitz-Gerald, G.F., and Grimshaw, R.H.J., 1979, "A note on the uniqueness of small-amplitude water waves travelling in a region of varying depth", *Proc. Camb. Phil. Soc.*, Vol. 86, pp. 511-519.

Massel, S.R., 1989, *Hydrodynamics of Coastal Zones*, Elsevier, Amsterdam.

Massel, S.R., 1993, "Extended Refraction-diffraction equation for surface waves", *Coastal Eng.*, Vol. 19, pp. 97-126.

Massel, S.R., 1996, *Ocean Surface Waves: Their Physics and Prediction*, Advanced Series on Ocean Engineering, Vol. 11, World Scientific, Singapore.

Mei, C.C., 1978, "Numerical methods in water wave diffraction and radiation", *Annual Rev. Fluid Mech.*, Vol. 10, pp. 393-416.

Mei, C.C., 1983, *The Applied Dynamics of Ocean Surface Waves*, World Scientific, Singapore.

Porter, D. and Chamberlain, P.G., 1997, "Linear wave scattering by two-dimensional topography", Ch. 2 in: Hunt, J.N. (Editor), *Gravity Waves in water of Finite Depth*, Computational Mechanics Publications, Southampton.

Simon, M.J. and Ursell, F., 1984, "Uniqueness in linearized two-dimensional water-wave problems", *J. Fluid Mech.*, Vol. 148, pp. 137-154.

Stoker, J.J., 1957, *Water Waves*, Interscience, New York.

Vainberg, B.R., and Maz'ja, V.G., 1973, "On the problem of the steady state oscillations of a fluid layer of variable depth", *Trans. Moscow Math. Soc.*, Vol. 28, pp. 56-73.

Wehausen, J.V., and Laitone, E.V., 1960, "Surface Waves", In: Flugge, W. (Editor), *Handbuch der Physik*, Vol. IV/3, pp. 446-778, Springer-Verlag, Berlin.

NEW MAPPED WAVE INFINITE ELEMENT AND DIFFRACTION OF WAVES BY ELLIPTICAL CYLINDERS OF VARYING ASPECT RATIO

Jacqueline A. Bettess
Information Technology Service
University of Durham
Science Laboratories, South Road, Durham, DH1 3LE, UK

Peter Bettess
School of Engineering
University of Durham
Science Laboratories, South Road, Durham, DH1 3LE, UK

1. Introduction

The paper describes the extension of existing mapped infinite wave theory, (or *unconjugated* infinite element theory) to deal with non-circular meshes of infinite elements. The method is illustrated by results for the diffraction of waves by an elliptical cylinder. The waves are governed by Helmholtz equation, and the Sommerfeld radiation condition. The two main aspects are:

(i) The extension of the existing theory of infinite wave elements (which should also be applicable to wave envelope, or conjugated elements) to the case when the elements are not placed strictly radially. This involves modifications to the element theory, which are given below.

(ii) An example of the results for an elliptical cylinder diffraction problem, and comparison with analytical solution.

2. Background Wave Theory

We consider a body of fluid of varying depth, with a free surface, in which is placed some object. Waves are incident from some direction. They impinge upon the object and are scattered by it, and radiate away to infinity. The general arrangement is shown in many publications and will not be repeated.

2.1 The governing equation

The waves are governed by the shallow water wave equation,

$$\frac{\partial}{\partial x}\left(h\frac{\partial \eta}{\partial x}\right)+\frac{\partial}{\partial y}\left(h\frac{\partial \eta}{\partial y}\right)=\frac{\partial^2 \eta}{\partial t^2}\frac{1}{g} \tag{1}$$

In equation (1), η is the wave elevation, x and y are the cartesian co-ordinates in the horizontal plane of the undisturbed free surface, (z is taken as vertically upwards), g is acceleration due to gravity, $h(x,y)$ is the water depth and t is time. This equation will not be derived here.

T.L. Geers (ed.), IUTAM Symposium on Computational Methods for Unbounded Domains, 33–42.

For periodic problems, the time dependence of the variable η, and other variables, can be written as

$$\eta(x,y,t) = R_e\eta(x,y)\exp(i\omega t) \tag{2}$$

where R_e denotes the real part of a complex number. In equation (2), ω is the angular frequency, and T, the wave period is given by $2\pi/\omega$. Other important parameters are the wave length, λ, and the wave number, $k = 2\pi/\lambda$. The *dispersion* relation

$$\omega^2 = gk \tanh kh \tag{3}$$

relates the frequency to the acceleration due to gravity, g, depth, h and wavenumber, k. In the frequency domain, for constant depth, the wave equation, (1), becomes Helmholtz equation, (4).

$$\frac{\partial^2\eta}{\partial x^2} + \frac{\partial^2\eta}{\partial y^2} + k^2\eta = \nabla^2\eta + k^2\eta = 0 \tag{4}$$

2.2 Incident and Scattered Waves

The wave can be separated into an *incident* wave and a *scattered* wave. This gives

$$\eta_{total} = \eta_{incident} + \eta_{scattered} \tag{5}$$

By 'scattered' we mean waves that have been scattered, diffracted or reflected from a fixed object in the water. The incident wave gives rise to a line integral on a boundary in the problem, which may be the boundary of an object which is diffracting the waves, or some other boundary.

2.3 Radiation Conditions

After impinging on the cylinder, the scattered waves radiate away, towards infinity. The boundary condition to be applied to these outgoing waves for periodic problems was given by Sommerfeld and Rellich. A summary of further information about this radiation condition is given in [Bettess, 1992]. For our case the radiation condition is written

$$\lim_{r\to\infty} \sqrt{r}\left(\frac{\partial\eta}{\partial r} + ik\eta\right) = 0 \tag{6}$$

The theory of finite elements for such problems is well understood, and is described, in for example [Bettess, Emson and Chiam, 1984].

3. Theory of Mapped Periodic Infinite Elements

We now deal with the application of the mapped infinite element to periodic wave problems.

3.1 Zienkiewicz mapped infinite element

The theory for this mapping is well known, but for reference will be briefly summarised here. Further details are given in [Bettess, 1992], [Astley *et al.*, 1991], [Bettess and Zienkiewicz, 1977], [Bettess *et al.*, 1984], [Zienkiewicz and Bettess, 1975], [Zienkiewicz *et al.*, 1978], [Zienkiewicz and Bettess, 1978], [Zienkiewicz *et al.*, 1981], [Zienkiewicz *et al.*, 1983], [Zienkiewicz *et al.*, 1985].

The mapping between the finite element domain and the mapped infinite element domain can be summarised as follows:

$$r = \frac{A}{1-\xi} \quad \text{and} \quad \xi = 1 - \frac{A}{r} \tag{7}$$

Two-dimensional exterior wave problems have solutions in which the wave amplitude decays like $1/\sqrt{r}$, and this can be acommodated in the mapped infinite element theory.

3.2 Introducing the wave component

In two-dimensional exterior domains the asymptotic solution to Helmholtz's equation can be described by a series of combined Hankel and trigonometric functions, the simplest solution to Helmholtz's equation being $H_0(kr)$. For large r the zeroth-order Hankel function oscillates roughly like $\cos(kr) + i\sin(kr)$ while decaying in magnitude as $r^{-1/2}$. We will use a series of terms $1/r$, $1/r^2$ etc., generated by the mapping, multiplied by $r^{1/2}$ and the periodic component $\exp(ikr)$ in order to model the $r^{-1/2}$ decay. The shape function is therefore the original finite element shape function, multiplied by an exponential term, giving

$$N(\xi,\eta) = P(\xi,\eta) r^{1/2} \exp(ikr) \tag{8}$$

On substituting the mapping expression, 7, the shape function becomes

$$N(\xi,\eta) = P(\xi,\eta)\left(\frac{A}{1-\xi}\right)^{1/2} \exp\left[\frac{ikA}{1-\xi}\right] \tag{9}$$

where $P(\xi,\eta)$ is the shape function of the parent finite element. To ensure continuity, with finite elements a factor $\exp(-ikA/2)$ is introduced so that the shape function becomes

$$N(\xi,\eta) = P(\xi,\eta)\left(\frac{A}{1-\xi}\right)^{1/2} \exp\left[\frac{-ikA}{2}\right] \exp\left[\frac{ikA}{1-\xi}\right] = C.P(\xi,\eta)\left(\frac{A}{1-\xi}\right)^{1/2} \exp\left[\frac{ikA}{1-\xi}\right] \tag{10}$$

where

$$C = \exp\left[\frac{-ikA}{2}\right] \tag{11}$$

and its derivatives are

$$\frac{\partial N}{\partial \xi} = C\left(\frac{A}{1-\xi}\right)^{1/2} \exp\left[\frac{ikA}{1-\xi}\right]\left[\frac{\partial P}{\partial \xi} + PikA\frac{1}{(1-\xi)^2} + \frac{1}{2(1-\xi)}\right] \tag{12}$$

$$\frac{\partial N}{\partial \eta} = C\frac{\partial P}{\partial \eta}\left(\frac{A}{1-\xi}\right)^{1/2} \exp\left[\frac{ikA}{1-\xi}\right] \tag{13}$$

The functional corresponding to Helmholtz's equation in three dimensions is given by

$$F = \int_\Omega \left[\left(\frac{\partial \phi}{\partial x}\right)^2 + \left(\frac{\partial \phi}{\partial y}\right)^2 - k^2\phi^2\right] d\Omega - \int_\Gamma \phi \frac{\partial \phi}{\partial n} d\Gamma \tag{14}$$

The evaluation of the integral 14 can be carried out using the special integration schemes, details of which are given in [Bettess, 1992], [Zienkiewicz *et al.*, 1981] and [Zienkiewicz *et al.*, 1985]. So far, these integration formulas give accurate and reliable results.

4. Elliptical Meshes

In the results given to date, in [Zienkiewicz et al., 1985], it has been assumed that the mapping parameter, A, in equation (7), is a constant, for the problem, which implies that all the meshes are circular, and radiate from a single pole. In solving wave diffraction problems where the aspect ratio of the problem is not close to unity, this can lead to large meshes of finite elements*. This has been criticised as a defect of the method by [Burnett, 1994], who has developed a special infinite element in ellipsoidal co-ordinates, which removes this difficulty. His infinite element is, however, non-conforming with the finite elements. In this paper, we extend the theory of the infinite elements described above so that the mapping parameter, A, is a function of η. This gives one additional term in the shape function derivative, and is thus simple to program. The efficacy is then shown by comparisons between the new and old versions of the infinite element with the analytical solution for elliptical cylinder diffraction.

5. New Mapped Wave Infinite Element

The first step is to write $A = A(\eta)$. This allows different infinite elements to have different 'poles', and different rates of decay, while remaining conforming, as shown in Figure 1. Within the infinite element, the parameter A is interpolated in the η direction. The values of A are determined on the lines $\eta = -1, 0, +1$, (for a quadratic element). On each line the x and y co-ordinates are used to determine A, from the mapping function, (7). Let these values be labelled A_1, A_2 and A_3. Then in the interior of the infinite element, at a location η, the value of A, and its derivatives, are given by

$$A = \sum_{i=1}^{3} N_i A_i \qquad \text{and} \qquad \frac{\partial A}{\partial \eta} = \sum_{i=1}^{3} \frac{\partial N_i}{\partial \eta} A_i \tag{15}$$

It is also convenient to define a variable, $\alpha = 2kA$. Now the only alteration to the shape function derivative is an additional term in the η derivative in equation (13), which becomes

$$\frac{\partial N}{\partial \eta} = C \left[\frac{\partial P}{\partial \eta} + i\frac{1}{4}\left(\frac{1+\xi}{1-\xi}\right) P(\xi,\eta)\frac{\partial \alpha}{\partial \eta} \right] \left(\frac{A}{1-\xi}\right)^{1/2} \exp\left[\frac{ikA}{1-\xi}\right] \tag{16}$$

Because the integration abscissæ are dependent upon A, it is now necesary to evaluate the abscissæ and weights for all the different values of A used in the analysis, instead of only once. Fortunately, this is not a large computational overhead and does not affect the solution times to a large extent. It would however, be good to find a way to avoid this extra computation. The rest of the evaluation of the element matrix and its use in the program, remains unchanged. Techniques similar to that presented here, in which the mapping parameter varies, although with a different notation, have been used in *wave envelope* infinite elements, by [Astley *et al.*, 1994] and [Cremers *et al.*, 1994].

6. Results

Figure 1. illustrates an elliptical cylinder of aspect ratio 2, with a relatively coarse mesh of twelve finite elements surrounded by twelve infinite elements. The product of the wave number, k, and the semi-major axis, a, is $ka = 2$ and the waves are incident along the x axis from $x = -\infty$, $\theta_i = 0\,\text{deg}$.

---

* It would also be possible to have A varying between elements, but this makes the infinite elements non-conforming with each other, and seems to cause significant errors, see later.

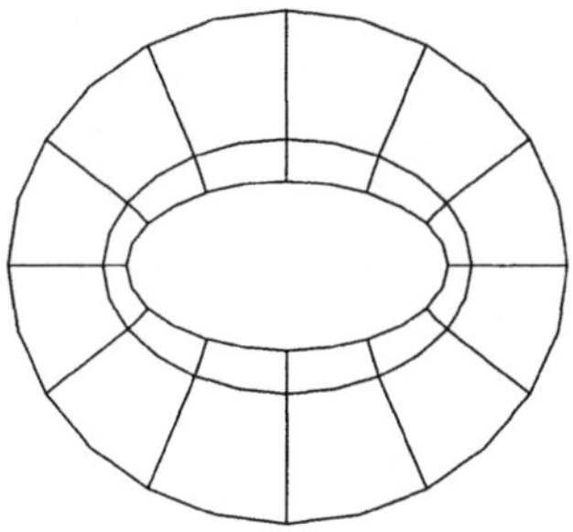

Aspect ratio 1:2

Figure 1 Elliptical mesh of finite and infinite elements

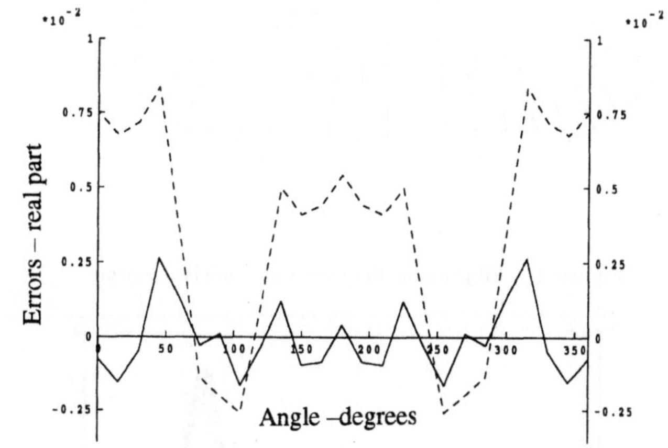

Figure 2 Errors relative to incident wave amplitude in real part of wave elevation as a function of angle around cylinder

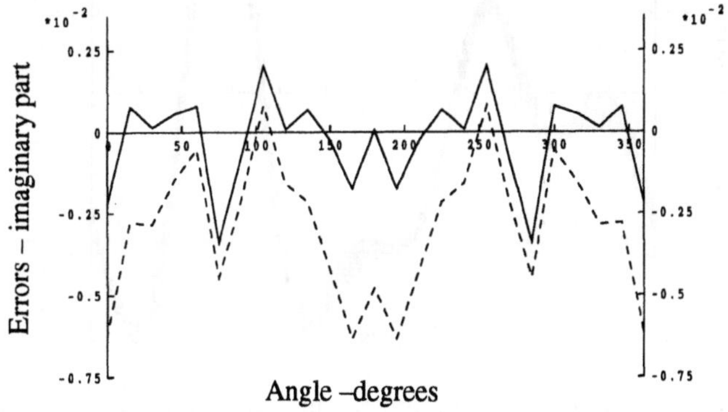

Figure 3 Errors relative to incident wave amplitude in imaginary part of wave elevation as a function of angle around cylinder

Figures 2 and 3 show the real and imaginary parts of the errors for the mesh above and with the value of A calculated by two different methods. The dotted line shows the errors when A is constant for each element, and the solid line shows the errors when A varies within the infinite elements according to the theory presented.

The fine mesh shown in Figure 4 is for an ellipse of aspect ratio 1:10 and contains 9 elements radially and 60 circumferentially. The product of the wave number, k, and the semi-major axis, a, is $ka = 4$ and the waves are incident from $-\infty$, at an angle of $\theta_i = 30$ deg.

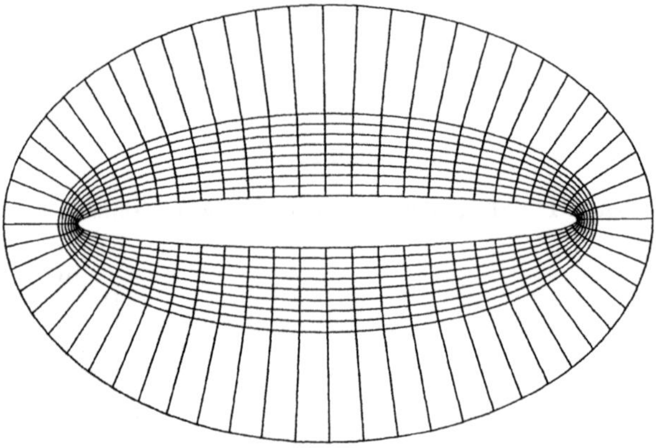

Aspect ratio 1:10

Figure 4 Elliptical mesh of finite and infinite elements

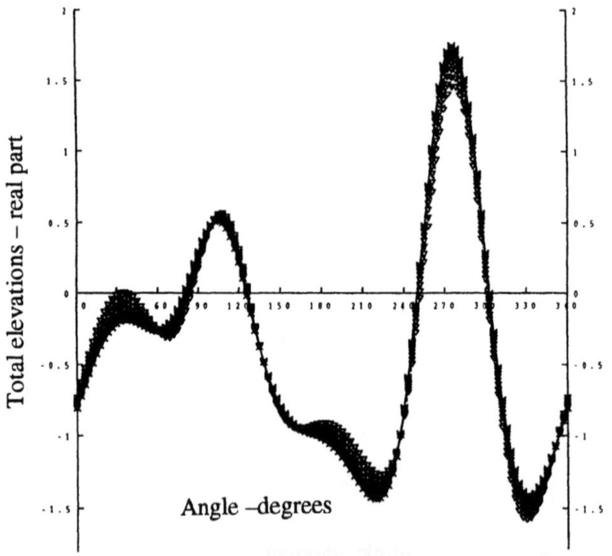

Figure 5 Real component of the wave elevations on ellipse as a function of angle around cylinder

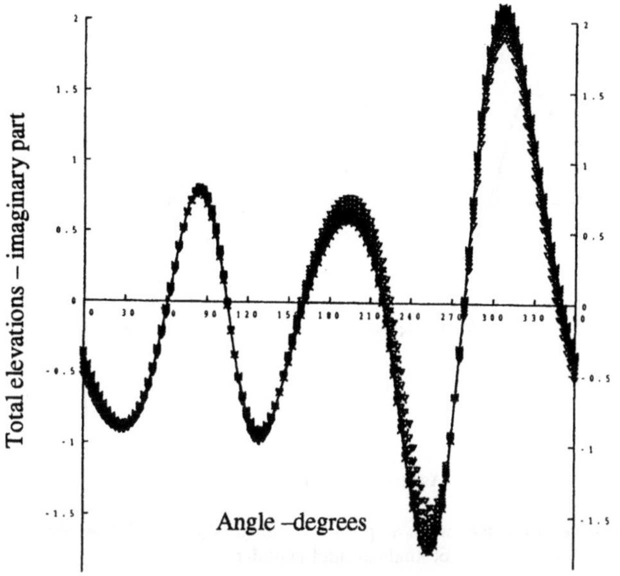

Figure 6 Imaginary component of the wave elevations on ellipse as a function of angle around cylinder

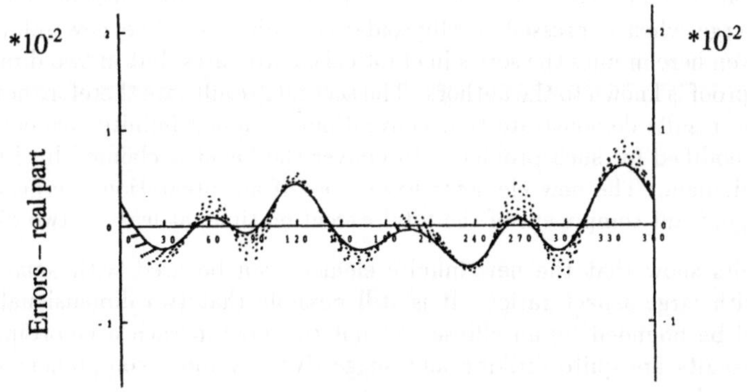

Aspect ratio 1:10

Figure 7 Errors relative to incident wave amplitude in real part of wave elevation as a function of angle around cylinder

Figures 5 and 6 show the real and imaginary components of the wave elevation on the surface of the cylinder. The solid line is on the surface of the cylinder with the dotted values being on concentric ellipses radiating outwards from the surface. Figures 7 and 8 show the errors in the real and imaginary part of the numerical solution around the ellipse. Again, the solid line is on the surface of the cylinder with the dotted values being on concentric ellipses radiating outwards from the surface. The errors on the ellipse surface can be seen to have a maximum value of about 1.2%.

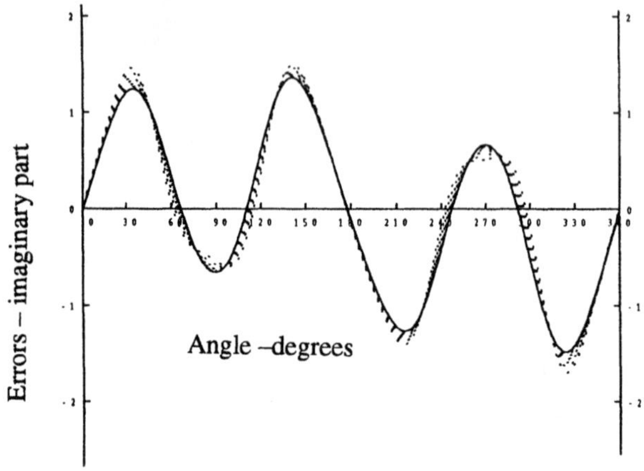

Aspect ratio 1:10

Figure 8 Errors relative to incident wave amplitude in imaginary part of wave elevation as a function of angle around cylinder

7. Conclusion

In absolute terms the results are very good, the highest error being only 1.2%. Burnett refers to a so far unpublished proof that solutions to similar three dimensional diffraction problems converge when expressed in ellipsoidal co-ordinates. The new infinite element formulation given here mimics the series in elliptical co-ordinates, but in two dimensions, for which no such proof is known to the authors. The accurate results are therefore not altogether surprising. The results demonstrate that conventional mapped infinite element techniques can easily be modified for such problems, to deliver the benefits claimed by Burnett, of a much smaller domain. The new elements have more of an integration overhead, although this is not a significant component of the total execution time, at least in two dimensions.

The results show that the new infinite element can be used with *some* confidence in problems with large aspect ratios. It is still possible that two dimensional diffraction problems could be bounded by an ellipse, yet not converge in such a co-ordinate system. However the results are quite striking and suggestive. A more comprehensive study of convergence is underway.

8. References

Bettess, P., 1992, *Infinite Elements*, Penshaw Press, Sunderland, U.K., ISBN 0-9518806-0-8.

Astley, R.J., Bettess, P. and Clark, P.J., 1991, Letter to the editor concerning Zienkiewicz *et al.* (1985), *Int. Journ. Num. Meth. Engng.*, Vol. 32, pp 207 - 209.

Astley, R. J. Macaulay G. J. and Coyette J. P. , 1994, 'Mapped Wave Envelope Elements for Acoustical Radiation and Scattering', *Journal of Sound and Vibration*, **170**(1), pp 97 - 118.

Cremers, L. Fyfe K. R. and Coyette J. P. , 1994, 'A Variable Order Infinite Acoustic Wave Envelope Element', *Journal of Sound and Vibration*, **171**(4), pp 483 - 508.

Bettess, P. and Zienkiewicz, O.C., 1977, "Diffraction and refraction of surface waves using finite and infinite elements", *Int. J. Num. Meth. Eng.*, Vol. 11, pp. 1271-1290.

Figure 9 shows the contours of the wave elevations.

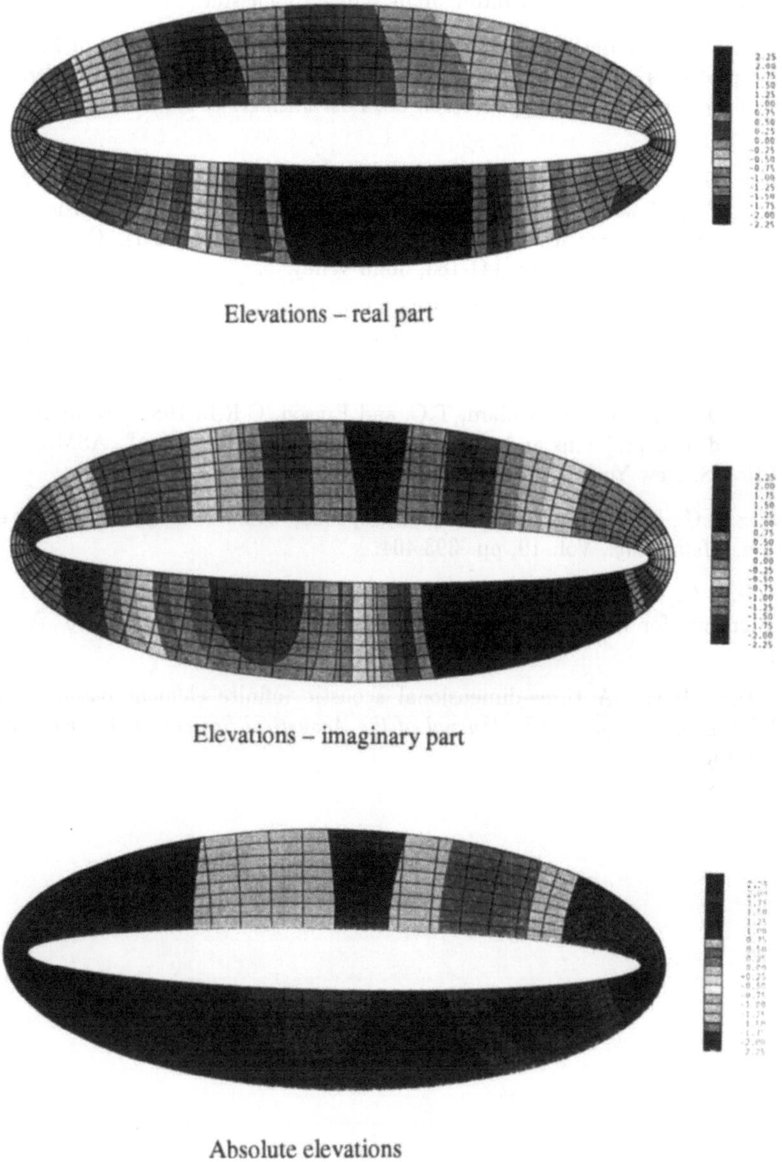

Aspect ratio 1:10

Figure 9 Contour plot of wave elevations

Bettess, P., Emson, C.R.I. and Chiam, T.C., 1984, "A New Mapped Infinite Element for Exterior Wave Problems", Chapter 17 of *Numerical Methods in Coupled Systems*, eds R. W. Lewis, P. Bettess and E. Hinton, John Wiley, Chichester.

Zienkiewicz, O.C. and Bettess, P., 1975, "Infinite elements in the study of fluid structure interaction problems", *Proc. 2nd Int. Symp. on Comp. Methods Appl. Sci.*, Versailles, also published in *Lecture Notes in Physics*, Vol. 58, Eds. J.Ehlers *et al.*, Springer-Verlag, Berlin, 1976.

Zienkiewicz, O.C., Bettess, P. and Kelly, D.W., 1978, "The Finite Element Method for Determining Fluid Loadings on Rigid Structures: Two- and Three- Dimensional Formulations", Chapter 4 of *Numerical Methods in Offshore Engineering*, Eds. O. C. Zienkiewicz, R. W. Lewis and K. G. Stagg, pp 141-183, John Wiley.

Zienkiewicz, O.C. and Bettess, P., 1978, "Fluid Structure Interaction ", Chapter 5 of *Numerical Methods in Offshore Engineering*, Eds. O. C. Zienkiewicz, R. W. Lewis and K. G. Stagg, John Wiley.

Zienkiewicz, O.C., Bettess, P., Chiam, T.C. and Emson, C.R.I., 1981, "Numerical methods for unbounded field problems and a new infinite element formulation", ASME, *AMD*, Vol. 46, pp. 115-148, New York.

Zienkiewicz, O.C., Emson, C.R.I. and Bettess, P., 1983, "A novel boundary infinite element", *Int.J. Num. Meth. Eng.*, Vol. 19, pp. 393-404.

Zienkiewicz, O.C., Bando, K., Bettess, P., Emson, C.R.I. and Chiam, T.C., 1985, "Mapped Infinite Elements for Exterior Wave Problems", *Int. J. Num. Meth. Eng.*, Vol. 21, pp. 1229-1251.

Burnett, D.S., 1994, "A three-dimensional acoustic infinite element based on a prolate spheroidal multipole expansion", *Journal of the Acoustical Society of America*, Vol. 95(5), pp. 2798-2816.

ABSORBING BOUNDARIES FOR ACOUSTIC WAVE PROPAGATION PROBLEMS[1]

Jacobo Bielak†, Loukas F Kallivokas†, and Richard C. MacCamy‡
†*Computational Mechanics Laboratory,*
Department of Civil and Environmental Engineering
‡*Department of Mathematical Sciences*
Carnegie Mellon University, Pittsburgh, PA 15213, USA

1. Introduction

In this article we present a summary of recent efforts at Carnegie Mellon University devoted to the development of efficient and accurate finite element procedures for the analysis in the frequency and time domains of large-scale acoustic and other wave propagation problems in unbounded domains.

One goal of our work has been to develop a methodology that will allow one to solve transient and steady-state harmonic radiation and scattering problems using standard tools created originally for problems defined over interior domains. We favor finite element methods for their ability to efficiently resolve multiscale phenomena, the ease with which they handle traction boundary conditions, and their firm theoretical foundation.

In order to apply the finite element methodology to exterior problems, it is necessary to truncate the computational domain by means of a fictitious boundary on which one imposes an approximate radiation, or absorbing boundary, condition to minimize spurious reflections. A desirable feature of this absorbing boundary is that it be readily implementable into existing software and that the errors incurred be sufficiently small to maintain the desired accuracy of the calculations for a given application. Clearly, different applications demand different degrees of approximation. This point will be illustrated later by means of examples.

In the next section we consider a canonical exterior Neumann problem in acoustics and describe a family of absorbing boundary conditions of increasing accuracy. A parametric study is then conducted to assess the actual accuracy of the various approximations and to examine additional errors that arise from their finite element implementation. In a subsequent section we describe briefly an elastic wave propagation problem that arises in earthquake ground motion modeling in large closed basins and illustrate how for the particular application considered a low order absorbing boundary condition is sufficient for obtaining satisfactory results.

2. Exterior Neumann Problem in Acoustics

We consider the prototype problem of a closed cavity in an infinite acoustic fluid, whose boundary is subjected to a prescribed acceleration, as shown in Fig. 1a.

The mathematical problem consists in finding the pressure p such that

$$\ddot{p}(\boldsymbol{x},t) = c^2 \Delta p(\boldsymbol{x},t), \quad \boldsymbol{x} \in \Omega, \quad t \geq 0, \tag{1a}$$

$$p_\nu(\boldsymbol{x},t) = f_N(\boldsymbol{x},t), \quad \boldsymbol{x} \in \Gamma, \quad t > 0 \tag{1b}$$

[1]Portions of this article have appeared in *Comp. Methods Appl. Mech. Engrg.*, 147 (1997) 235–262, and 152 (1998) 85–102; reprinted with permission from Elsevier Science.

T.L. Geers (ed.), IUTAM Symposium on Computational Methods for Unbounded Domains, 43–51.

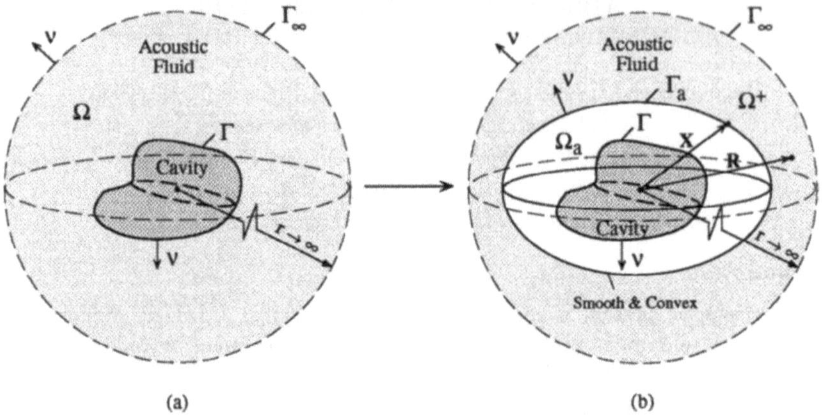

Figure 1: (a) Model of cavity surrounded by infinite acoustic fluid; (b) Reduced model with finite fluid region Ω_a and absorbing boundary.

$$\lim_{r\to\infty} r(p_r + \frac{1}{c}\dot{p}) = 0, \quad \text{and} \tag{1c}$$

$$p(\boldsymbol{x},0) = 0, \quad \dot{p}(\boldsymbol{x},0) = 0, \quad \boldsymbol{x} \in \bar{\Omega}. \tag{1d}$$

In these equations c is the velocity of wave propagation, and f_N is a prescribed function. Condition (1c) is the Sommerfeld radiation condition, and as indicated by (1d) the system is taken to be initially at rest.

The main difficulty associated with the solution of (1) is the need to ensure that the radiation condition (1c) is satisfied at infinity. Instead of dealing with an infinite domain one truncates the exterior region by introducing an artificial boundary Γ_a that contains Γ in its interior, as shown in Fig. 1b. For the solution p to coincide with that of the original problem within the truncated region Ω_a, it is necessary to specify an appropriate boundary condition on Γ_a that will ensure that the outgoing waves crossing Γ_a are undisturbed. The exact boundary condition, however, is nonlocal both in space and in time, and is difficult to describe explicitly.

We use concepts from geometrical optics to derive approximate, local, boundary conditions on Γ_a. Here we merely state our results. Further details are given by Kallivokas et al [1997a] in these proceedings; a detailed derivation of the absorbing boundary conditions, and a full description of the numerical results may be found in Kallivokas et al [1997b].

3. Absorbing Boundaries

By introducing an asymptotic expansion for the Laplace transform of the pressure p in terms of increasing powers of the inverse of the Laplace transform variable, one can construct successive approximations for the normal derivative p_ν of the pressure in terms of time and tangential derivatives of p. The first three approximations are:

$$\text{0th order}: \; p_\nu = -\frac{1}{c}\dot{p}, \tag{2a}$$

$$\text{1st order}: \; p_\nu = -\frac{1}{c}\dot{p} + H\,p, \tag{2b}$$

$$\begin{aligned}\text{2nd order}: \; \dot{p}_\nu + \gamma\,p_\nu = & -\frac{1}{c}\ddot{p} + \left(H - \frac{\gamma}{c}\right)\dot{p} + H\,\gamma\,p \\ & + \frac{c}{2}\left\{\frac{1}{\sqrt{h}}\left\{\left[\frac{1}{\sqrt{h}}(h_{22}\,p_u - h_{12}\,p_w)\right]_u + \left[\frac{1}{\sqrt{h}}(-h_{12}\,p_u + h_{11}\,p_w)\right]_w\right\}\right. \\ & \left. + (H^2 - K)p\right\}.\end{aligned} \tag{2c}$$

In these equations H and K are the mean and Gaussian curvatures of Γ_a, h_{ij} is the i, j component of the Eucledian metric tensor on Γ_a, and γ is an arbitrary constant introduced for stability. γ, which represents numerical damping, must satisfy the condition

$$\gamma \geq \gamma_{cr} = \frac{c}{2}\left(-\frac{H^2 - K}{H}\right)_{max}.$$

The first condition, (2a), is the well known plane wave approximation (PWA) [Mindlin and Bleich, 1953]. For the particular case of a sphere, (2b) reduces to the also well-known exact condition for radially symmetric waves [Bayliss and Turkel, 1980], and (2c) coincides with the second-order Bayliss and Turkel [1980] condition provided one chooses $\gamma = \gamma_0 = c/a$, where a is the radius of the sphere.

4. Numerical Examples

To assess the accuracy of the individual approximations, we consider the Neumann problem over a sphere with a concentric spherical absorbing boundary, as shown in Fig. 2.

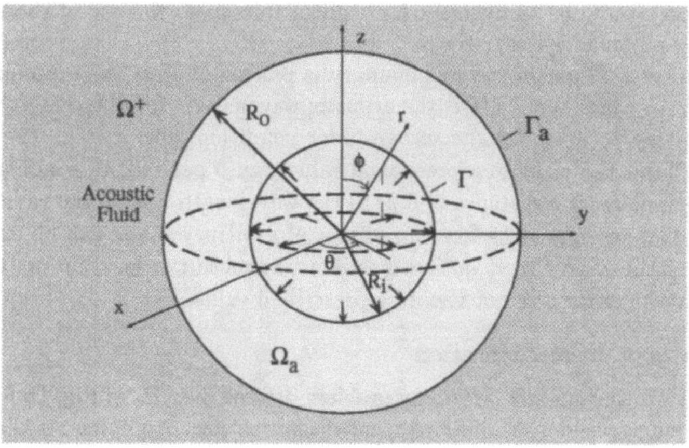

Figure 2: Model for studying radiation from a spherical cavity.

For a prescribed steady-state radial acceleration field on the inner boundary of the cavity of the form:

$$p_r = \rho\, A_0\, P_n^{(m)}(\cos\phi) \cos m\theta\, e^{i\omega t}, \quad \text{at } r = R_i, \tag{3}$$

where A_0 is the amplitude of the prescribed acceleration field and $P_n^{(m)}$ is the associated Legendre function of the first kind, degree n and order m, it can be shown that the exact solution can be expressed as:

$$p = \rho\, A_0\, R_i\, Q_{ex}(r/R_i, kR_i, n)\, P_n^{(m)}(\cos\phi)\, \cos m\theta, \tag{4}$$

in which Q_{ex} represent the radial component of p with frequency measured through the wave number $k = \omega/c$. One can also obtain the exact solutions to the approximate problems corresponding to the three conditions (2). This can be written in the form:

$$p_{app} = \rho\, A_0\, R_i\, Q_{app}^{(j)}(r/R_i, R_o/R_i, kR_i, n)\, P_n^{(m)}(\cos\phi)\, \cos m\theta, \tag{5}$$

in which Q_{app}^{j} denotes the radial component of the solution, and the superscript j refers to the order of the approximation. The accuracy of the various approximations can then be measured by the relative error E, defined by:

$$E = \frac{|Q_{ex} - Q_{app}^{(j)}|}{|Q_{ex}|}, \quad \text{for } j = 0, 1, 2. \tag{6}$$

To illustrate how E varies with the location of the absorbing boundary for various modes n, Fig. 3 depicts E versus R_o/R_i (Fig. 2) for the first six modes (n = 1,2,...,6), and for a fixed frequency of excitation, corresponding to the various boundary conditions. Three different values of γ are considered for the second-order boundary: $\gamma = 0$, $\gamma = \gamma_0$, and $\gamma = \gamma_{opt}$, where γ_{opt} is such that E is minimum for a fixed n, and wavenumber kR_i. Results for $n = 0$ are not included since all conditions, except for the zeroth order, are exact. Notice that for $n = 1$ the second-order conditions are also exact. In general, the errors decrease with the order of the approximation. Conversely, for a given tolerance, the use of the second-order boundary requires a smaller buffer zone.

In Fig. 4, the error E on the cavity boundary is plotted against the normalized annulus size $(R_o - R_i)/\lambda$, in which $\lambda = 2\pi/k$ is the acoustic wavelength, for different values of n and kR_i. All the calculations are for the second-order condition with $\gamma = \gamma_0$. It can be seen that in order to limit the error to a prescribed value, say 5 percent, it is sufficient to place the absorbing boundary at a distance of 0.24 of the wavelength λ from the cavity boundary. Since any wave system emanating from a radiator of arbitrary shape can be expressed as a series of terms of the form (3), it is then possible to choose the location of the absorbing boundary so that the error will not exceed a prescribed value.

5. Finite Element Implementation

We now return to the general Neumann problem defined over Ω_a in Fig. 1b for a cavity of arbitrary shape and consider its finite element discretization. Applying standard Galerkin ideas to the wave equation leads to a term that involves the normal derivative of the pressure on Γ and on Γ_a. Hence, the 0th- and 1st-order approximations (2a) and (2b) can be readily incorporated into the formulation by replacing p_ν by the right sides of (2a) and (2b). The

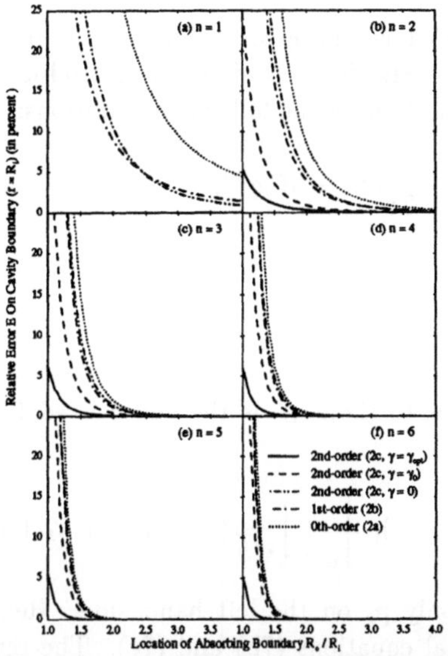

Figure 3: Relative error on the cavity boundary as a function of the location of the absorbing boundary R_o/R_i; $kR_i = 0.5$; various harmonics; various absorbing conditions.

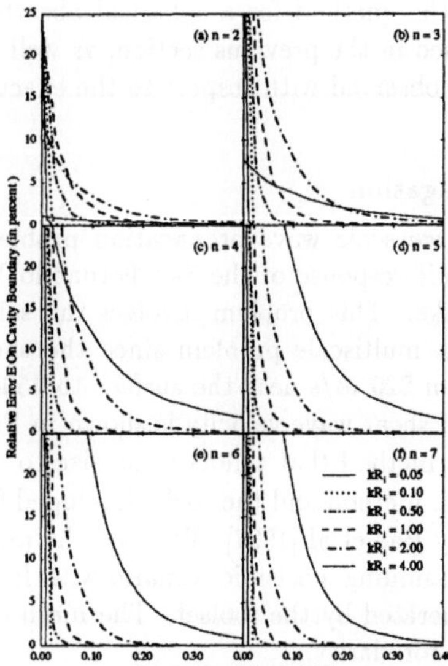

Figure 4: Relative errors on the cavity boundary as function of the annulus size $(R_o - R_i)/\lambda$ due to (2c) with $\gamma = \gamma_0$; various harmonics; various frequencies kR_i.

2nd-order approximation (2c), however, involves both p_ν and its time derivative $\dot{p}_\nu$. To make (2c) amenable to standard integration techniques we introduce two auxiliary variables $q^{(1)}$ and $q^{(2)}$ and replace this condition by the following conditions:

$$-p_\nu = \frac{1}{c}\dot{p} - H\,p - \frac{c}{2\gamma}\mathcal{J}q^{(1)} - \frac{c}{2\gamma}(H^2 - K)q^{(2)}, \tag{7a}$$

$$\mathcal{J}p - \mathcal{J}q^{(1)} - \frac{1}{\gamma}\mathcal{J}\dot{q}^{(1)} = 0, \tag{7b}$$

$$p - q^{(2)} - \frac{1}{\gamma}\dot{q}^{(2)} = 0, \tag{7c}$$

which are entirely equivalent to (2c). $\mathcal{J}$ is the generalized Beltrami-Mitchell differential operator, defined by:

$$\mathcal{J}\cdot = \frac{1}{\sqrt{h}}\left\{\left[\frac{1}{\sqrt{h}}\left(h_{22}\,(\cdot)_u - h_{12}\,(\cdot)_w\right)\right]_u + \left[\frac{1}{\sqrt{h}}\left(-h_{12}\,(\cdot)_u + h_{11}\,(\cdot)_w\right)\right]_w\right\}. \tag{8}$$

Equation (7a) involves only p_ν on the left hand side. Hence, it can be implemented readily, as can the additional equations (7b) and (7c). The resulting spatially discretized equations are of the form:

$$\boldsymbol{M}\ddot{\boldsymbol{P}} + \boldsymbol{C}\dot{\boldsymbol{P}} + \boldsymbol{K}\boldsymbol{P} = \boldsymbol{F}, \tag{9}$$

in which the mass, damping, and stiffness matrices, $\boldsymbol{M}$, $\boldsymbol{C}$, and $\boldsymbol{K}$, are sparse and symmetric.

We have implemented this method for a spherical absorbing boundary and solved the canonical examples discussed in the previous section, as well as related transient problems. No visible differences were observed with respect to the exact, closed form, solutions of the corresponding problems.

6. Elastic Wave Propagation

As an example of a large scale wave propagation problem in an unbounded medium we consider next the seismic response of the San Fernando Valley to an aftershock of the 1994 Northridge Earthquake. This problem involves the solution of Navier's equation of elastodynamics. This is a multiscale problem since the shear wave velocity in the San Fernando Valley varies from 220 m/s near the surface to 4500 m/s for rock at depth. The near-surface distribution of shear wave velocity is shown in Fig. 5.

We use a finite element method that tailors mesh size to the local wavelength of propagating waves. Details of the method and the tools developed for implementing it on parallel computers are described by Bao et al [1997]. We used the mesh generator contained in our toolset to create a mesh, assuming a seismic scenario with frequencies up to 1.6 Hz. Figure 6 depicts the elements generated by the toolset. The mesh comprises 77 million elements, but many fewer are shown for clarity.

As can be seen from the figure, the density of nodes is highest in the softest soil, where the shortest wavelengths occur. Simple variations of the zeroth order (PWA) boundary conditions were used both on the side and bottom boundaries. Because the computational domain extends into the rock material in all directions, we expected that the ground motion within the basin would be much stronger than in the vicinity of the absorbing boundaries.

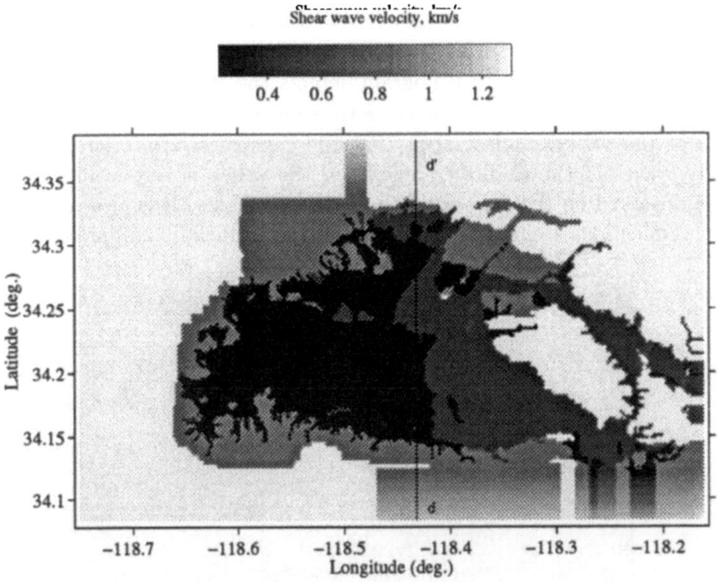

Figure 5: Near-surface distribution of shear-wave velocity in the San Fernando Valley; actual depth is 1m.

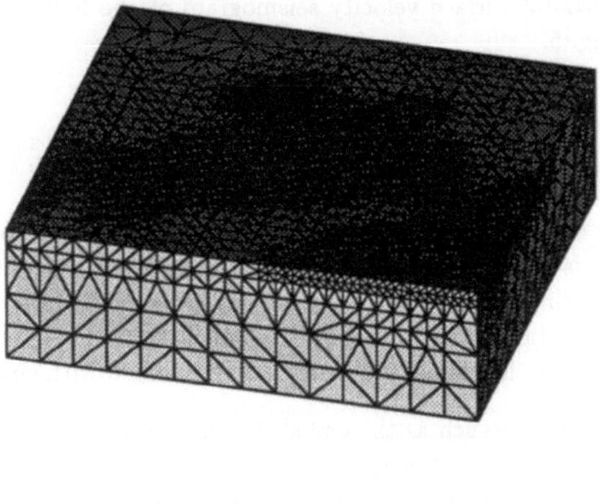

Figure 6: Tetrahedral element mesh of the San Fernando Valley. Maximum tetrahedral aspect ratio is 5.5. Again, for illustration purposes, the mesh shown is much coarser than those used for simulation.

Thus, small errors caused by spurious wave reflections would not affect the basin ground motion. Fig. 7 shows the E-W surface velocity component along the d-d' axis shown in Fig. 5; the column on the left of the seismogram depicts the shear wave velocity profile of the basin along the same axis. While it is clear that longer durations are associated with the deeper parts of the valley, it also seems that the constructive interference of surface and trapped body waves in the shallower regions of the valley is responsible for the stronger amplification observed on the surface overlying those regions. It is also noteworthy that no spurious wave reflections seem to be generated at the artificial boundaries.

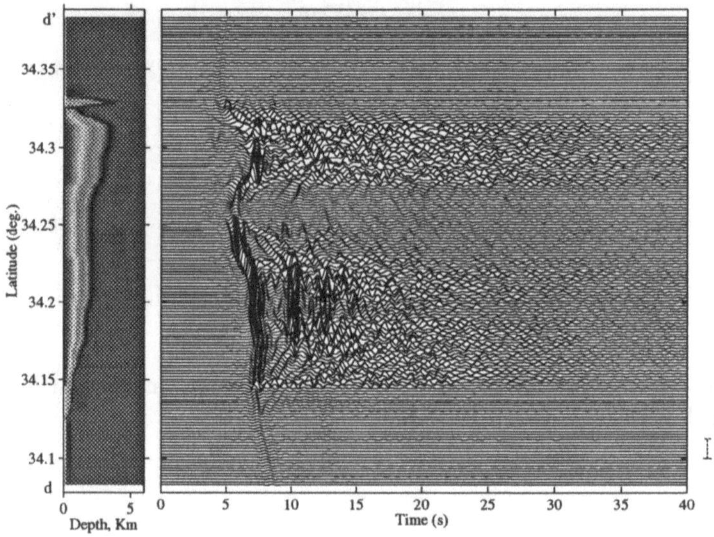

Figure 7: Horizontal surface velocity seismogram of the E-W component along the d-d′ axis shown in Fig. 5.

7. Conclusions

Several conclusions may be drawn from this study:

- The finite element implementation of the three absorbing boundary conditions discussed in this paper permits one to retain the familiar form of the discretized equations of motion with their sparsity and symmetry intact. The entire procedure lends itself to easy incorporation into existing finite element codes for interior problems. Transient and steady-state harmonic problems can be handled with equal ease.

- Parametric studies such as the one sketched here of the accuracy of various members of a family of absorbing boundary conditions can be used for choosing the location and particular member required to attain a prescribed accuracy.

- Only the spherical high-order artificial boundary has been implement to date in our work. For slender bodies, further economy of the computational domain can be expected from the use of an ellipsoidal or some other elongated absorbing boundary.

- Simple absorbing boundaries suffice for some applications.

8. Acknowledgment

Support for the work on elastic wave propagation was provided by the National Science Foundation's Grand Challenges in High Performance Computing and Communications program, under grant CMS-9318163. Computing services were provided by the Supercomputing Center under grant BCS-960001P.

9. References

[1] Bao, H.,Bielak J., Ghattas O., Kallivokas L. F., O'Hallaron D. R., Shewchuk J. R., and Xu J., 1998, "Large-Scale Simulation of Elastic Wave Propagation in Heterogeneous Media on Parallel Computers," *Comp. Methods Appl. Mech. Engrg.*, **152**, pp 85–102

[2] Bayliss A. and Turkel E., 1980, "Radiation Boundary Conditions for Wave-Like Equations," *Comm. Pure Appl. Math.*, **33**(6), pp 707–725

[3] Mindlin, R. D. and Bleich H. H., 1953, "Response of an Elastic Cylindrical Shell to a Transverse, Step Shock Wave, *ASME J. Appl. Mech.*, **20**, pp 189–195

[4] Kallivokas, L. F., Bielak J., and MacCamy R. C., 1997a, "Absorbing Boundaries of Arbitrary Shape for the Three-Dimensional Wave Equation," in *Proc. IUTAM Conference on Computational Methods for Unbounded Domains*

[5] Kallivokas, L. F., Bielak J., and MacCamy R. C., 1997b, "A Simple Impedance-Infinite Element for the Finite Element Solution of the Three-Dimensional Wave Equation in Unbounded Domains," *Comp. Methods. Appl. Mech. Engrg.*, **147**, pp 235–262

MULTIPOLE-BASED 3-D INFINITE ELEMENTS: AN ELLIPSOIDAL ACOUSTIC ELEMENT AND A SPHERICAL ELECTROMAGNETIC ELEMENT

David S. Burnett and Richard L. Holford

Bell Laboratories, Lucent Technologies, Whippany, NJ 07981, USA

1. Introduction

In previous papers [Burnett, 1994; Holford, submitted 1997; Burnett and Holford, accepted 1997] the authors presented new 3-D time-harmonic prolate and oblate spheroidal acoustic infinite elements, based on new prolate and oblate spheroidal multipole expansions, for modeling acoustic fields in unbounded domains. Here we present the development of an ellipsoidal infinite element, which is the logical generalization of those elements. This development is also based on a new multipole expansion as well as a new system of ellipsoidal coordinates. Both the new coordinate system and the new element include the previous spheroidal coordinate systems and spheroidal elements as limiting cases. Therefore all the previously reported performance data for the spheroidal elements apply to this ellipsoidal element when used in spheroidal form.

Since the three axes of an ellipsoid can be chosen independently, an ellipsoid can circumscribe any structural shape at least as closely, and generally more closely, than a prolate or oblate spheroid. The resulting reduction in size of the finite computational domain will result in even greater computational speeds than those already reported for the spheroidal elements. Since this infinite element provides maximum computational efficiency for structures of all shapes, software element libraries would need only this one element for all problems in unbounded domains.

Finally, we show how multipole-based infinite elements may be developed for electromagnetism.

2. Ellipsoidal coordinates

The basis for our development of an ellipsoidal acoustic infinite element is a multipole expansion for the pressure field in a region exterior to an ellipsoid surrounding the structure of interest. In order to derive a suitable system of coordinates for effecting this, we begin by orienting Cartesian coordinates (x, y, z) with the x-axis aligned with the longest dimension of the structure and the z-axis aligned with the shortest. Then, choosing the origin of coordinates at the center of the circumscribing ellipsoid, the latter will be given by the equation

$$\frac{x^2}{a^2} + \frac{y^2}{b^2} + \frac{z^2}{c^2} = 1 \tag{1}$$

where 2a, 2b, and 2c are the major, intermediate, and minor axes of the ellipsoid, respectively,

T.L. Geers (ed.), IUTAM Symposium on Computational Methods for Unbounded Domains, 53–62.

such that $a \geq b \geq c$. In the plane y = 0, the cross-section of the ellipsoid given by Eq. (1) is an ellipse with foci at x = ± f, where $f^2 = a^2 - c^2$. In the plane z = 0, the cross-section is an ellipse with foci at x = ± g, where $g^2 = a^2 - b^2$. And in the plane x = 0, the cross-section is an ellipse with foci at y = ± h, where $h^2 = b^2 - c^2$. We note that $f > g$, and that $h^2 = f^2 - g^2$, so that the positions of only two sets of foci are independent. We now introduce ellipsoidal coordinates (r, θ, ϕ) as follows:

$$
\begin{aligned}
x &= r(1 - q\cos^2\theta)^{1/2}\cos\phi\,, \\
y &= (r^2 - g^2)^{1/2}\sin\theta\sin\phi\,, \qquad (2) \\
z &= (r^2 - f^2)^{1/2}\cos\theta\,(1 - p\cos^2\phi)^{1/2}\,,
\end{aligned}
$$

where

$$
f^2 = a^2 - c^2\ ,\ g^2 = a^2 - b^2\ ,\ p = g^2/f^2\ ,\ q = 1 - p\,. \qquad (3)
$$

The ranges for the variables r, θ, and ϕ are given by $r \geq f$, and either $0 \leq \theta \leq 2\pi,\ 0 \leq \phi \leq \pi$, or $0 \leq \theta \leq \pi,\ 0 \leq \phi \leq 2\pi$.

The ellipsoid given by Eq. (1) corresponds to the surface r = a. In general, the surface r = constant is a confocal ellipsoid, and the surfaces θ = constant and φ = constant are confocal hyperboloids (of one and two sheets, respectively). These three sets of quadrics are mutually orthogonal; see Fig. 1. We note that the radial variable r is the same as the variable ρ used by [Hobson, 1931], but the introduction of the angular variables θ and φ appears to be new.

When two of its axes are equal, the ellipsoid becomes a surface of revolution; i.e., a spheroid. Infinite elements for these surfaces have been described in [Burnett, 1994] and [Burnett and Holford, op. cit.]. These special cases are: (i) b=c, yielding a prolate spheroid; and (ii) a=b, yielding an oblate spheroid. In case (i), Eqs. (3) reduce to those given in Eqs. (2.13) of Burnett and Holford, with x <-> z and θ <-> φ; and in case (ii), Eqs. (3) reduce to those given in Eqs. (2.18) of the same reference.

3. Ellipsoidal multipole expansion

We shall now derive the multipole expansion in ellipsoidal coordinates that forms the basis for the infinite element representation to be presented in Section 4. This derivation is modeled after that for spheroidal coordinates previously given by the authors.

Let S be the ellipsoid specified by Eq. (1), forming an artificial surface surrounding a structure that is vibrating with angular frequency ω. Let $p(r)e^{i\omega t}$ be the acoustic field generated in the exterior of S, where $p(r)$ satisfies the Helmholtz equation and the Sommerfeld radiation condition:

$$
\nabla^2 p + k^2 p = 0\ ,\quad \frac{\partial p}{\partial R} + ikp = o(1/R)\ as\ R \to \infty,
$$

respectively, with k (=ω/c) the acoustic wavenumber, and $R = |r|$. We shall call any such function p, which is assumed to be at least twice differentiable on and exterior to S, a "radiation function" in the exterior of S. With respect to the ellipsoidal coordinates defined by Eqs. (2) and (3), for which S is the surface r = a, we can now state the following theorem:

EXPANSION THEOREM. *Let p(r) be a radiation function in the region exterior to the ellipsoid S (r = a). Then, p (r) can be expanded in ellipsoidal coordinates (r, θ, φ) in the form*

$$p(r) = e^{-ikr} \sum_{n=1}^{\infty} \frac{G_n(\theta,\phi;k)}{r^n} , \tag{4}$$

where the series converges for r > a, and converges absolutely and uniformly in r, θ, and φ in any region r>a+ε(ε>0). The series may be differentiated any number of times, and the resulting series all converge absolutely and uniformly in r>a+ε.

The proof takes as its starting point the Helmholtz integral representation, which expresses the field in the exterior of S in terms of its values, and those of its normal derivative, on S:

$$p(r) = \frac{1}{4\pi}\iint_S \left\{ p(r') \frac{\partial}{\partial n'}\left[\frac{e^{-ik\rho}}{\rho}\right] - \frac{e^{-ik\rho}}{\rho}\frac{\partial p(r')}{\partial n'} \right\} d\sigma', \tag{5}$$

where n' is the outward normal to S, $\rho=|r-r'|$, and $d\sigma'$ is the element of area at position r' on S. As we did for spheroids, we now introduce ellipsoidal coordinates given by Eqs. (2) and (3) into the definition of the distance function ρ and show that $e^{-ik(\rho-r)}/\rho$ is an analytic function of the complex variable $\zeta = 1/r$ in the interior of the circle $|\zeta| = a^{-1}$, for all points r' on S, provided the zeros of ρ, considered as a function of ζ, lie *outside* this circle. (This has been proven mathematically in Holford [op. cit.] for the special case when S is a spheroid. For general ellipsoids, we have verified this condition numerically for values of a:b:c varying from 3:2:1 to 100:10:1.) It then follows directly from Eq. (5) that $p(r)e^{ikr}$ is also analytic in $|\zeta| < a^{-1}$, and may therefore be expanded as a power series in $\zeta = 1/r$, as in Eq. (4). The complete proof of the expansion theorem follows using elementary arguments.

4. Ellipsoidal acoustic infinite element

4.1 Global geometry: shape and orientation of elements

The infinite element is shown in Fig. 2. One face of the element, the "base", must attach to, and conform to the shape of, an ellipsoidal surface of ellipsoidal radius r_1 (previously denoted by a in Sections 2 and 3) surrounding the structure, called the "infinite element ellipsoid". The base may be a curvilinear quadrilateral or triangle; it need not conform to the θ,φ coordinate lines. Any type of FE representation may be used over the base.

The infinite element ellipsoid uniquely determines three pairs of foci, which in turn define a system of confocal ellipsoids and hyperboloids (cf. Fig. 1) that are used in the construction of the rest of the element. The side faces of the element are the loci of confocal hyperbolas emanating from the sides of the base. A multipole of order $m[p=(a_1/r+\cdots+a_m/r^m)e^{-ikr}]$ requires m layers of nodes that are on confocal ellipsoids of ellipsoidal radii $r_1, r_2, \cdots, r_m$. The nodal pattern is identical on each ellipsoid. The value $m = 2$ corresponds to a dipole, $m = 3$ to a quadrupole, etc. Finally, there is an outer ellipsoidal face, $\hat{S}$, at ellipsoidal radius $\hat{r}$, that recedes to infinity in the theoretical development. Since the base, intermediate nodal layers, and outer face conform to confocal ellipsoids and the side faces are the loci of confocal hyperbolas, the element is a right cylinder in r, θ, φ-space.

Conventional ξ,η coordinates are defined over the element cross section, i.e., on ellipsoids, by the mapping

$$\theta(\xi,\eta) = \sum_{\nu=1}^{n} \theta_\nu \chi_\nu^a(\xi,\eta)\,, \qquad \phi(\xi,\eta) = \sum_{\nu=1}^{n} \phi_\nu \chi_\nu^a(\xi,\eta), \tag{6}$$

where n is the number of nodes in the quadrilateral or triangle, θ_ν and ϕ_ν are the ellipsoidal angular coordinates of the νth node, and $\chi_\nu^a(\xi,\eta)$ are interpolation polynomials.

If the geometry for the finite-size elements is generated using a conventional polynomial mapping (rather than blending functions), there will be a discontinuity (in ξ,η coordinates and therefore in the dependent variable, pressure) between the infinite elements and adjacent finite-size elements that are on the inside of the infinite element ellipsoid. The finite-size elements will not conform exactly to the shape of the ellipsoid, whereas the infinite elements do conform exactly. This does not cause a problem because the error due to the geometric discontinuity converges to zero, as the mesh is h or p-refined, at the same rate as does the (well-known) error at the boundary of a curved domain when using conventional polynomial-mapped elements. Thus, there is asymptotic continuity in pressure across the infinite element ellipsoid.

4.2 Finite element representation of pressure

The scattered and/or radiated pressure in the infinite element is represented as follows,

$$p(\xi,\eta,r) = \sum_{j=1}^{N} \psi_j(\xi,\eta,r) P_j\,, \tag{7}$$

where

$$\psi_j(\xi,\eta,r) = \psi_\nu^a(\xi,\eta)\,\psi_\mu^r(r) \qquad \nu = 1,2,\ldots,n\,;\mu = 1,2,\cdots,m\,;\, n\times m = N \tag{8}$$

Here $\psi_\nu^a(\xi,\eta)$ are "angular" shape functions that interpolate p over ellipsoidal surfaces confocal to the infinite element surface, and $\psi_\mu^r(r)$ are "radial" shape functions that interpolate p along hyperbolic rays.

The angular shape functions $\psi_\nu^a(\xi,\eta)$ are conventional 2-D interpolation-polynomial shape functions, e.g., Lagrange, serendipity, or hierarchic, which provide C^0 continuity between the sides of adjacent infinite elements.

The radial shape functions $\psi_\mu^r(r)$ use a truncated form of the radial part of the multipole expansion of Sec. 3, viz., an mth order multipole expansion:

$$\psi_\mu^r(r) = e^{-ik(r-r_\mu)} \sum_{\mu'=1}^{m} \frac{h_{\mu\mu'}}{(kr)^{\mu'}} \qquad \mu = 1,2,\cdots,m \;\; (m \geq 2)\,. \tag{9}$$

Eq. (9) represents m m^{th} degree polynomials in the reciprocal variable $1/r$ in the finite interval $1/r \in (0\,,1/r_1]$, with the constant terms omitted because $\psi_\mu^r(r) \to 0$ as $1/r \to 0$. Hence there are m coefficients $h_{\mu\mu'}$ in each polynomial, which are determined by the requirement, for convergence, of asymptotic C^0 continuity (cf. end of Sec. 4.1) between the base of the infinite element and adjacent finite-size elements on the inside of the infinite element ellipsoid. This is achieved by transforming the polynomials to Lagrange interpolation polynomials in $1/r$ with m interpolation points at $1/r_1,\ldots,1/r_m$, using the interpolation property

$\psi^r_\mu(r_{\mu'}) = \delta_{\mu\mu'}$, which yields m sets of m linear algebraic equations, $hS = I$, where $S_{ij} = (kr_j)^{-i}$ and I is the identity matrix. Therefore, $h = S^{-1}$. This procedure defines m layers of nodes (with n nodes on each layer) lying on ellipsoids of ellipsoidal radii $r_1, r_2, \cdots r_m$. The resulting Lagrange polynomials in $1/r$ are, in effect, "interpolation multipoles" in r over a semi-infinite interval.

4.3 Differential geometry of ellipsoidal coordinates

The elipsoidal covariant base vectors are derived from Eqs. (2). From these we develop expressions for: the metric tensor (which is diagonal, so the coordinates are orthogonal); scale factors; volume Jacobian, J_V; surface Jacobian, J_S; and gradient operator, ∇. Both Jacobians reduce to the two spheroidal forms [Burnett, 1994, and Burnett and Holford, op. cit.].

4.4 Stiffness, mass, and radiation-damping matrices

The remaining mathematical development follows, step by step, the same procedure as for the spheroidal elements [Ibid.]; only the algebraic details differ. We briefly describe these steps and then present the final equations. A full, detailed development will be published separately [Burnett and Holford, submitted 1997].

The element matrices are derived by starting with a finite-size element, i.e., with the outer face on an ellipsoid of ellipsoidal radius $\hat{r}$ (cf. Fig. 2), and then taking the limit as $\hat{r} \rightarrow \infty$. Thus, applying the Bubnov-Galerkin weighted residual method over a single element yields

$$\lim_{\hat{r} \rightarrow \infty} \int\int_{V^{(e)}}\int (B\nabla^2 p + \omega^2 \rho p)\ \psi_i\, dV = 0 \qquad i = 1,2,\cdots,N \tag{10}$$

using $k^2 = \omega^2 \rho / B$. The first integral is converted into a surface integral and another volume integral using the identity $\psi_i\ \nabla^2 p = \nabla \cdot (\psi_i \nabla p) - \nabla \psi_i \cdot \nabla p$ and the divergence theorem. Substituting Eq. (7) into the volume integrals yields the stiffness matrix, K and mass matrix, M. Substituting the Sommerfeld radiation condition and balance of linear momentum into the surface integral yields the radiation-damping matrix, C, and particle displacement vector, D. Thus, Eq. (10) becomes the following element matrix equation,

$$(K + i\omega C - \omega^2 M)P = D \tag{11}$$

where

$$K_{ij} = \lim_{\hat{r} \rightarrow \infty} \int\int_{V^{(e)}}\int B\nabla\psi_i \cdot \nabla\psi_j J_V dr d\theta d\phi$$

$$M_{ij} = \lim_{\hat{r} \rightarrow \infty} \int\int_{V^{(e)}}\int \rho\psi_i\psi_j\ J_V dr d\theta d\phi \tag{12}$$

$$C_{ij} = \rho c \lim_{\hat{r} \rightarrow \infty} \int\int_{S^{(\hat{r})}} \psi_i\psi_j\ J_S d\theta d\phi$$

Equation (8) is substituted for ψ_i and ψ_j, introducing primes to distinguish the indices ν and μ that are associated with the index i:

$$\psi_i(\xi,\eta,r) = \psi^a_{\nu'}(\xi,\eta)\,\psi^r_{\mu'}(r) \quad ; \quad \psi_j(\xi,\eta,r) = \psi^a_{\nu}(\xi,\eta)\,\psi^r_{\mu}(r)\,. \tag{13}$$

Performing these substitutions, we then make two observations: (i) all the integrands are sums of terms, each of which is a product of an angular function and a radial function [the angular shape functions are implicitly functions of θ, ϕ because of the coordinate mapping in Eqs. (6)]; and (ii) since the base of the element conforms to an ellipsoid and the side faces are the loci of confocal hyperbolas, the element is a right cylinder in r,θ,ϕ–space (or r,ξ,η-space). Therefore, the 3-D volume integrals for K_{ij} and M_{ij} separate into products of 2-D angular integrals, $A^{(i)}_{\nu'\nu}$, and 1-D radial integrals, $R^{(i)}_{\mu'\mu}$; and the 2-D surface integral for C_{ij} separates into a product of a 2-D angular integral and a radial function. (The D_i integrals are ignored because they are zero in virtually all practical applications.)

The angular integrals require no further algebraic manipulation. For each of the radial integrals we substitute Eq. (9) and perform various algebraic operations, separating out undefined oscillatory terms that all cancel, leaving only well-defined integrals.

The final form of the element matrix equation for the ellipsoidal infinite element is as follows:

$$(\boldsymbol{K}^{\infty} - \omega^2 \boldsymbol{M}^{\infty})\,\boldsymbol{P} = \boldsymbol{D}\,, \tag{14}$$

where

$$\begin{aligned} K^{\infty}_{ij} &= B\left[A^{(1)}_{\nu'\nu}R^{(1)}_{\mu'\mu} + A^{(2)}_{\nu'\nu}R^{(2)}_{\mu'\mu} - \varepsilon_1^2 A^{(3)}_{\nu'\nu}R^{(3)}_{\mu'\mu}\right] \\ M^{\infty}_{ij} &= \rho r_1^2\left[A^{(1)}_{\nu'\nu}R^{(4)}_{\mu'\mu} - \varepsilon_1^2 A^{(4)}_{\nu'\nu}R^{(2)}_{\mu'\mu} + \varepsilon_1^4 A^{(5)}_{\nu'\nu}R^{(3)}_{\mu'\mu}\right] \end{aligned} \tag{15}$$

and B is bulk modulus, ρ is density, and $\varepsilon_1 = f/r_1$. The indices i and j are related to ν', μ' and ν,μ, respectively, as in Eqs. (8) and (13). The vector $\boldsymbol{D}$ is zero in virtually all practical applications.

The angular integrals are as follows:

$$\begin{aligned} A^{(1)}_{\nu'\nu} &= \iint_{\sigma^{(e)}} \frac{P^2+Q^2-1}{PQ}\,\psi^a_{\nu'}\psi^a_{\nu}\,d\theta d\phi \\ A^{(2)}_{\nu'\nu} &= \iint_{\sigma^{(e)}} \left[\frac{Q}{P}\frac{\partial\psi^a_{\nu'}}{\partial\theta}\frac{\partial\psi^a_{\nu}}{\partial\theta} + \frac{P}{Q}\frac{\partial\psi^a_{\nu'}}{\partial\phi}\frac{\partial\psi^a_{\nu}}{\partial\phi}\right] d\theta d\phi \\ A^{(3)}_{\nu'\nu} &= \iint_{\sigma^{(e)}} \left[\frac{Q(1-P^2)}{P}\frac{\partial\psi^a_{\nu'}}{\partial\theta}\frac{\partial\psi^a_{\nu}}{\partial\theta} + PQ\frac{\partial\psi^a_{\nu'}}{\partial\phi}\frac{\partial\psi^a_{\nu}}{\partial\phi}\right] d\theta d\phi \\ A^{(4)}_{\nu'\nu} &= \iint_{\sigma^{(e)}} \frac{(1+Q^2-P^2)(P^2+Q^2-1)}{PQ}\,\psi^a_{\nu'}\,\psi^a_{\nu}\,d\theta d\phi \\ A^{(5)}_{\nu'\nu} &= \iint_{\sigma^{(e)}} \frac{Q(1-P^2)(P^2+Q^2-1)}{P}\,\psi^a_{\nu'}\,\psi^a_{\nu}\,d\theta d\phi \end{aligned} \tag{16}$$

where $P = (1-p\cos^2\phi)^{1/2}$, $Q = (1-q\cos^2\theta)^{1/2}$, $p = g^2/f^2$, $q = 1-p$, and $\sigma^{(e)}$ is the "ellipsoidal cross section" of the element, i.e., any confocal ellipsoidal surface inside the element that is bounded by the side faces. The integrals may be evaluated numerically by transforming them to the local ξ, η-coordinates defined in Eqs. (6), using $d\theta d\phi = Jd\xi d\eta$. All five integrals can be numerically integrated using Gaussian rules since the integrands are smooth and bounded. The derivatives of ψ_ν^a are evaluated in the conventional manner, using the inverse of the Jacobian matrix.

The radial integrals are as follows:

$$R^{(1)}_{\mu'\mu} = L_{\mu'\mu}\left[\frac{b_2}{2i}e^{-i2\zeta} + \sum_{\beta=0}^{2} mb_{\beta+2}I_\beta\right], \qquad R^{(2)}_{\mu'\mu} = L_{\mu'\mu}\sum_{\beta=2}^{2m} c_\beta\, J_\beta$$

$$R^{(3)}_{\mu'\mu} = L_{\mu'\mu}\zeta^2\sum_{\beta=2}^{2m} c_\beta\, J_{\beta+2}, \qquad R^{(4)}_{\mu'\mu} = L_{\mu'\mu}\frac{1}{\zeta^2}\left[\frac{c_2}{2i}e^{-i2\zeta} + \sum_{\beta=0}^{2m-2} c_{\beta+2}\, J_\beta\right] \tag{17}$$

where $\zeta = kr_1$, $L_{\mu'\mu} = e^{ik(r_{\mu'}+r_\mu)}/k$, $b_\beta = \sum_{\gamma=1}^{\beta-1} a_{\mu'\gamma}\, a_{\mu,\beta-\gamma}$, $c_\beta = \sum_{\gamma=1}^{\beta-1} h_{\mu'\gamma}\, h_{\mu,\beta-\gamma}$, $a_{\mu\alpha} = -ih_{\mu\alpha} - (\alpha-1)\,h_{\mu,\alpha-1}$, and $h_{\mu\alpha}$ are determined as in Sec. 4.2.

Also,

$$I_\beta = \int_{r_1}^{\infty} \frac{FG - \delta_{\beta 0}}{(kr)^\beta}\; e^{-i2kr}\, k\, dr\ , \quad \beta \ge 0 \tag{18}$$

$$J_\beta = \int_{r_1}^{\infty} \left[\frac{1}{FG(kr)^\beta} - \delta_{\beta 0}\right] e^{-i2kr} k\, dr\ , \quad \beta \ge 0 \tag{19}$$

where $\delta_{\beta 0}$ is the Kronecker Delta (= 1 if $\beta = 0$; = 0 if $\beta > 0$), $F = (1-f^2/r^2)^{1/2}$, and $G = (1-g^2/r^2)^{1/2}$. The I_β and J_β integrals are not independent, but are related by the expression

$$I_\beta = J_\beta - (1+p)\zeta^2\varepsilon_1^2 J_{\beta+2} + p\zeta^4\varepsilon_1^4 J_{\beta+4}\ , \quad \beta \ge 0 \tag{20}$$

Perhaps the easiest way to evaluate the J_β (and I_β) integrals is to transform them to Fourier sine and cosine transforms with the change of variable $\chi = kr-\zeta$, since Fourier routines may be found in many mathematics software packages.

The radial integrals, $R^{(i)}_{\mu'\mu}$, $i = 1, 2, 3, 4$ are identical for every infinite element in a mesh because they are independent of angular variables and are along identical radial paths, i.e., hyperbolas emanating from the same ellipsoid. Their computational cost is insignificant and they only need to be evaluated once for a given problem. Consequently, the numerical integration required to generate K^∞ and M^∞ for each infinite element involves only the evaluation of the 2-D angular integrals, making these 3-D elements as cheap to generate as 2-D elements. In addition, since the frequency dependence of the element is contained only in the radial integrals, generation of the infinite element mesh during a frequency sweep is essentially free after the first frequency.

Finally, it is noted that in the limiting cases of prolate and oblate spheroidal coordinates, the above expressions for K_{ij}^{∞} and M_{ij}^{∞} are identical to those reported in Burnett and Holford [op. cit.].

5. Spherical Electromagnetic Infinite Element

The motivation for this development is the recognition that a wide class of practical EM radiation and scattering problems, e.g., wireless telecommunication and radar, are in the atmosphere (air), for which the EM properties are essentially homogeneous and isotropic in the open regions surrounding objects.

In source-free, homogeneous, isotropic regions a general time-harmonic EM field may be decomposed [Felsen and Marcuvitz, 1994] into a TE- (or H-) mode and a TM- (or E-) mode relative to spherical coordinates, with the radial (r) and angular (θ, ϕ) directions defined as the longitudinal and transverse directions, respectively.

Let

$$\boldsymbol{E}(\boldsymbol{r}) = \boldsymbol{E}'(\boldsymbol{r}) + \boldsymbol{E}''(\boldsymbol{r}) \quad , \quad \boldsymbol{H}(\boldsymbol{r}) = \boldsymbol{H}'(\boldsymbol{r}) + \boldsymbol{H}''(\boldsymbol{r}) \tag{21}$$

where single primes denote TM-mode and double primes denote TE-mode. From the decomposition theorem we have the following relations:

$$\begin{aligned} \boldsymbol{E}'(\boldsymbol{r}) &= \nabla \times \nabla \times [\boldsymbol{r}\Pi'(\boldsymbol{r})] \quad , \quad \boldsymbol{E}''(\boldsymbol{r}) = -j\omega\mu \, \nabla \times [\boldsymbol{r}\Pi''(\boldsymbol{r})] \\ \boldsymbol{H}'(\boldsymbol{r}) &= j\omega\varepsilon \, \nabla \times [\boldsymbol{r}\Pi'(\boldsymbol{r})] \quad , \quad \boldsymbol{H}''(\boldsymbol{r}) = \nabla \times \nabla \times [\boldsymbol{r}\Pi''(\boldsymbol{r})] \end{aligned} \tag{22}$$

where $\Pi'(\boldsymbol{r})$ and $\Pi''(\boldsymbol{r})$ are two independent *scalar* potentials, known as the Debye (or Hertz) potentials. Each potential satisfies the scalar Helmholtz equation, i.e.,

$$(\nabla^2 + k^2)\Pi'(\boldsymbol{r}) = 0 \quad , \quad (\nabla^2 + k^2)\Pi''(\boldsymbol{r}) = 0 \tag{23}$$

as well as the Sommerfeld radiation condition. In other words, $\Pi'(\boldsymbol{r})$ and $\Pi''(\boldsymbol{r})$ are each governed by the *acoustic* wave equation.

Thus, in source-free, homogeneous, isotropic regions, general radiated and/or scattered EM fields may be represented by a pair of spherical acoustic infinite elements outside the minimum sphere circumscribing an object [Burnett and Liabianca, in preparation].

These Debye potential infinite elements can be coupled to conventional vector-based EM finite elements inside the sphere by the standard Galerkin operations on both sides of the sphere since all fields exist on both sides of the sphere. A Bubnov formulation, e.g., would yield symmetric, off-diagonal coupling matrices.

We anticipate that the many advantages that have been realized by multipole-based infinite elements in acoustics will also be realizable in electromagnetism.

6. Advantages of multipole-based infinite elements

1. Computationally very fast. Has been found to be several hundred times faster than the boundary element method for the same accuracy [Burnett, op. cit.].
2. Broadband. There are no critical frequencies as with the BEM.

3. Numerically well conditioned. Has been tested up to $ka = 100$ [Burnett and Holford, op. cit.], with no indication of round-off error.
4. Symmetric element matrices.
5. Low cost element matrix generation. The cost is approximately the same as for a 2-D element.
6. Conventional FEA code architecture.
7. Compatible with *hp*-adaptive technology. The FE representation is a polynomial in all directions.
8. Extendable to other forms of wave propagation in unbounded regions. Has recently been extended to electromagnetism [Burnett and Labianca, op. cit.].

7. Conclusions

We have presented the theoretical foundations for ellipsoidal acoustic infinite elements. A new system of ellipsoidal coordinates, comprising two angular and one radial coordinate, has been introduced. This system includes prolate and oblate spheroidal coordinates as limiting cases; hence, this element includes the authors' previously reported spheroidal elements as special cases. Proof of convergence of the underlying ellipsoidal multipole expansion has been outlined and mathematical expressions for the stiffness and mass matrices have been presented. Numerical results for ellipsoidal elements used in ellipsoidal form are not presented here, but results in spheroidal form may be found in Burnett [1994] and Burnett and Holford [accepted 1997]. We have also shown how this infinite element technology can be extended to electromagnetism.

References

Burnett, D.S., 1994, "A three-dimensional acoustic infinite element based on a prolate spheroidal multipole expansion", *J. Acoust. Soc. Am.*, Vol. 96, pp. 2798-2816.

Burnett, D.S., and Holford, R.L., "Prolate and oblate spheroidal acoustic infinite elements", *Comput. Methods Appl. Mech. Engrg.*, accepted 1997.

Burnett, D.S., and Holford, R.L., "An ellipsoidal acoustic infinite element", *Comput. Methods Appl. Mech. Engrg.*, submitted 1997.

Burnett, D.S., and Labianca, F.M., "A 3-D electromagnetic infinite element for wave propagation in unbounded domains", in preparation.

Felsen, L.B., and Marcuvitz, N., 1994, *Radiation and Scattering of Waves*, IEEE Press.

Hobson, E.W., 1931, *The Theory of Spherical and Ellipsoidal Harmonics*, Cambridge University Press, Cambridge, U.K., Ch.11.

Holford, R.L., "A multipole expansion for the acoustic field exterior to a prolate or oblate spheroid", *J. Acoust. Soc. Am.*, submitted 1997.

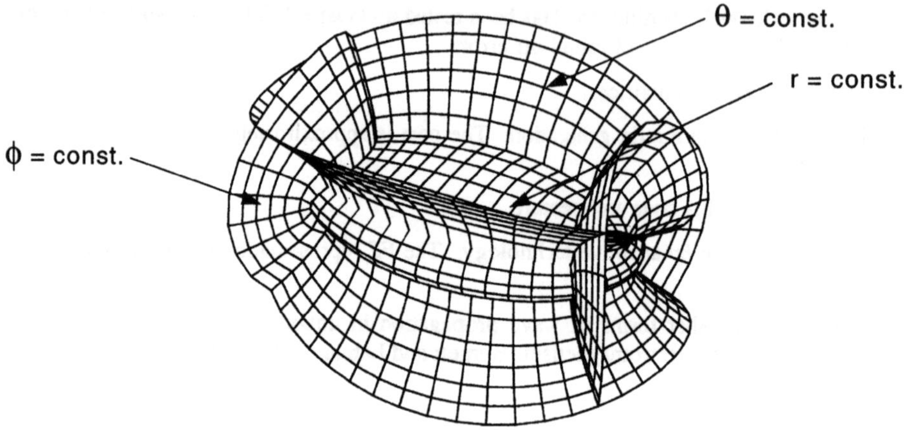

Figure 1. Ellipsoidal coordinates

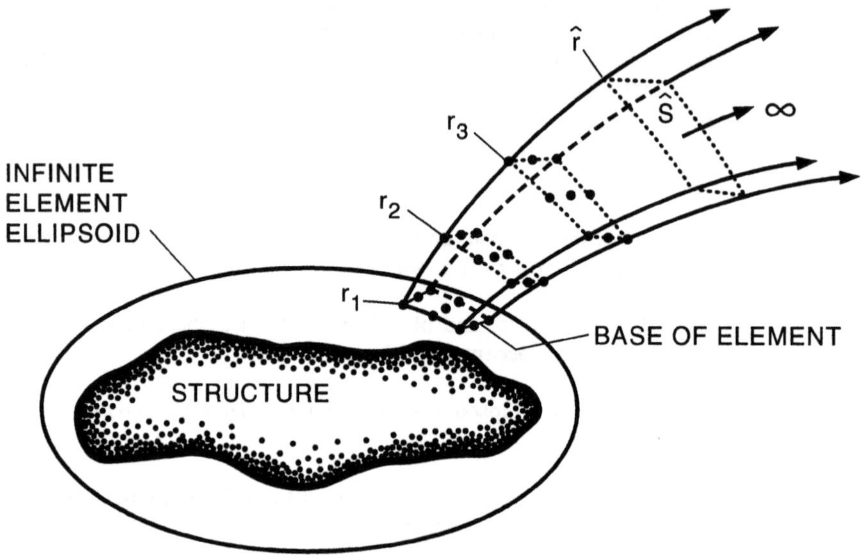

Figure 2. The ellipsoidal infinite element

A PRACTICAL COMPARISON OF FINITE ELEMENT AND BOUNDARY ELEMENT FORMULATIONS FOR MODELING SOUND RADIATION FROM ELASTIC PLATES

Jean-Pierre Coyette and Luc Cremers

LMS-Numerical Technologies
Leuven, Belgium

1. Introduction

Modeling sound radiation from vibrating structures usually relies, at least in the 'low' frequency range, on boundary integral representations of the acoustic field. Either direct and indirect integral representations are available for that purpose and form the mathematical basis of related acoustic boundary element (BE) methods [Ciskowski and Brebbia, 1991]. Such formulations have been used for a long time by the acoustic community. Model size's reduction and *a priori* verification of the Sommerfeld radiation condition are presented as key characteristics of BE methods. On the other hand, domain methods (like finite element (FE) methods) require an appropriate treatment of the radiation condition, involve usually a larger amount of data (as implied by the related volume mesh) but are characterized by sparse matrices and reduced calculation times.

The present paper deals with a practical comparison of BE and FE methods as they could be applied to sound radiation from a baffled elastic plate subjected to a point load. Such an application is representative of a wider class of vibro-acoustic problems where a time-harmonic mechanical excitation is driving a structure which in turns radiates sound in the surrounding fluid. The numerical treatment of this problem is performed by combining a structural FE model (and his related modal representation) with an acoustic FE or BE model. The acoustic FE model relies on variable order infinite wave envelope elements [Astley, Macaulay and Coyette, 1994; Cremers, Fyfe and Coyette, 1994] for handling approximately the Sommerfeld radiation condition. Different classes of infinite wave envelope elements (spherical and spheroidal) of variable order [SYSNOISE, 1996] have been used for that purpose. A comparison of results associated to different models is presented for this particular benchmark problem. Both field values (acoustic pressures) and integrated values (radiated power) are involved in this comparison. They show the capability of infinite element formulations to resolve the acoustic field in the vicinity of a vibrating structure.

2. Benchmark definition

The problem considered is related to sound radiation from an baffled elastic plate (Figure 1) excited by a point load. This problem is extracted from a set of benchmarks

T.L. Geers (ed.), IUTAM Symposium on Computational Methods for Unbounded Domains, 63–72.

published by SFM (Société Française de Mécanique) as a first guide for validating vibro-acoustic softwares [Valor, 1996].

The following data are used for the elastic plate (length a = 1 m, length b = 1 m, thickness t = 0.01 m, Young modulus $E = 2.10^{+11}$ Pa, Poisson ratio $\nu = 0.3$, density $\rho = 7800$ kg/m^3, structural loss factor $\eta = 10^{-2}$). The plate is simply supported along the four edges and is surrounded by a rigid (plane) baffle. The acoustic fluid (on one side of the plate) is the air with conventional material properties (density $\rho = 1.2$ kg/m^3 and speed of sound c = 340 m/s). A point load of amplitude F = 1 N (with a time dependence like $e^{+i\omega t}$ where ω is the circular frequency) is applied at location ($x_1 = 0.2$ m, $x_2 = -0.3$ m).

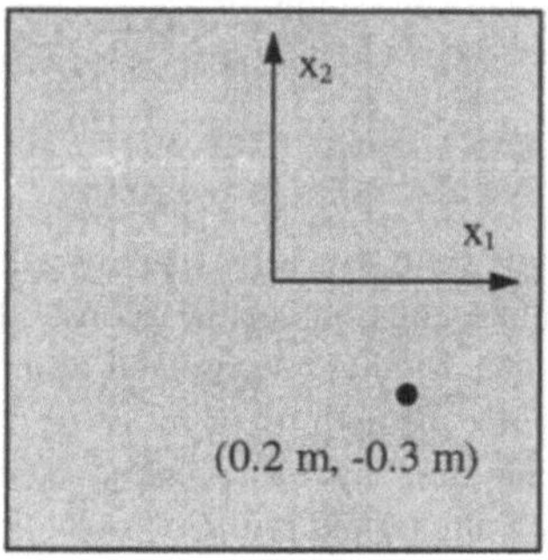

Figure 1: Elastic plate problem.

3. Numerical models

The problem has been modeled using two particular coupled methods. Both methods refer to a conventional structural FE model for the plate. The first method combines this structural FE model with an acoustic BE model (based on Rayleigh's integral representation for baffled problems) while the second one combines the structural FE model with an acoustic FE model where the 'far' field is modeled by infinite wave envelope elements [Astley, Macaulay and Coyette, 1994]. The coupled effects are taken into account in the usual way [Morand and Ohayon, 1995] so that the discrete coupled system can be formulated as:

$$\begin{bmatrix} Z_S(\omega) & C \\ \rho\omega^2 C^T & Z_A(\omega) \end{bmatrix} \begin{pmatrix} U \\ P \end{pmatrix} = \begin{pmatrix} F_S \\ F_A \end{pmatrix} \qquad (1)$$

where Z_S and Z_A are the structural and acoustic impedance matrices, C is the geometrical coupling matrix, F_S and F_A are the structural and acoustical load vectors, U and P are the vectors of nodal displacements (or modal participation factors) and nodal pressures. Details are omitted here but can be found in references listed at the end of the paper.

3.1 Structural FE model

The FE structural mesh is represented in Figure 2 (100 QUAD4 thin shell elements). A reduced modal basis involving the first 25 eigenmodes has been selected. The related eigenfrequencies and modal orders along x_1- and x_2-directions are listed in Table 1. The first 4 modes are represented in Figure 2

Mode	Hz	Mode	Hz	Mode	Hz	Mode	Hz	Mode	Hz
(1,1)	48.140	(2,1)	120.35	(3,1)	240.70	(4,1)	409.19	(5,1)	625.82
(1,2)	120.35	(2,2)	192.56	(3,2)	312.91	(4,2)	481.40	(5,2)	698.03
(1,3)	240.07	(2,3)	312.91	(3,3)	433.26	(4,3)	601.75	(5,3)	818.38
(1,4)	409.19	(2,4)	481.40	(3,4)	601.75	(4,4)	770.24	(5,4)	986.87
(1,5)	625.82	(2,5)	698.03	(3,5)	818.38	(4,5)	986.87	(5,5)	1203.5

Table 1: Elastic plate - First 25 eigenfrequencies.

Figure 2: Elastic plate - FE mesh (100 QUAD4, 121 nodes).

3.2 Acoustic BE model

The acoustic BE model relies on Rayleigh's integral representation (as valid for a baffled plane structure). The related BE mesh is identical to the structural FE mesh shown in Figure 2. This acoustic BE model (coupled to the above structural FE model) produces a solution nearly identical (at least in the considered frequency range) to the available reference solution [Valor, 1996].

3.3 Acoustic FE models

Various acoustic FE models have been used for this benchmark. All these models rely on a conventional pressure formulation for the acoustic fluid and a suitable volume mesh in the vicinity of the vibrating structure. This particular 'inner' domain correspond to either a box (Figure 4) for meshes A1 and A2 (Figure 6) or a semi-ellipsoid (Figure 5) for meshes A3 and A4 (Figure 7). The models differ by the nature of infinite wave envelope elements selected for the 'outer' field. 'Spherical' infinite elements are used in conjunction with meshes A1 and A2 while 'oblate spheroidal' infinite elements are selected for meshes A3 and A4. It should be emphasized that meshes A1 and A2 violate the 'circumscribing' sphere condition while the external ellipsoidal boundary surface of meshes A3 and A4 is enclosing totally the vibrating structure.

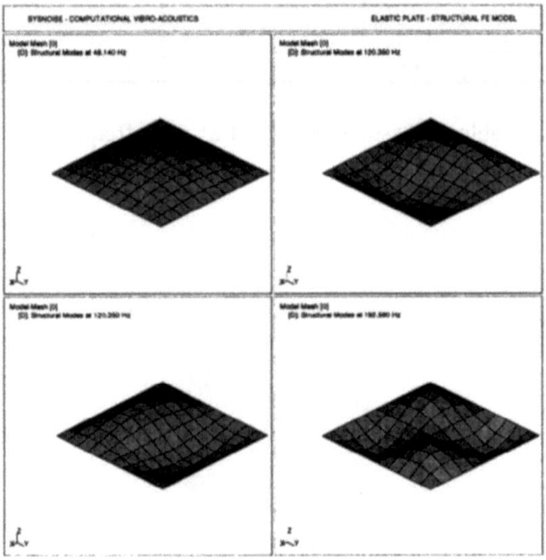

Figure 3: Elastic plate - First 4 mode shapes.

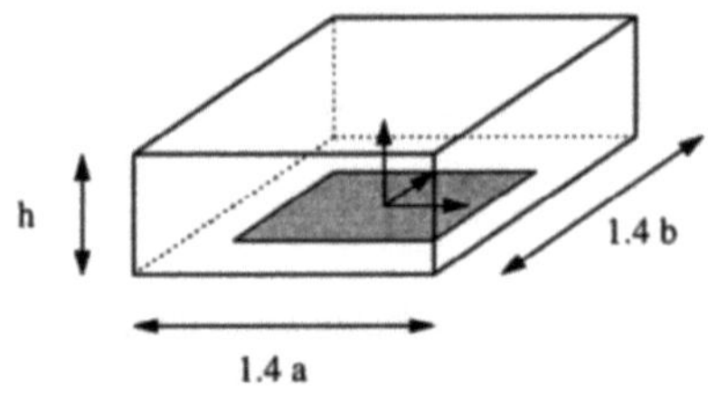

Figure 4: 'Inner' acoustic domain related to meshes A1 (h=0.5 a) and A2 (h=0.3 a).

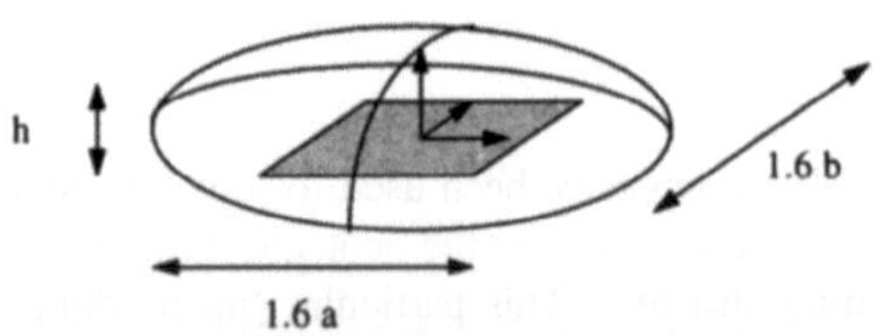

Figure 5: 'Inner' acoustic domain related to meshes A3 (h=0.25 a) and A4 (h=0.10 a).

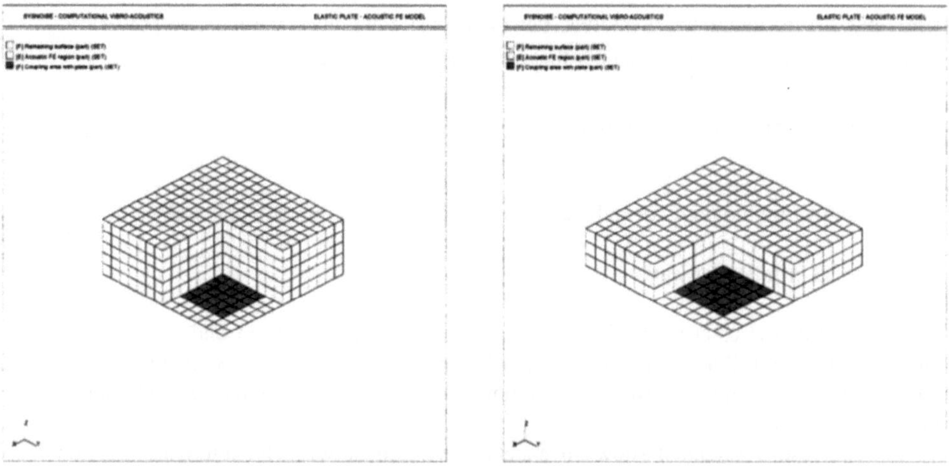

Figure 6: Acoustic FE meshes A1 (left; 1350 nodes and 980 elements) and A2 (right; 900 nodes and 588 elements).

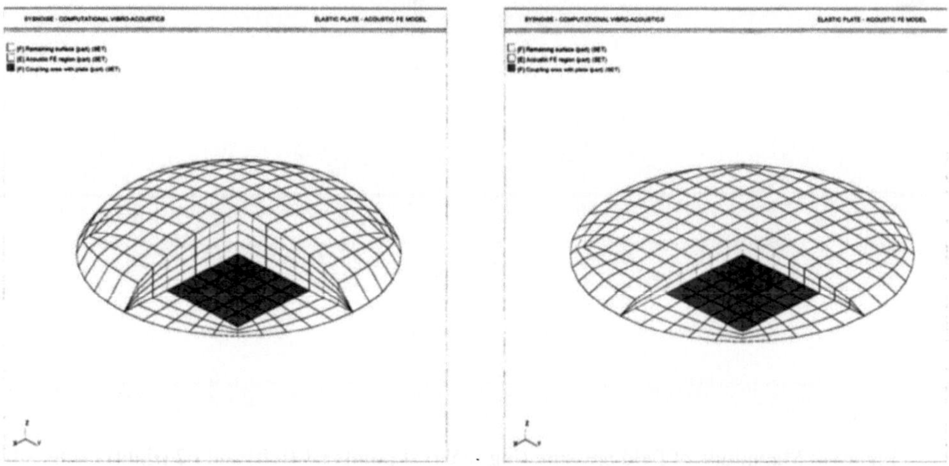

Figure 7: Acoustic FE meshes A3 (left; 1045 nodes and 880 elements) and A4 (right; 643 nodes and 440 elements).

4. Numerical results

The acoustic response has been evaluated for the different models in the frequency range 10-300 Hz using a step of 1 Hz. For each model, attention has been paid to single field values (pressure at two field points located at (P1: x_1=0, x_2=0, x_3=0) and at (P2: x_1=0, x_2=0,

x_3=1) but also to integrated quantities. For the acoustic engineer, the radiated power W_{rad} is such a meaningful integrated quantity that can be evaluated from:

$$W_{rad}(\omega) = \int_S \mathrm{Re}\big(p(\mathbf{x},\omega)v_n^*(\mathbf{x},\omega)\big)\,dS(\mathbf{x}) \tag{2}$$

where p and v_n are the pressure and the normal velocity along the radiating surface S.

4.1 Spherical infinite elements

Field pressures at points P1 and P2 as obtained using mesh A1 and spherical infinite elements of order 1, 2 and 3 are presented in Figure 8. Similar results obtained with mesh A2 are presented in Figure 9. The radiated power as obtained using meshes A1 and A2 is presented in Figure 10.

The results obtained show an excellent agreement with the reference solution for elements of order greater or equal to 2 along the radial direction.

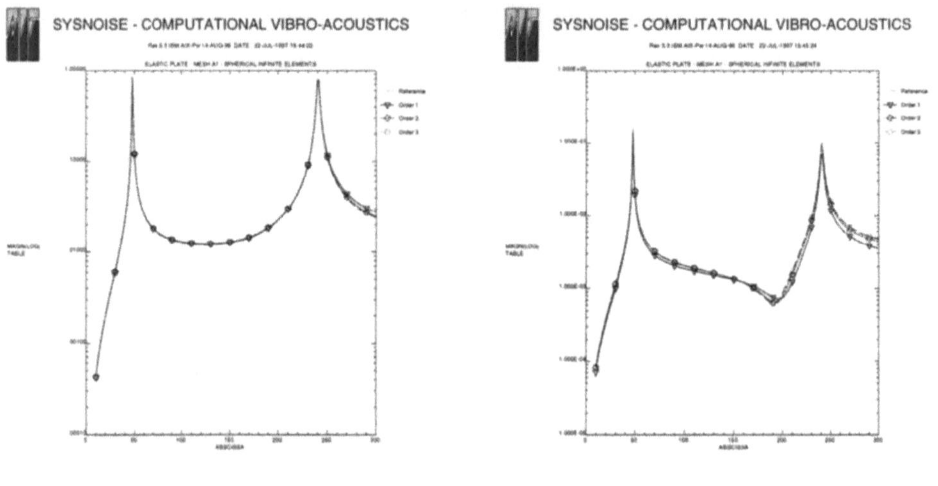

Figure 8: Magnitude of field pressure (log scale) at points P1 (left) and P2 (right) using mesh A1 (*spherical* infinite elements of order 1,2 and 3) compared to reference solution (SFM, 1996).

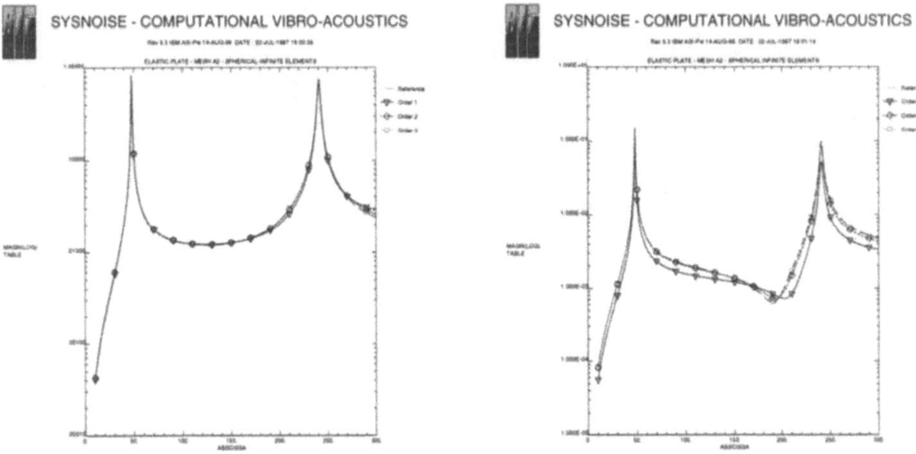

Figure 9: Magnitude of field pressure (log scale) at points P1 (left) and P2 (right) using mesh A2 (*spherical* infinite elements of order 1,2 and 3) compared to reference solution (SFM, 1996).

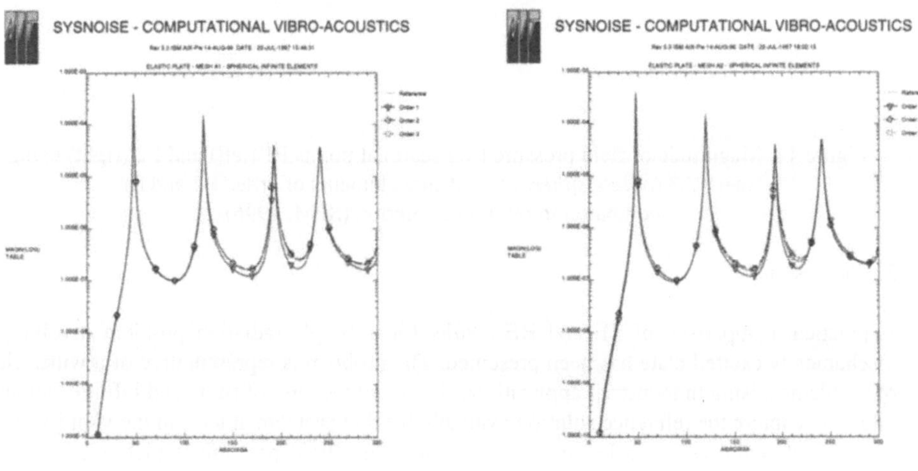

Figure 10: Radiated power (log scale) using mesh A1 (left) and A2 (right) (*spherical* infinite elements of order 1,2 and 3) compared to reference solution (SFM, 1996).

4.2 Oblate spheroidal infinite elements

In the same way, field pressures at points P1 and P2 as obtained using mesh A3 and oblate spheroidal infinite elements of order 1, 2 and 3 are presented in Figure 11. Similar results obtained with mesh A4 are presented in Figure 12. The radiated power as obtained using meshes A3 and A4 is presented in Figure 13.

As it can be seen from inspection of Figures 11 and 12, the convergence is still good along the radiating surface but deteriorates at greater distance.

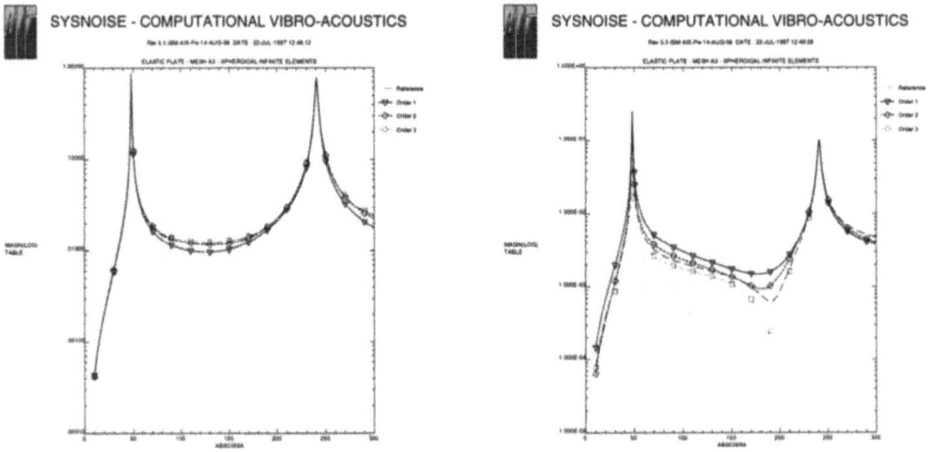

Figure 11: Magnitude of field pressure (log scale) at points P1 (left) and P2 (right) using mesh A3 (*oblate spheroidal* infinite elements of order 1,2 and 3) compared to reference solution (SFM, 1996).

5. Conclusion

A practical comparison of FE and BE results for a simple radiation problem involving a mechanically excited plate has been presented. The problem is representative of a wider class of problems arising in industrial applications. The combined use of finite and infinite elements allows to capture the reference solution available for that problem at least in the vicinity of the radiating structure as it can be seen from single value (field pressure at plate's center) and integrated value (radiated power) predictions. These good performances have been obtained for both spherical and oblate spheroidal infinite elements. On the other hand, spheroidal elements are leading (at least for the considered case) to a slower convergence that should be investigated in more details.

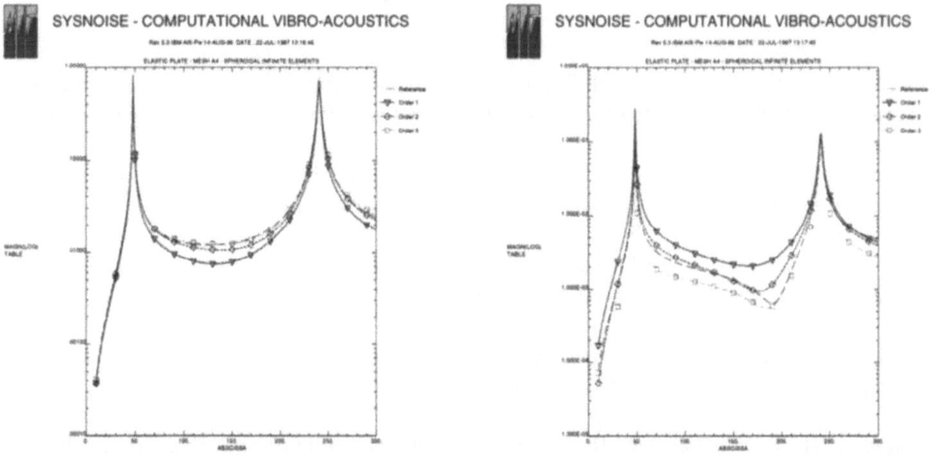

Figure 12: Magnitude of field pressure (log scale) at points P1 (left) and P2 (right) using mesh A4 (*oblate spheroidal* infinite elements of order 1,2 and 3) compared to reference solution (SFM, 1996).

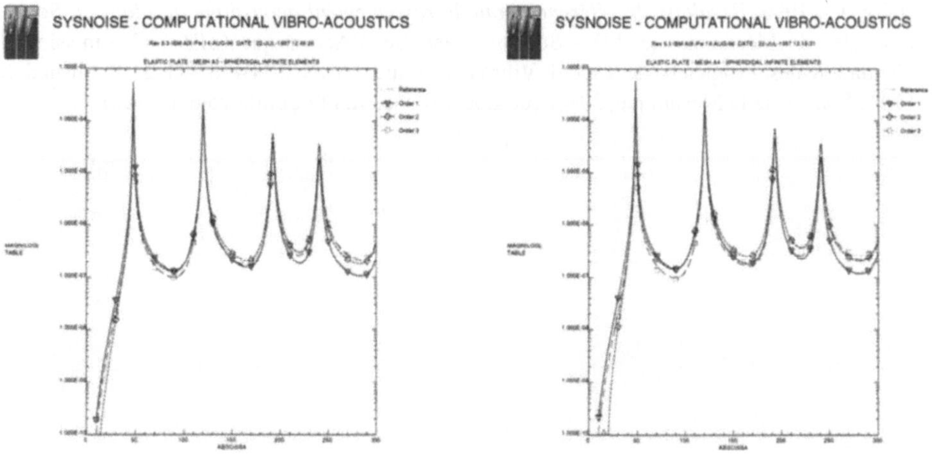

Figure 13: Radiated power (log scale) using mesh A3 (left) and A4 (right) (*oblate spheroidal* infinite elements of order 1,2 and 3) compared to reference solution (SFM, 1996).

6. References

Astley, R.J., Macaulay, G.J. and Coyette, J.P., 1994, "Mapped wave envelope elements for acoustical radiation and scattering", *Journal of Sound and Vibration*, Vol. 170, pp. 97-118.

Cremers, L., Fyfe, K.R. and Coyette, J.P., 1994, "A variable order infinite acoustic wave envelope element", *Journal of Sound and Vibration*, Vol. 171, pp. 483-508.

Burnett, D.S., 1994, "A three-dimensional acoustic infinite element based on prolate spheroidal multipole expansion", *J. Acoust. Soc. America*, Vol. 96, pp. 2789-2816.

Ciskowski, R.D. and Brebbia, C.A., 1991, *Boundary element methods in acoustics*, Computational Mechanics Publications, Boston.

Morand, H.J.P. and Ohayon, R., 1995, *Fluid-structure interaction*, John Wiley & Sons, Chichester.

SYSNOISE Revision 5.3, 1996, User's Manual, LMS-Numerical Technologies, Leuven (Belgium).

Valor, C., 1996, *Résultats de référence pour le rayonnement acoustique de plaques*, Société Française de Mécanique (SFM) - Société Française d'Acoustique (SFA), Commission de Validation des Progiciels de Calcul Vibroacoustique, ISBN 2-904983-11-2 (Published by SFM, Maison de la Mécanique, 39-41 rue Louis Blanc, 92400 Courbevoie, France).

Approximation of Transient Hydrodynamics on Unbounded Domains Using Rational Functions

Christopher J. Damaren

Department of Mechanical Engineering
University of Canterbury
Private Bag 4800
Christchurch, New Zealand

1 Introduction

Consider a floating body, possibly flexible, which floats on the surface of an infinitely deep ocean of infinite extent in the presence of surface waves. We are concerned with the case where the body motions and wave heights are small and the fluid medium is incompressible, of constant density, inviscid, and irrotational. The goal is determination of a finite-dimensional, linear time-invariant (FDLTI) representation for the mapping from (transient) wave motion to body motion. Such a description is a basis for simulation but is also a necessary prelude to control system design.

The problem has long been studied in the frequency domain in terms of steady-state harmonic solutions [John, 1950], and the connection to time-domain transient solutions noted by Cummins [1962] and Wehausen [1971]. The latter emphasized that inverse Fourier transformation of harmonic solutions was but one possibility; alternatively, the problem can be formulated directly in the time domain using a time-varying transient Green's function. These functions are known from the work of Finkelstein [1957] and have been used by Yeung [1982], Newman [1985], Beck and Liapis [1987], and Pot and Jami [1991] who studied the radiation problem for various cylinders and spheres.

The transient motion of a two-dimensional cylinder has been studied by Ursell [1964] and Maskell and Ursell [1970] where properties of the added mass and damping coefficients were used to infer asymptotic properties of the temporal solution. The full solution was obtained numerically using inverse Fourier transformation. None of the above approaches yield an explicit representation or *realization* of the transient body response to radiation and diffraction forces which is the subject of this work.

A Green's function approach is used here to determine the corresponding steady-state harmonic solutions for the radiation and diffraction problems. It is proposed to fit the frequency responses associated with the 'radiation impedance' and wave exciting forces with appropriate analytic functions. These are chosen to possess certain key properties which the corresponding exact solutions are known to have. By using rational functions for the approximation, the transformation from the frequency domain to the

T.L. Geers (ed.), IUTAM Symposium on Computational Methods for Unbounded Domains, 73–82.

time domain is facilitated. Similar approximations have been used by Yu and Falnes [1995] who conducted the approximation using impulse responses in the time domain. Our method is illustrated for both two-dimensional and three-dimensional problems using a floating cylinder and a thin flexible plate. Applications of the method to the design of control systems for optimal suppression of wave-induced motion are indicated.

2 Problem Formulation

The motion equations for the floating body $\mathcal{B}$ are assumed to be described by

$$\sigma(\mathbf{r})\ddot{\mathbf{w}} + \mathcal{K}\mathbf{w}(\mathbf{r},t) = \mathbf{f}(\mathbf{r},t) \quad (1)$$

where $\mathbf{r} = [x\ y\ z]^T$, $\mathbf{w} = [w_1\ w_2\ w_3]^T$, $\sigma(\mathbf{r})$ is the mass density, $\mathcal{K}$ is a linear stiffness operator, and $\mathbf{f}$ represents the external forces. The z-axis is vertically upwards. A spatial discretization of the form

$$\mathbf{w}(\mathbf{r},t) = \sum_{\alpha=1}^{N} \boldsymbol{\psi}_\alpha(\mathbf{r})q_\alpha(t) = \boldsymbol{\Psi}(\mathbf{r})\mathbf{q}(t) \quad (2)$$

is employed where $\boldsymbol{\Psi} = \text{row}\{\boldsymbol{\psi}_\alpha\}$ and $\mathbf{q} = \text{col}\{q_\alpha\}$. Included in $\boldsymbol{\Psi}$ are the six rigid-body motions. The ensuing discrete motion equations can be determined using a variational formulation of (1) which yields

$$\boldsymbol{M}\ddot{\mathbf{q}} + \boldsymbol{K}\mathbf{q} = \boldsymbol{f}(t) \quad (3)$$

where $M_{\alpha\beta} = \int_{\mathcal{B}} \boldsymbol{\psi}_\alpha^T \boldsymbol{\psi}_\beta \sigma\, dV$, $f_\alpha = \int_{\mathcal{B}} \boldsymbol{\psi}_\alpha^T \mathbf{f}\, dV$, and $K_{\alpha\beta}$ is calculated using the strain energy bilinear form corresponding to $\int_{\mathcal{B}} \boldsymbol{\psi}_\alpha^T \mathcal{K} \boldsymbol{\psi}_\beta\, dV$. Let $\mathbf{n}(\mathbf{r})$ denote the components of the outward normal to the wetted portion of $\mathcal{B}$, $\mathcal{S}_b$, and define $n_\alpha(\mathbf{r}) = \mathbf{n}^T\boldsymbol{\psi}_\alpha$.

The motion of the fluid is governed by the velocity potential $\Phi(\mathbf{r},t)$ which, in $\mathcal{V} = \{z \leq 0 \backslash \mathcal{B}\}$ satisfies

$$\nabla^2\Phi = 0,\ \mathbf{r} \in \mathcal{V};\quad \frac{\partial^2\Phi}{\partial t^2} = -g\frac{\partial\Phi}{\partial z},\quad \mathbf{r} \in \mathcal{S}_f;\quad \lim_{z\to-\infty} \frac{\partial\Phi}{\partial z} = 0; \quad (4)$$

$$\frac{\partial\Phi}{\partial n} = \mathbf{n}^T\dot{\mathbf{w}} = \sum_{\alpha=1}^{N} n_\alpha(\mathbf{r})\dot{q}_\alpha\ ,\quad \mathbf{r} \in \mathcal{S}_b. \quad (5)$$

Here $\mathcal{S}_f = \{z = 0\backslash\mathcal{B}\}$ denotes the free surface and g is the acceleration due to gravity. We make the standard decomposition

$$\Phi = \Phi_I + \Phi_S + \sum_{\alpha=1}^{N} \phi_\alpha * \dot{q}_\alpha \quad (6)$$

where Φ_I describes the incident wave field and $*$ is the temporal convolution product. The scattered (Φ_S) and radiated fields (ϕ_α) satisfy (4) with

$$\frac{\partial\Phi_S}{\partial n} = -\frac{\partial\Phi_I}{\partial n},\quad \frac{\partial\phi_\alpha}{\partial n} = n_\alpha\delta(t),\quad \mathbf{r} \in \mathcal{S}_b. \quad (7)$$

In addition, Φ_S and ϕ_α are bounded as are their first derivatives as $R = \sqrt{x^2+y^2} \to \infty$. A further decomposition of ϕ_α is possible which separates memoryless effects from nonmemoryless ones in the convolution in (6) [Cummins, 1962]. Initial conditions are required to complete the problem and are discussed later.

For simplicity, the incident wave field is assumed to consist of transient waves propagating in the x-direction, *i.e.*,

$$\Phi_I = \frac{1}{2\pi}\int_{-\infty}^{\infty} \widetilde{\phi}_I e^{-j(k'a'+\omega t)} A(\omega)\, d\omega, \quad \widetilde{\phi}_I = -\frac{jg}{\omega} e^{kz+jk'x}, \quad k=\omega^2/g, \quad k' = k\,\mathrm{sgn}(\omega), \tag{8}$$

and a' is an appropriately chosen constant whose selection is described below. The free surface elevation satisfies

$$\eta(x,y,t) = -\frac{1}{g}\frac{\partial \Phi_I}{\partial t}\bigg|_{z=0} = \frac{1}{2\pi}\int_{-\infty}^{\infty} e^{jk'(x-a')} A(\omega) e^{-j\omega t}\, d\omega \tag{9}$$

so that we can make the identification $A(\omega) = \widetilde{A}(t)$ where $A(t) = \eta(a', y, t)$ and $\widetilde{(\cdot)}$ denotes the Fourier transform:

$$\widetilde{\Phi}(\mathbf{r}) = \int_{-\infty}^{\infty} \Phi(\mathbf{r}, t) e^{j\omega t}\, dt \tag{10}$$

Using Bernoulli's equation, the (linearized) component of $\boldsymbol{f}(t)$ stemming from fluid forces is

$$\boldsymbol{f}_h(t) = \mathrm{col}\left\{ \int_{S_b} \left(\rho \frac{\partial \Phi}{\partial t} + \rho g z\right) n_\alpha\, dS \right\} = \boldsymbol{f}_D(t) + \boldsymbol{f}_R(t) - \boldsymbol{K}_s \mathbf{q}(t). \tag{11}$$

Here, ρ is the fluid density and $\boldsymbol{K}_s = \mathrm{matrix}\{K_{s,\alpha\beta}\} = \boldsymbol{K}_s^T \geq \mathbf{O}$ is the matrix of hydrostatic restoring coefficients; see Newman (1977) for the rigid-body case. The diffraction and radiation forces satisfy $\boldsymbol{f}_D = \mathrm{col}\{f_{D\alpha}\}$, $\boldsymbol{f}_R = \mathrm{col}\{f_{R\alpha}\}$, where

$$f_{D\alpha} = \rho \int_{S_b} \left[\frac{\partial \Phi_I}{\partial t} + \frac{\partial \Phi_S}{\partial t}\right] n_\alpha\, dS, \quad f_{R\alpha} = \rho \sum_{\beta=1}^{N} \int_{S_b} \left[\frac{\partial}{\partial t}(\phi_\beta * \dot{q}_\beta)\, n_\alpha\right] dS. \tag{12}$$

The problem is to now relate $A(t)$ to $\mathbf{q}(t)$ in an explicit manner.

3 Frequency Domain Modeling

Taking the Fourier transform of the force expressions in (12) gives

$$\widetilde{f}_{D\alpha} = \underbrace{-j\omega\rho \int_{S_b} \left[\widetilde{\phi}_I + \widetilde{\phi}_S\right] n_\alpha\, dS e^{-jk'a'}}_{\widetilde{X}_\alpha(\omega)} A(\omega), \quad \widetilde{f}_{R\alpha} = -\sum_{\beta=1}^{N} \underbrace{\left[j\omega\rho \int_{S_b} \left(\widetilde{\phi}_\beta n_\alpha\right) dS\right]}_{\widetilde{H}_{\alpha\beta}(\omega)} \widetilde{\dot{q}}_\beta \tag{13, 14}$$

where we have written $\widetilde{\Phi}_I = \widetilde{\phi}_I e^{-jk'a'} A(\omega)$ and $\widetilde{\Phi}_S = \widetilde{\phi}_S e^{-jk'a'} A(\omega)$. Transformation of the boundary value problem in (4) leads to

$$\nabla^2 \widetilde{\Phi} = 0, \ \ \mathbf{r} \in \mathcal{V}; \qquad \frac{\partial \widetilde{\Phi}}{\partial z} = k\widetilde{\Phi}, \ \ \mathbf{r} \in \mathcal{S}_f; \qquad \lim_{z\to-\infty} \frac{\partial \widetilde{\Phi}}{\partial z} = 0. \tag{15}$$

Using (7), the potentials $\widetilde{\phi}_\alpha$ and $\widetilde{\phi}_S$ satisfy the above with $\widetilde{\phi}_{\alpha,n} = n_\alpha$, $\widetilde{\phi}_{S,n} = -\widetilde{\phi}_{I,n}$ on $\mathcal{S}_b$. In addition they satisfy an outgoing radiation condition as $R \to \infty$. Both potentials can be obtained as solutions to a Fredholm integral equation of the form

$$-2\pi\widetilde{\phi}(\mathbf{r}) + \int_{S_b} \frac{\partial G(\mathbf{r}, \boldsymbol{\xi})}{\partial n_\xi} \widetilde{\phi}(\boldsymbol{\xi}) dS_\xi = \int_{S_b} G(\mathbf{r}, \boldsymbol{\xi}) \frac{\partial \widetilde{\phi}(\boldsymbol{\xi})}{\partial n_\xi} dS_\xi \tag{16}$$

where $G(\mathbf{r}, \boldsymbol{\xi})$ is the Green's function which satisfies Laplace's equation and the free surface, bottom, and radiation conditions [Thorne, 1953]. In the sequel, such an approach in the guise of source distributions is employed to determine $\widetilde{\phi}_\alpha$ and $\widetilde{\phi}_s$ at discrete points on $\mathcal{S}_b$ for several values of k. Problems associated with irregular frequencies are handled using the modifications to $G(\mathbf{r},\boldsymbol{\xi})$ outlined by Ursell [1981].

It is conventional to describe $\widetilde{H}_{\alpha\beta}$ in terms of the added mass and damping coefficients: $\lambda_{\alpha\beta}(\omega) = Re\{\widetilde{H}_{\alpha\beta}\}$, $\mu_{\alpha\beta}(\omega) = -\omega^{-1}Im\{\widetilde{H}_{\alpha\beta}\}$. It can shown using energy considerations that $\mathbf{\Lambda} = \text{matrix}\{\lambda_{\alpha\beta}\}$ is symmetric and positive-semidefinite and $\mathbf{\Lambda} \to \mathbf{O}$ as $\omega \to 0$ and $\omega \to \infty$. The added mass matrix$\{\mu_{\alpha\beta}\}$ is positive-definite in both the low- and high-frequency limits. The values of $\lambda_{\alpha\beta}$, $\mu_{\alpha\beta}$, and $\widetilde{X}_\alpha$ are readily determined from (13) using numerical approximations to the corresponding potential functions.

Assuming that the motions $A(t)$ and $\mathbf{q}(t)$ start at rest at $t=0$, the radiation and diffraction forces can be written in the time domain as

$$f_{D\alpha}(t) = \int_0^t X_\alpha(t-\tau)A(\tau)\,d\tau, \quad f_{R\alpha}(t) = -\sum_{\beta=1}^{N}\int_0^t H_{\alpha\beta}(t-\tau)\dot{q}_\beta(\tau)\,d\tau \tag{17}$$

where $X_\alpha(t)$ and $H_{\alpha\beta}(t)$ are inverse Fourier transforms of the quantities in (14). It is known that $\widetilde{H}_{\alpha\beta}(\omega)$ is an analytic function in a region containing the upper half of the complex ω plane thus yielding a causal convolution operator above for the radiation forces. Arguing on physical grounds, we take a' in (13) to satisfy $a' \le \inf_{\mathbf{r}\in\mathcal{S}_b}[1\ 0\ 0]^T\mathbf{r}$ (the x-coordinate of the wetted portion of the body first exposed to the incoming wave) so that $X_\alpha(t)$ is causal and hence $\widetilde{X}_\alpha(\omega)$ is analytic for $Im\{\omega\} > 0$. Discussion provided in [Falnes, 1995] suggests that the 'diffraction operator' X_α can be made only approximately causal by such a device.

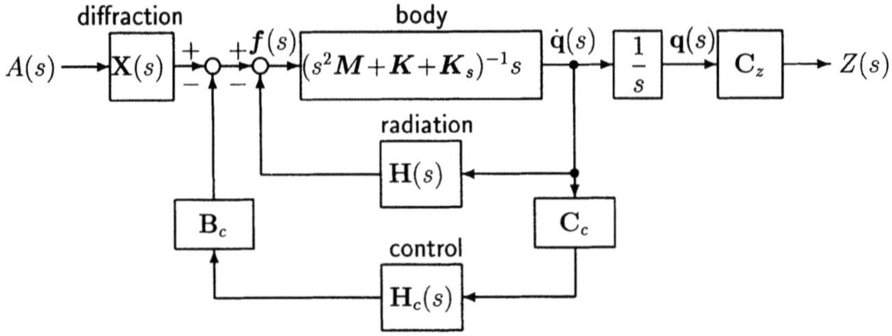

Figure 1: Block Diagram of Combined Radiation/Diffraction Problem

Our purposes are better served using the Laplace transform $\mathcal{L}\{\cdot\}$. Setting $X_\alpha(s) = \mathcal{L}\{X_\alpha(t)\}$, $H_{\alpha\beta}(s) = \mathcal{L}\{H_{\alpha\beta}(t)\}$, and taking the transform of (17) gives

$$\mathbf{f}_D(s) = \mathbf{X}(s)A(s)\,, \quad \mathbf{f}_R(s) = -\mathbf{H}(s)\dot{\mathbf{q}}(s) \tag{18}$$

Note that $\mathbf{X}(j\omega) = \text{col}\{\widetilde{X}^*_\alpha(\omega)\}$ and $\mathbf{H}(j\omega) = \text{matrix}\{\widetilde{H}^*_{\alpha\beta}(\omega)\}$. The block diagram in Figure 1 illustrates the relationship between the wave height and the body motion. The transfer matrix describing the body is a rational (matrix) function of s and gives rise to the constant-coefficient differential equations in (3). We seek similar descriptions for $\mathbf{H}(s)$ and $\mathbf{X}(s)$. The block $\mathbf{H}_c(s)$ corresponds to a control system designed to minimize the effect of $A(t)$ on $Z(t) = \mathbf{C}_z\mathbf{q}(t)$. The quantity $\mathbf{C}_c\dot{\mathbf{q}}(t)$ represents rate measurements available to the controller and $\mathbf{B}_c$ describes the (spatial) distribution of control forces on the body. If the entire diagram is summarized by $Z(s) = T(s)A(s)$ and we define $\|Z\|_2^2 = \int_0^\infty Z^2(t)\,dt$ then a suitable control objective is to select $\mathbf{H}_c(s)$ to minimize $\sup \|Z\|_2/\|A\|_2$ where the supremum is taken over all $A(t)$ with $\|A\|_2 < \infty$, *i.e.*, minimize the L_2-induced norm of the mapping from wave height to body response as measured by $Z(t)$. Efficient algorithms exist for the synthesis of such controllers when the controlled system is described in a FDLTI fashion.

4 Approximation by Analytic Functions

It can be established that $\mathbf{H}(s)$ satisfies the following conditions for $Re\{s\} > 0$: (i) $\mathbf{H}(s)$ is analytic; (ii) $\mathbf{H}(s)$ is real for real s; (iii) $\frac{1}{2}[\mathbf{H}(s) + \mathbf{H}^H(s)] \geq \mathbf{O}$. These properties constitute the definition of a positive real (PR) function (Zemanian, 1965) and are known to be equivalent to the passivity of an LTI system, *i.e.*, in the present case

$$-\int_0^T \boldsymbol{f}_R^T(t)\dot{\mathbf{q}}(t)\,dt \geq 0, \;\; \forall T \geq 0$$

which states that radiation is a dissipative process. It will be further assumed that $\mathbf{H}(s)$ and $\mathbf{X}(s)$ are continuous on the axis $s = j\omega$ which excludes the possibility of isolated poles on the imaginary axis. The radiation impedance can be further written as

$$\mathbf{H}(s) = \mathbf{H}_r(s) + s\boldsymbol{M}_\infty\,, \quad \boldsymbol{M}_\infty = \text{matrix}\{\lim_{\omega\to\infty} \mu_{\alpha\beta}(\omega)\}$$

$\mathbf{H}_r(s)$ continues to enjoy the positive real property but it is strictly proper, *i.e.*, $\mathbf{H}_r(s) \to \mathbf{O}$ as $s \to \infty$ since $\mathbf{\Lambda} \to \mathbf{O}$ in this limit. It is assumed that $\boldsymbol{M}_\infty$ is known; this calculation is feasible using a special limiting form of the Green's function.

Since $\mathbf{X}(s)$ and $\mathbf{H}_r(s)$ are analytic in a region containing the open right-half plane, they can be uniformly approximated in compact regions of this region by rational functions with poles in the left-half plane (this is Runge's Theorem — see [Rudin, 1987]). We propose to make such an approximation on a finite part of the imaginary axis then extend the function to the right-half plane (RHP) by analytic continuation. Standard properties of rational positive real functions can be found in [Tao & Iaonnou, 1988]. In the scalar case, it is interesting to note that all rational PR functions have necessarily relative degree one so that rational approximations to the added damping and mass coefficients satisfy $\lambda_{\alpha\alpha} \propto \omega^{-2}$ and $[\mu_{\alpha\alpha} - \mu_{\alpha\alpha}(\infty)] \propto \omega^{-2}$ as $\omega \to \infty$. This is at odds with known asymptotics for these coefficients but the discrepancy can be pushed to arbitrarily high frequency by increasing the order of the approximation.

The low-frequency asymptotics are similarly mismatched. For example, in the case of a two-dimensional heaving cylinder, Ursell [1964] has shown that near $s = 0$, $H_r(s)$ (suitably nondimensionalized) behaves like

$$H_r(s) = \frac{8}{\pi^2}\left[\frac{-2s\log s + (\frac{3}{2} - 2\log 2 - \gamma)s\cdots}{1 + \frac{4}{\pi}s^2\log s + \cdots}\right] \tag{19}$$

and hence has a logarithmic branch point at $s = 0$. Much of the hydrodynamical literature has emphasized the role of the acceleration to force map ($H_r(s)/s$) which does not possess the positive real property. The impedance $H_r(s)$ is not only positive real but partially avoids the problem at $s = 0$. Unlike $H_r(s)/s$ on the imaginary axis, $H_r(j\omega)$ is continuous at $\omega = 0$; Mergelyan's extension to Runge's theorem [Rudin, 1987] permits uniform approximation by rational functions in a compact region where the function is analytic inside but upon whose boundary the function is merely continuous.

To fix ideas, consider a single degree of freedom $q(t)$, say the heaving motion of a cylinder in two dimensions. The simplest rational positive real function, without poles on the imaginary axis, and satisfying $Re\{\widehat{H}_r(\jmath\omega)\} = 0$ in the high- and low-frequency limits is one of second order:

$$\widehat{H}_r(s) = \frac{bs}{s^2 + 2\zeta\Omega s + \Omega^2} \tag{20}$$

with $b > 0$, $\zeta > 0$, and $\Omega > 0$ which corresponds to the following relationship between $\dot{q}(t)$ and $f_R(t)$: $f_R(t) = -M_\infty \ddot{q}(t) + f_r(t)$ with $f_r = -\dot{q}_r$, $\ddot{q}_r + 2\zeta\Omega\dot{q}_r + \Omega^2 q_r = b\dot{q}$. In general, we propose to approximate $H_r(s)$ as

$$\widehat{H}_r(s) = \frac{b(s)}{a(s)} = \frac{\sum_{i=1}^{n-1} b_i s^{n-i}}{s^n + \sum_{i=1}^{n} a_i s^{n-i}} \tag{21}$$

which can be written as a sum of terms of the form of (20) when n is even. Given the values of the added mass and damping coefficients at l discrete frequencies ω_i, we propose to minimize the following function with respect to the a_i and b_i:

$$\mathcal{J} = \tfrac{1}{2}\sum_{i=1}^{l} \left|\widehat{H}_r(\jmath\omega_i) - H_r(\jmath\omega_i)\right|^2, \quad H_r(\jmath\omega_i) = \lambda(\omega_i) + \jmath\omega_i(\mu_{\alpha\beta}(\omega_i) - \mu_{\alpha\beta}(\infty)) \tag{22}$$

This is not in the form of a standard least-squares problem but can be approximated by the series of problems

$$\mathcal{J}^{(m)} = \tfrac{1}{2}\sum_{i=1}^{l} \left|b^{(m)}(j\omega_i) - H_r(j\omega_i)a^{(m)}(j\omega_i)\right|^2 \left|W^{(m)}(j\omega_i)\right|^2, \quad m = 1, \ldots, M \tag{23}$$

which is quadratic in the b_i, a_i. Here, $W^{(m)}(s)$ is a weighting function which we take as 1 for $m = 1$ and $W^{(m)}(s) = 1/a^{(m-1)}(s)$, $m = 2, 3, \cdots, M$. In carrying out the optimization, it may so happen that $a^{(M)}(s)$ has poles in the RHP, leading to an unstable approximation. This is remedied by carrying out one further optimization with a prescribed denominator for $H_r(s)$ containing the stable poles of $a^{(M)}(s)$. Typically the resulting approximations are positive real since the data possess this property. If not, it is possible to further approximate $\widehat{H}_r(s)$ by the closest positive real function with the same poles.

The constituents of $\mathbf{X}(s)$, $X_\alpha(s)$, can be approximated in the same way but the positive real and low frequency conditions are not required. Ideally, the same denominator polynomial $a_D(s)$ is used for each $\widehat{X}_\alpha(s)$ so that

$$\widehat{X}_\alpha(s) = \frac{b_{D\alpha}(s)}{a_D(s)} = \frac{\sum_{i=1}^{n} b_{\alpha i} s^{n-i}}{s^n + \sum_{i=1}^{n} a_{di} s^{n-i}} \tag{24}$$

Then, the first mapping in (18) can be realized in the the time domain as

$$\boldsymbol{f}_D(t) = \mathbf{C}_d \mathbf{x}_d, \quad \dot{\mathbf{x}}_d = \mathbf{A}_d \mathbf{x}_d + \mathbf{B}_d A(t) \tag{25}$$

where

$$\mathbf{C}_d = \text{matrix}\{b_{\alpha i}\}, \quad \mathbf{B}_d = [1\ 0\ \cdots 0\ 0]^T, \quad \mathbf{A}_d = \begin{bmatrix} -a_{d1} & -a_{d2} & \cdots & -a_{d,n-1} & -a_{dn} \\ 1 & 0 & \cdots & 0 & 0 \\ 0 & 1 & \cdots & 0 & 0 \\ \vdots & \vdots & \ddots & \vdots & \vdots \\ 0 & 0 & \cdots & 1 & 0 \end{bmatrix}$$

The denominator $a_D(s)$ can be the product of individual ones optimised for each $X_\alpha(s)$ and the $b_{\alpha i}$ can be obtained using one final least squares optimization for each $\widehat{X}_\alpha$ using the composite $a_D(s)$.

In the multivariable case, we propose to represent $\mathbf{H}_r(s)$ as

$$\widehat{\mathbf{H}}_r(s) = \sum_{k=1}^{n'} \mathbf{W}_k \frac{s}{s^2 + 2\zeta_k \Omega_k s + \Omega_k^2}, \quad \zeta_k > 0, \;\; \Omega_k > 0, \;\; \mathbf{W}_k = \mathbf{W}_k^T. \tag{26}$$

Every rational matrix PR function without poles on the imaginary axis, which is strictly proper, and has a real part vanishing in the low-frequency limit admits such a representation. The ζ_k and Ω_k are determined by applying the scalar approach outlined above to the diagonal terms $H_{r,\alpha\alpha}(s)$. With these parameters then fixed, the quantities $W_{k,\alpha\beta}$ are determined by approximating $H_{r,\alpha\beta}(s)$ with fixed poles according to (22). The resulting matrices $\mathbf{W}_k$ can be factored as $\mathbf{W}_k = \mathbf{B}_{rk}\mathbf{B}_{rk}^T$ where $\mathbf{B}_{rk}$ has full column rank which we denote by m_k. The advantage of (26) with this factorization lies in its ability to be realized in a matrix second-order form:

$$\boldsymbol{f}_r(t) = -\mathbf{B}_r\dot{\mathbf{q}}_r, \quad \ddot{\mathbf{q}}_r + \mathbf{D}_r\dot{\mathbf{q}}_r + \mathbf{K}_r\mathbf{q}_r = \mathbf{B}_r^T\dot{\mathbf{q}} \tag{27}$$

where $\mathbf{B}_r = \text{row}\{\mathbf{B}_{rk}\}$, $\mathbf{D}_r = \text{block diag}\{2\zeta_k\Omega_k\mathbf{1}_{m_k}\}$, $\mathbf{K}_r = \text{block diag}\{\Omega_k^2\mathbf{1}_{m_k}\}$, and $\mathbf{1}_{m_k}$ is the $m_k \times m_k$ identity matrix. Hence, there are $\sum m_k$ coordinates in $\mathbf{q}_r$. When the radiation forces are written as $\boldsymbol{f}_R(t) = -\boldsymbol{M}_\infty\ddot{\mathbf{q}} + \boldsymbol{f}_r(t)$ and combined with (3), (11), (25), and (27), we have

$$\begin{bmatrix} \boldsymbol{M}+\boldsymbol{M}_\infty & \mathbf{O} \\ \mathbf{O} & \mathbf{1} \end{bmatrix} \begin{bmatrix} \ddot{\mathbf{q}} \\ \ddot{\mathbf{q}}_r \end{bmatrix} + \begin{bmatrix} \mathbf{O} & \mathbf{B}_r^T \\ \mathbf{B}_r & \mathbf{D}_r \end{bmatrix} \begin{bmatrix} \dot{\mathbf{q}} \\ \dot{\mathbf{q}}_r \end{bmatrix} + \begin{bmatrix} \boldsymbol{K}+\boldsymbol{K}_s & \mathbf{O} \\ \mathbf{O} & \mathbf{K}_r \end{bmatrix} \begin{bmatrix} \mathbf{q} \\ \mathbf{q}_r \end{bmatrix} = \begin{bmatrix} \mathbf{C}_d \\ \mathbf{O} \end{bmatrix} \mathbf{x}_d \tag{28}$$

with $\dot{\mathbf{x}}_d = \mathbf{A}_d\mathbf{x}_d + \mathbf{B}_d A(t)$. This model is of the desired form and the unforced problem ($A(t) \equiv 0$) is characterized by symmetric mass, damping, and stiffness matrices. The corresponding eigenproblem permits the determination of the 'hydroelastic modes' which is one way of further reducing the order of the model.

5 Numerical Examples

Consider the two-dimensional heaving motion of a semicircular cylinder of radius a. The added mass and damping coefficients obtained with an integral equation approach with 100 constant source panels are shown in Figure 2a. The diffraction force coefficient is shown in Figure 2b where a value of $a' = -2a$ was used to describe the wave height. Fits to both functions are shown for various values of n. A single second order approximation captures the added mass and damping coefficients quite well and for $n = 4$ there is little discernible difference. The corresponding time behaviour of the free decay from a displaced initial configuration is shown in Figure 2c for $n = 4$ and $n = 10$. Our results show excellent agreement with those of Yeung [1982]. A similar approach was employed for the heaving motion of a (three-dimensional) hemisphere and similar agreement was obtained between the panel solutions and approximations obtained using the same number of functions as the cylinder case. The free decay results were in good agreement with those of Beck and Liapis [1987].

Now consider a thin floating plate with dimensions $a \times b \times c$ ($a/b = 2.5$, $a/c = 30$)

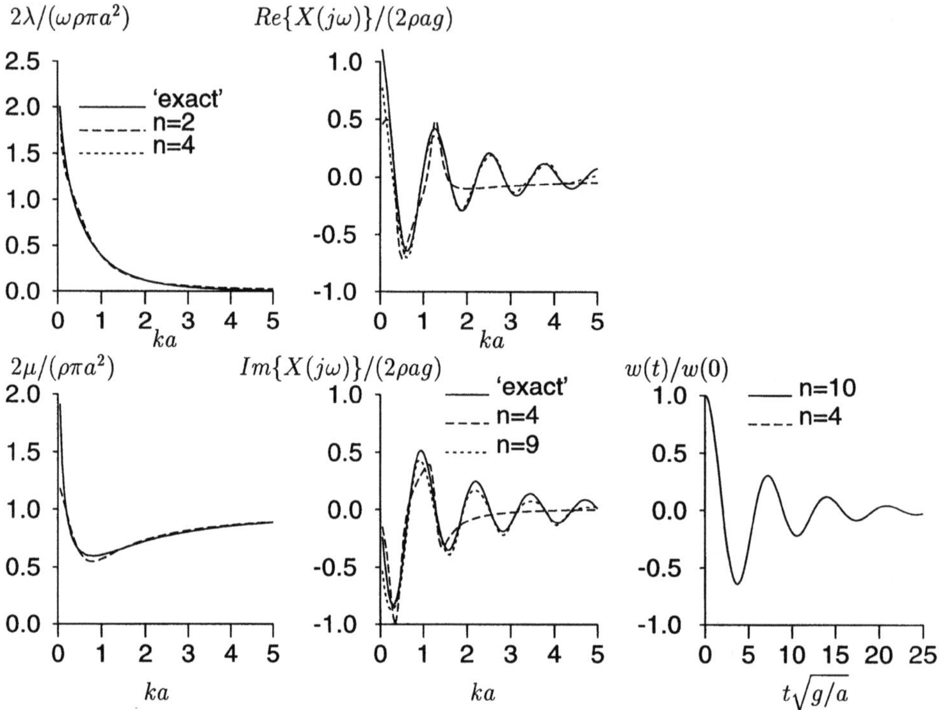

(a) Radiation Coefficients (b) Diffraction Coefficients (c) Free Decay in Heave

Figure 2: Results for Two-Dimensional Heave of a Cylinder

and constant mass density σ. The stiffness operator is $\boldsymbol{K} = D\nabla^4$ where D is the modulus of rigidity. The strain energy expression is well known in this case and was used in conjunction with a 10×10 array of Legendre polynomials to discretize (1). The eigenproblem associated with $\boldsymbol{M}$ and $\boldsymbol{K}$ was solved to obtain the vibration modes relative to the polynomial basis. These then comprised the $\boldsymbol{\psi}_\alpha$ for the fluid problem with $\alpha = 1, 2, 3$ corresponding to the rigid-body modes in heave, pitch, and roll. The fluid problem was solved for the first 9 modes with an array of 30×12 constant source panels on the wetted plate surface.

Some of the radiation and diffraction force coefficients are depicted in Figure 3. A fourth order approximation like (21) was used to initially fit each $H_{\alpha\alpha}(s)$, $\alpha = 1 \ldots 9$, which furnished 18 appropriate values for the the Ω_k and ζ_k in (26). The entries in each $\mathbf{W}_k$ were then obtained by fitting each $H_{\alpha\beta}(s)$ with the denominators fixed. The resulting approximations are shown in Figure 3 for rigid-body heave, the first vibration mode ($\alpha = 4$), and the radiation coupling between them. The approximations are very good and are typical of that obtained for other modes. The diffraction forces were obtained using a value of $a' = -2a$ with $n = 8$ for each $\widehat{X}_\alpha$. The resulting approximation in (25) was of 66th order after unstable poles were removed. Given in Figure 3 are the ensuing approximations for $\alpha = 1$, 4, and 8.

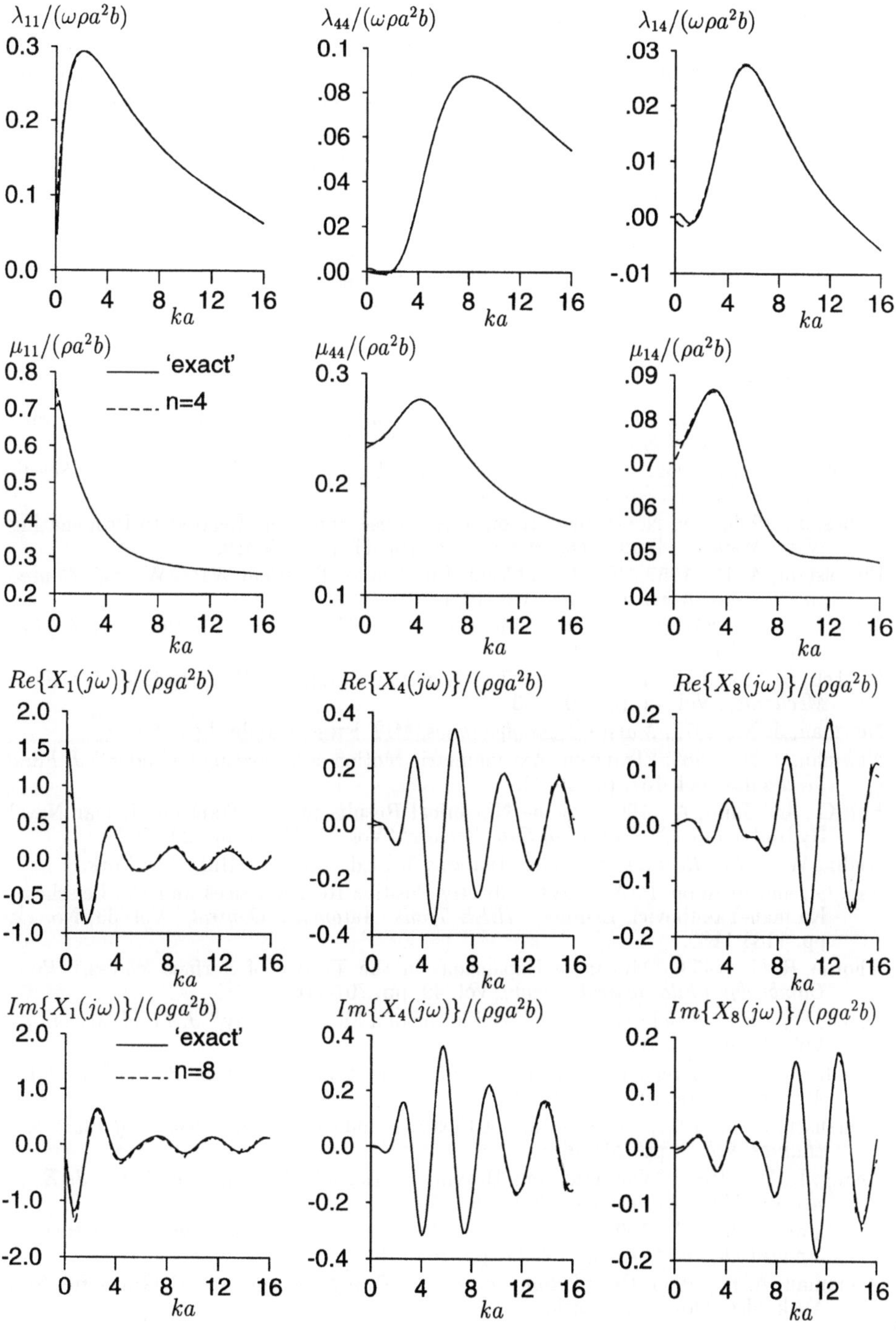

Figure 3: Radiation and Diffraction Force Coefficients for a Thin Floating Plate

6 Concluding Remarks

The approach presented here has the ability to approximate transient solutions arbitrarily closely and provides an alternative methodology upon which comparison with other techniques can be made. It is also based on frequency-domain solutions for which well established procedures already exist. Most important is the explicit characterization of the dynamics relating a transient sea to the body motions in terms of linear constant-coefficient matrix differential equations. This is particularly important when one contemplates the problem of control system design for large floating structures. The success of the method suggests the possibility of using rational approximations for the Green's function itself so that the motion equations may be formulated directly in the time domain.

References

Beck, R. F., and Liapis, S., 1987, "Transient Motions of Floating Bodies at Zero Forward Speed," *Journal of Ship Research*, Vol. 31, No. 3, pp. 164-176.

Cummins, W. E., 1962, "The Impulse Response Function and Ship Motions," *Schiffstechnik*, Vol. 47, No. 9, pp. 101-109.

Falnes, J., 1995, "On Non-Causal Impulse Response Functions Related to Propagating Water Waves," *Applied Ocean Research*, Vol. 17, pp. 379-389.

Finkelstein, A. B., 1957, "The Initial Value Problem for Transient Water Waves," *Comm. on Pure and Applied Maths.*, Vol. 10, pp. 511-522.

John, F., 1950, "On the Motion of Floating Bodies II.," *Comm. on Pure and Applied Maths.*, Vol. 3, pp. 45-101.

Maskell, S. J., and Ursell, F., 1970, "The Transient Motion of a Floating Body," *J. Fluid Mechanics*, Vol. 44, pp. 303-313.

Newman, J. N., 1977, *Marine Hydrodynamics*, MIT Press, Cambridge, MA.

Newman, J. N., 1985, "Transient Axisymmetric Motion of a Floating Cylinder," *J. Fluid Mechanics*, Vol. 157, pp. 17-33.

Pot, G., and Jami, A., 1991, "Some Numerical Results in 3-D Transient Linear Naval Hydrodynamics," *Journal of Ship Research*, Vol. 35, No. 4, pp. 295-303.

Rudin, W., 1987, *Real and Complex Analysis*, 3rd Ed., McGraw-Hill, New York.

Tao, G., and Ioannou, P. A., 1988, "Strictly Positive Real Matrices and the Lefschetz-Kalman-Yakubovich Lemma," *IEEE Trans. Automatic Control*," Vol. 33, No. 12, pp. 1183-1185.

Thorne, R. C., 1953, "Multipole Expansions in the Theory of Surface Waves," *Proc. Cambridge Philosophical Society*, Vol. 49, pp. 707-716.

Ursell, F., 1964, "The Decay of the Free Motion of a Floating Body," *J. Fluid Mechanics*, Vol. 19, pp. 305-319.

Ursell, F., 1981, "Irregular Frequencies and the Motion of Floating Bodies," *J. Fluid Mechanics*, Vol. 105, pp. 143-156.

Wehausen, J. V., 1971, "The Motion of Floating Bodies," *Annual Review of Fluid Mechanics*, Vol. 3, pp. 237-268.

Yeung, R. W., 1982, "The Transient Heaving Motion of Floating Cylinders," *J. Engineering Mathematics*, Vol. 16, pp. 97-119.

Yu, Z., and Falnes, J., 1995, "State-Space Modelling of a Vertical Cylinder in Heave," *Applied Ocean Research*, Vol. 17, pp. 265-275.

Zemanian, A. H., 1965, *Distribution Theory and Transform Analysis*, McGraw-Hill, New York; also, Dover publications.

PROOF OF CONVERGENCE FOR THE COUPLED FINITE/INFINITE ELEMENT METHODS FOR HELMHOLTZ EXTERIOR BOUNDARY-VALUE PROBLEMS

Leszek Demkowicz and Frank Ihlenburg

Texas Institute for Computational and Applied Mathematics
The University of Texas at Austin
Austin, TX 78712

1 Introduction

The note outlines the idea of coupled Finite Element (FE) - Infinite Element (IE) approximations for the exterior Helmholtz equation and summarizes the convergence analysis presented in [6]. The main difficulty in analyzing the convergence of approximations to the Helmholtz equation in *exterior* domains lies in two facts:

- The sesquilinear form corresponding to the variational formulation is not positive-definite [1],
- The weighted Sobolev space for exterior domains *is not* compactly imbedded in the L^2-space and therefore the standard asymptotic convergence argument [4] cannot be applied.

We show how to circumference both problems.

The presentation is concluded with some preliminary considerations for an analogous concept for Maxwell's equations.

2 Approximation by finite/infinite elements

Given a bounded domain Ω with boundary Γ, we wish to solve the Helmholtz equation in the exterior of Ω. The equation is accompanied by a Neumann boundary condition on Γ and

[1] The corresponding proof for the Laplace equation, based on a Poincare inequality, is elementary, see e.g. [7]

T.L. Geers (ed.), IUTAM Symposium on Computational Methods for Unbounded Domains, 83–93.

the Sommerfeld radiation condition at ∞^2.

$$\begin{cases} -\Delta u - k^2 u &= 0 \qquad \text{in } \Omega^e = \mathbb{R}^3 - \bar{\Omega} \\ \dfrac{\partial u}{\partial n} &= g \qquad \text{on } \Gamma \\ \dfrac{\partial u}{\partial r} - iku &= O(\frac{1}{r^2}) \quad \text{as } r \to \infty \end{cases} \tag{2.1}$$

Here k is the wave number, (r, θ, ϕ) denote the standard spherical coordinates with an origin inside of Ω and n stands for the outward normal unit for domain Ω.

To approximate the problem, we introduce first a truncating sphere S surrounding domain Ω (in practice as close as possible)[3]. The sphere S partitions the exterior domain Ω^e into domain Ω_S and domain Ω_S^e exterior to the sphere. By a simple scaling argument, we can always assume that S has a unit radius. Next we mesh the truncated domain Ω_S with a FE grid supporting a standard C^0-approximation. In particular we can use hp meshes, with *locally* variable element size h and order of approximation p, to efficiently capture possible irregularities of the solution (scattering form corners, edges, resonating cavities). For the sake of an argument we shall assume that the mesh matches the geometry of the truncating sphere perfectly, i.e. that the *exact geometry* elements are used. Finally, we construct a spectral-type approximation in the exterior spherical domain Ω_S^e, matching the FE mesh on the sphere. If $e_m(\hat{\boldsymbol{x}}), \hat{\boldsymbol{x}} \in S$ is a typical FE basis function defined on the unit sphere, the corresponding approximation is defined as:

$$u(r, \hat{\boldsymbol{x}}) = \sum_{n=1}^{N} \frac{e^{ikr}}{r^n} \sum_{m=1} u_{nm} e_m(\hat{\boldsymbol{x}}) \tag{2.2}$$

The "shape" functions of radius r correspond to the classical expansion formula for the exact solution *valid outside* the smallest sphere surrounding domain Ω. In this way we satisfy the *approximability condition*, i.e. make sure that, as the number N of terms in the radial direction increases, there is a chance for resolving the exact solution. Note also that functions (2.2) satisfy automatically the radiation condition.

Now comes a variational formulation. Since the domain is unbounded, we multiply $(2.1)_1$ with complex conjugate[4] of a test function v and integrate first only over a truncated domain Ω_γ^e extending from boundary Γ to a sphere S_γ with radius γ. Standard integration by parts gives:

$$\int_{\Omega_\gamma^e} [\boldsymbol{\nabla} u \boldsymbol{\nabla} \bar{v} - k^2 u \bar{v}] - \int_{S_\gamma} \frac{\partial u}{\partial n} \bar{v} = \int_\Gamma g \bar{v} \tag{2.3}$$

Next we want to pass with $\gamma \to \infty$.

[2]The Neumann boundary condition, corresponding to the rigid scattering problem, can be replaced with a Dirichlet or a local impendance boundary condition without any essential changes in the presented results.

[3]In practical applications prolate or oblate spheroids are preferred [2].

[4]An alternative development including the original version of Burnett is possible using unconjugated test functions [2, 8].

Option 1. If the integrals are to be interpreted in the Lebesgue sense, the test functions cannot be approximated with (2.2). Instead, we have to use

$$v(r,\hat{\boldsymbol{x}}) = \sum_{n+1}^{N} \frac{e^{ikr}}{r^{n+2}} \sum_{m=1}^{M} v_{nm} e_m(\hat{\boldsymbol{x}}) \tag{2.4}$$

The S_γ boundary term in (2.3) dissappears in the limit. The resulting Petrov-Galerkin approximation corresponds precisely to the Leis functional setting based on the *weighted Sobolev spaces* [10, 7], see also [1].

Option 2. We start with the domain decomposition,

$$\int_{\Omega^e_\gamma} [\nabla u \nabla \bar{v} - k^2 u\bar{v}] = \int_{\Omega_s} [\,] + \int_{\Omega^e_S} [\,] \tag{2.5}$$

and interprete the integral over Ω^e_S in the Cauchy Principal Value sense,

$$⨍_{\Omega^e_S} [\nabla u \nabla \bar{v} - k^2 u\bar{v}]\, d\Omega^e_S := \int_S \lim_{\gamma\to\infty} \int_1^\gamma [\nabla u \nabla \bar{v} - k^2 u\bar{v}]\, r^2 dr dS \tag{2.6}$$

Upon a closer inspection, we find out that the Lebesgue non-integrable contributions corresponding to term e^{ikr}/r in expansion (2.2), used now for *both* u and v, cancel out.

Additionally, we build the Sommerfeld radiation condition into the formulation by replacing $\partial u/\partial n$ in the S_γ-term with $iku + \phi$, where $\phi = O(1/r^2)$ is an unknown function. Upon passing with $\gamma \to \infty$, the term with ϕ disappears and we get a conjugated analog of Burnett's formulation [2]

$$⨍_{\Omega^e} [\nabla u \nabla \bar{v} - k^2 u\bar{v}] - ik \lim_{\gamma\to\infty} \int_{S_\gamma} u\bar{v}\, dS_\gamma = \int_\Gamma g\bar{v} \tag{2.7}$$

Note that the second term involves only the first order terms ($n = 1$) in expansion (2.2). Finally, notice that this time we deal with a Bubnov-Galerkin approximation as the discretization of both u and v is identical.

3 Convergence analysis

We begin by rewriting (2.1) as a coupled problem.

$$\left\{ \begin{array}{rll} -\Delta u - k^2 u &= 0 & \text{in } \Omega_S \\ \frac{\partial u}{\partial n} &= g & \text{on } \Gamma \\ -\Delta u_1 - k^2 u_1 &= 0 & \text{in } \Omega^e_S \\ \frac{\partial u_1}{\partial r} - iku_1 &= O(\frac{1}{r^2}) & \text{as } r \to \infty \\ u &= u_1 & \text{on } S \\ \frac{\partial u}{\partial n} &= \frac{\partial u_1}{\partial n} & \text{on } S \end{array} \right. \tag{3.1}$$

where n denotes the outward unit vector for Ω_S.

Multiplying $(3.1)_1$ by a test function $\bar{v}$, integrating over Ω_S, integrating by parts, using the Neumann boundary condition on Γ and the first compatibility condition on S, we arrive at:

$$\int_{\Omega_S} \nabla u \nabla \bar{v} - k^2 u\bar{v} - \int_S \frac{\partial u_1}{\partial n}\bar{v} = \int_\Gamma g\bar{v} \tag{3.2}$$

At this point we introduce the DtN operator[5] [9, 3] corresponding to the exterior spherical domain Ω_S^e.

$$G: \quad H^{\frac{1}{2}}(S) \ni u_1^0 \to \frac{\partial u_1}{\partial n}|_S \in H^{-\frac{1}{2}}(S) \tag{3.3}$$

Here u_1 denotes the solution of the Dirichet problem:

$$\begin{cases} -\Delta u_1 - k^2 u_1 &= 0 \qquad \text{in } \Omega_S^e \\ \dfrac{\partial u_1}{\partial r} - iku_1 &= O(\frac{1}{r^2}) \quad \text{as } r \to \infty \\ u_1 &= u_1^0 \qquad \text{on } S \end{cases} \tag{3.4}$$

Using the second compatibility condition on S, we end up with the final variational formulation in the form:

$$\int_{\Omega_S} \nabla u \nabla \bar{v} - k^2 u\bar{v} - \int_S (Gu)\bar{v} = \int_\Gamma g\bar{v} \tag{3.5}$$

involving the *non-local* term with the DtN operator.

Exactly the same procedure can be repeated for the approximate solution. In particular, one can think of an intermediate step where the solution of problem (3.4) is done using only N terms in the radial direction,

$$u(r, \hat{\boldsymbol{x}}) = \sum_{n=1}^{N} \frac{e^{ikr}}{r^n} u_n(\hat{\boldsymbol{x}}) \tag{3.6}$$

but with arbitrary functions $u_n \in H^1(S)$. This results in an *approximate DtN operator* G_N replacing G and a corresponding approximate solution u^N of:

$$\int_{\Omega_S} \nabla u^N \nabla \bar{v} - k^2 u^N \bar{v} - \int_S (G_N u)\bar{v} = \int_\Gamma g\bar{v} \tag{3.7}$$

With u_h^N denoting the ultimate FE/IE solution, we have the obvious estimate:

$$\|u - u_h^N\| \le \|u - u^N\| + \|u^N - u_h^N\| \tag{3.8}$$

where $\|\ \|$ denotes the $H^1(\Omega_S)$ norm. We shall restrict ourselves to estimating the first part of the error, corresponding to the IE approximation only.

[5]also known as the Poincare-Steklov operator or the Green operator of the first kind

THEOREM 1

Suppose, for a given data g, that the variational problem (3.5) has u unique solution $u \in H^1(\Omega_S)$ *with trace* $u|_S \in H^\alpha(S), \alpha \geq \frac{1}{2}$. *Assume additionally that the semidiscrete problem (3.7) satisfies the inf-sup condition with constant* $\gamma_N > 0$.

Then the following bound holds

$$\|u - u^N\| \leq T\gamma_N^{-1}\|G - G_N\|_{\mathcal{L}(H^\alpha, H^{-\frac{1}{2}})}\|u|_S\|_{H^\alpha(S)} \tag{3.9}$$

where T is the trace constant relating norms $\|u\|$ *and* $\|u|_S\|_{H^{\frac{1}{2}}}$. ∎

The message of Theorem 1 is standard. In order to guarantee the convergence, we need the stability of the discrete problem, represented by the discrete inf-sup constant γ_N, and the approximability expressed in terms of the convergence of operator G_N to G.

4 Spectral decomposition

Operators G and G_N admit standard spectral decomposition using spherical harmonics. Representing $u|_S$ and $v|_S$ in the standard form,

$$u(1, \hat{\boldsymbol{x}}) = \sum_{n=1}^{\infty} \sum_{m=-n}^{n} u_n^m Y_n^m(\hat{\boldsymbol{x}}), \quad \hat{\boldsymbol{x}} \in S \tag{4.1}$$

with $Y_n^m(\hat{\boldsymbol{x}})$ denoting the spherical harmonics, we obtain:

$$\int_S Gu\bar{v}\, dS = \sum_{n=1}^{\infty} \lambda_n \sum_{m=-n}^{n} u_n^m \bar{v}_n^m \tag{4.2}$$

where

$$\lambda_n = -k\frac{h_n'(k)}{h_n(k)} \tag{4.3}$$

with $h_n(z)$ denoting the Hankel function of the first kind of order n. The formula above results from the standard solution to the Dirichlet problem (3.4) using the separation of variables.

Lemma 1

The eigenvalues λ_n *have the following properties.*

(i) The imaginary parts of all λ_n *are strictly negative and*

$$\Im\lambda_n \to 0 \quad as\ n \to \infty \tag{4.4}$$

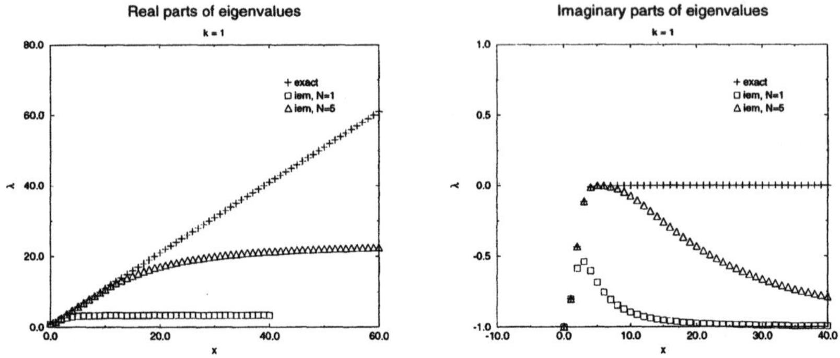

Figure 1: Eigenvalues for $k = 1$

(ii) The real parts of λ_n admit a bound:

$$\max\{1, n+1-k\} \leq \Re\lambda_n \leq n+1+k \tag{4.5}$$

Hence, for large n,

$$\lambda_n \approx n+1 \tag{4.6}$$

∎

An identical spectral decomposition can be introduced for the approximate DtN operator G_N. The "approximate" eigenvalues λ_n^N depend upon the variational formulation and number N of terms in the expansion.

Figures 1,2 present a comparison of the exact and typical approximate eigenvalues for wave numbers $k = 1, 20$ and number of terms $N = 1, 5$. The approximate eigenvalues correspond to the Leis formulation. Note that with N terms used, the first N eigenvalues *are always reproduced exactly*, i.e.,

$$\lambda_n = \lambda_n^N, \quad n = 1, \ldots, N \tag{4.7}$$

Based on the numerical experiments we postulate the following conjecture.

Conjecture 1
The discrete eigenvalues have the properties:
(1) The imaginary parts of all λ_n^N are negative and bounded with

$$\Im\lambda_n^N \to -k \qquad as\, n \to \infty\,. \tag{4.8}$$

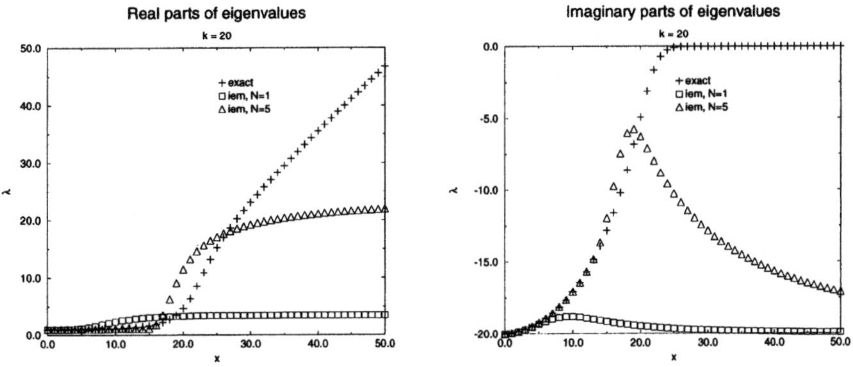

Figure 2: Influence of wavenumber ($k = 20$)

(2) The real parts of all λ_n^N form for each N an increasing sequence, converging to the limit

$$\lambda_*^N = \lim_{n\to\infty} \Re\lambda_n^N = N^2 + 4N + 1\,. \tag{4.9}$$

(3) The real parts are positive and bounded from below by one,

$$1 \leq \Re\lambda_n^N\,. \tag{4.10}$$

▮

Incidentally, the spectral decomposition and the properties of eigenvalues λ_i are crucial in establishing a simple proof of existence and uniqueness of solution to the reduced problem (3.5) and, consequently, to the original problem (2.1).

THEOREM 2

The reduced variational problem (3.5) has for all $g \in H^{-\frac{1}{2}}(\Gamma)$ a unique solution $u \in H^1(\Omega_S)$.

▮

Proof: Split the sesquilinear form $b(u,v)$ corresponding to variational formulation (3.7) into two parts:

$$\begin{aligned} a(u,v) &= \int_{\Omega_S} \nabla u \nabla \bar{v} + \sum_{n=1}^{\infty} \Re\lambda_n \sum_{m=-n}^{n} u_n^M \bar{v}_n^m \\ c(u,v) &= -k^2 \int_{\Omega_S} u\bar{v} - i \sum_{n=1} \Im\lambda_n \sum_{m=-n}^{n} u_n^M \bar{v}_n^m \end{aligned} \tag{4.11}$$

Properties of λ_n summarized in Lemma 1 imply that form $a(u,v)$ is H^1-coercive and (complex) symmetric, and form $c(u,v)$ is compact. Therefore, the Fredholm alternative holds, and it is sufficient to show uniqueness. But this follows from equating $b(u,u)$ to zero and using the strict negativeness of $\Im\lambda_n$. ∎

REMARK 1 The possibility of splitting the sesquilinear form into the symmetric positive definite part and the compact perturbation is also crucial in proving the convergence of the FE solution u_h^N to u^N, see e.g. [4]. ∎

The experimentally observed properties of the approximate eigenvalues λ_n^N allow to prove the stability result for the approximate reduced problem (3.7).

THEOREM 3

Assume that $\Re\lambda_n^N \geq 0$ and $\Im\lambda_n^N < 0$ for all n, N. Then

$$\inf_N \gamma_N > 0\,. \tag{4.12}$$

The discrete inf-sup constants are strongly bounded away from zero. ∎

Finally, we can estimate the error correponding to the IE approximation.

THEOREM 4

With the assumptions of Theorem 1, the error of the IEM solution can be estimated as

$$\|u - u_N\|_{H^1(\Omega_a)} \leq T\gamma_N^{-1}\,\frac{C}{N^{\alpha-3/2}}\,\|u\|_{H^\alpha(\Gamma_a)}, \qquad \alpha > \frac{3}{2}\,. \tag{4.13}$$

∎

Proof: According to Theorem 1, it is sufficient to estimate the difference between the exact and approximate DtN maps.

$$\begin{aligned}
\|G - G_N\|_{\mathcal{L}(H^\alpha, H^{-1/2})} &= \sup_{\|u\|_\alpha = 1}\left(\sup_{\|v\|_{1/2}=1} |\langle (G - G_N)u, v\rangle|\right) \\
&= \sup_{\|u\|_\alpha=1}\ \sup_{\|v\|_{1/2}=1}\left|\sum_{n=0}^{\infty}\frac{(\lambda_n - \lambda_n^N)}{(1+n^2)^{\frac{1+2\alpha}{4}}}\sum_{m=-n}^{n}(1+n^2)^{\alpha/2}u_{mn}(1+n^2)^{1/4}\bar{v}_{mn}\right| \\
&\leq \sup_{n\geq 0}\left|\frac{(\lambda_n - \lambda_n^N)}{(1+n^2)^{\frac{1+2\alpha}{4}}}\right|
\end{aligned} \tag{4.14}$$

Finally, since the first N eigenvalues are reproduced exactly, we have:

$$\sup_{n\geq 0}\left|\frac{(\lambda_n-\lambda_n^N)}{(1+n^2)^{\frac{1+2\alpha}{4}}}\right| \leq C \sup_{n\geq N+1}\frac{|\lambda_n-\lambda_n^N|}{n^{1/2+\alpha}} \leq C \sup_{n\geq N+1}\frac{1}{n^{\alpha-3/2}} \sup_{n\geq N+1}\frac{|\lambda_n-\lambda_n^N|}{n^2} \tag{4.15}$$

for some C. The first supremum on the right is obviously achieved for the lowest possible $n = N+1$. From Lemma 1 and Conjecture 1 it is easy to see that the second supremum is bounded. Thus

$$\|G - G_N\| \leq \frac{C}{N^{\alpha-3/2}} \tag{4.16}$$

and the statement follows from Theorem 1. ∎

We know from the standard regularity result for elliptic operators with constant coefficients that the solution u is analytic around the truncating sphere S. This indicates an unlimited rate of convergence and, ultimately, an exponential convergence. Indeed, we have the following result.

THEOREM 5
The approximate solution u^N of (3.7) converges exponentially to the exact solution u of (3.5) as $N \to \infty$. More precisely,

$$\|u - u_N\|_1 \leq C\, e^{-\kappa N} \tag{4.17}$$

for some positive constants C, κ. ∎

REMARK 2 Constant $\kappa = ln(1/r_0)$ where r_0 is the radius of the smallest sphere enclosing domain Ω. The further away the coupling sphere, the larger the ratio $1/r_o$. Hence the rate of convergence is expected to grow with the distance between the obstacle and the coupling surface. ∎

5 Maxwell's equations

Formally, the presented approach generalizes to Maxwell's equations:

$$\begin{cases} \nabla \times (\nabla \times \boldsymbol{E}) - k^2\boldsymbol{E} = \boldsymbol{0} & \text{in } \Omega^e = \omega - \bar{\Omega} \\ \boldsymbol{n} \times (\nabla \times \boldsymbol{E}) = \boldsymbol{g} & \text{on } \Gamma \\ \boldsymbol{e}_r \times (\nabla \times \boldsymbol{E}) - ik\boldsymbol{E} = O(\frac{1}{r^2}) & \text{as } r \to \infty \end{cases} \tag{5.1}$$

Here $\boldsymbol{E}$ denotes the electric field, k is the wave number and $\boldsymbol{e}_r$ is the unit vector in the radial direction. The variational formulation for the reduced problem takes on the form:

$$\int_{\Omega_S} (\nabla \times \boldsymbol{E}) \circ (\nabla \times \bar{\boldsymbol{F}}) - k^2 \boldsymbol{E} \circ \bar{\boldsymbol{F}} + ik \int_S \boldsymbol{G}(\boldsymbol{n} \times \boldsymbol{E}) \circ (\boldsymbol{n} \times \bar{\boldsymbol{F}}) = \int_\Gamma \boldsymbol{g} \circ \bar{\boldsymbol{F}} \tag{5.2}$$

where the DtN map $\boldsymbol{G}$ is now called the *electric to magnetic boundary component map* [3]. More precisely, map $\boldsymbol{G}$ sets tangential component of the $\boldsymbol{E}$-field on S into the corresponding tangential components of the magnetic field on S:

$$\boldsymbol{G}: \quad \boldsymbol{n} \times \boldsymbol{E} \to \frac{1}{ik} \boldsymbol{n} \times (\nabla \times \boldsymbol{E}) \tag{5.3}$$

The design of an infinite element approximation and the spectral representation of the DtN map are based on the classical series representation of the exact solution to Maxwell's equations outside sphere S [3].

$$\begin{aligned} \boldsymbol{E} &= \sum_{n=1}^{\infty} \sum_{m=-n}^{n} \quad a_n^m \nabla \times \{ r h_n(kr) Y_n^m(\hat{\boldsymbol{x}}) \hat{\boldsymbol{x}} \} \\ &\qquad + b_n^m \nabla \times \nabla \times \{ r h_n(kr) Y_n^m(\hat{\boldsymbol{x}}) \hat{\boldsymbol{x}} \} \\ &= \sum_{n=1}^{\infty} \sum_{m=-n}^{n} \quad -a_n^m h_n(kr) \hat{\boldsymbol{x}} \times \nabla_S Y_n^m(\hat{\boldsymbol{x}}) \\ &\qquad + b_n^m \{ n(n+1) \tfrac{h_n(kr)}{r} Y_n^m(\hat{\boldsymbol{x}}) \hat{\boldsymbol{x}} \} + [\tfrac{h_n(kr)}{r} + k h_n'(kr)] \nabla_S Y_n^m(\hat{\boldsymbol{x}}) \} \end{aligned} \tag{5.4}$$

Here $Y_n^m(\hat{\boldsymbol{x}}), \hat{\boldsymbol{x}} \in S$, denote the spherical harmonics, ∇_S is the gradient on the unit sphere, and $\mu_n \nabla_S Y_n^m(\hat{\boldsymbol{x}}), \mu_n \hat{\boldsymbol{x}} \times \nabla_S Y_n^m(\hat{\boldsymbol{x}})$ are the *vector spherical harmonics.* Here $\mu_n^{-1} = n(n+1)$ is a scaling factor. This leads to the following formula for the tangential components of the magnetic field on a sphere:

$$\begin{aligned} \tfrac{1}{ik} \hat{\boldsymbol{x}} \times (\nabla \times \boldsymbol{E}) &= \sum_{n=1}^{\infty} \sum_{m=-n}^{n} \quad \tfrac{E1_n^m}{ik} \{ h_n(kr)^{-1} [\tfrac{h_n(kr)}{r} - k h_n'(kr)] \mu_n \hat{\boldsymbol{x}} \times \nabla_S Y_n^m(\hat{\boldsymbol{x}}) \} \\ &\qquad ik E2_n^m \{ h_n(kr) [\tfrac{h_n(kr)}{r} - k h_n'(kr)]^{-1} \mu_n \nabla_S Y_n^m(\hat{\boldsymbol{x}}) \} \end{aligned} \tag{5.5}$$

where

$$\begin{aligned} \hat{\boldsymbol{x}} \times \boldsymbol{E} &= \sum_{n=1}^{\infty} \sum_{m=-n}^{n} \quad E1_n^m \{ \mu_n \nabla_S Y_n^m(\hat{\boldsymbol{x}}) \} \\ &\qquad E2_n^m \{ \mu_n \hat{\boldsymbol{x}} \times \nabla_S Y_n^m(\hat{\boldsymbol{x}}) \} \end{aligned} \tag{5.6}$$

The final modal representation of the DtN term in (5.2) reduces to:

$$\sum_{n=1}^{\infty} \left\{ (1 - \lambda_n) \sum_{m=-n}^{n} E1_n^m \bar{F2}_n^m - k^2 (1 - \lambda_n)^{-1} \sum_{m=-n}^{n} E2_n^m \bar{F1}_n^m \right\} \tag{5.7}$$

where λ_n denote the eigenvalues of the DtN operator for the Helmholtz equation.

Acknowledgement: The work has been supported by the Office of Naval Research under Contract N00014-89-J-1451.

References

[1] R. J. Astley, G. J. Macaulay and J. P. Coyette, "Mapped Wave Envelope Elements for Acoustical Radiation and Scattering", *Journal of Sound and Vibration*, vol. 170, no. 1, pp. 97-118, 1994.

[2] D. S. Burnett, "A Three-Dimensional Acoustic Infinite Element Based on a Prolate Spheroidal Multipole Expansion", *Journal of the Acoustical Society of America*, vol. 96, pp. 2798-2816, 1994.

[3] D. Colton and R. Kress, *Inverse Acoustic and Electromagnetic Scattering*, Springer Verlag, Berlin 1992.

[4] L. Demkowicz, "Asymptotic Convergence in Finite and Boundary Element Methods: Part1: Theoretical Results", *Computers and Mathematics with Applications*, vol. 27, no. 12, pp. 69-84, 1994.

[5] L. Demkowicz and K. Gerdes, "Convergence of the Infinite Element Methods for the Helmholtz Equation", TICAM Report 95-07, also in print in *Numerische Mathematik.*

[6] L. Demkowicz and F. Ihlenburg, "Analysis of a Coupled Finite-Infinite Element Method for Exteriori Helmholtz Problems", *TICAM Report 95-52*, in review in *Mathematics of Computation.*

[7] K. Gerdes and L. Demkowicz, "Solution of 3D-Laplace and Helmholtz Equation in Exterior Domains Using *hp* Infinite Elements", *Computer Methods in Applied Mechanics and Engineering*, Vol. 137, 239-274, 1996.

[8] K. Gerdes, "Solution of the 3D-Laplace and Helmholtz Equation in Exterior Domains of Arbitrary Shape Using *hp* Finite-Infinite Elements", Ph. Dissertation, The University of Texas at Austin, February 1996.

[9] M. J. Grote and J. B. Keller, "On Nonrefelecting Boundary Conditions", *Journal of Computational Physics*, Vol. 122, 231-243, 1995.

[10] R. Leis, *Initial Boundary Value Problems in Mathematical Physics*, Teubner, 1986.

USES OF THE BERENGER PML IN PSEUDOSPECTRAL METHODS FOR MAXWELL'S EQUATIONS

Tobin A. Driscoll
Bengt Fornberg
Department of Applied Mathematics
University of Colorado
Boulder, CO, 80309, USA.

1 Introduction

The Berenger perfectly matched layer (PML) has been very successful in the simulation of unbounded domains for the solution of Maxwell's equations by finite-difference time-domain (FDTD) techniques (Berenger [1994], Katz et al. [1994]). Compared to alternatives, the PML and its variations offer reflection coefficients that are usually orders of magnitude smaller.

The most common FDTD algorithm is due to Yee (Taflove [1995]) and employs second-order, staggered finite differences in both space and time variables. While such low-order methods are easy to implement and use, they suffer from large dispersion errors. As a result, a typical resolution requirement is at least 10 points per wavelength (PPW) (Taflove [1995]). For three-dimensional simulations involving electrically large objects, the storage and work needed to meet this standard can be huge.

Pseudospectral (PS), or spectral collocation, methods are frequently employed in the solution of differential equations. Their advantages include very low dispersive errors, and the resolution requirements can be as low as 2 PPW (periodic) or π PPW (nonperiodic). Thus it would be highly advantageous to use PS methods for the solution of Maxwell's equations. Some work has been done along these lines, including the incorporation of PML techniques (Yang et al. [1997]). In our numerical studies of two applications of PS methods to Maxwell's equations, we find that PML techniques work well.

We choose to work with the equations for transverse electric (TE) waves. In two dimensions, the equations for the PML are

$$\begin{aligned}
\epsilon\frac{\partial E_1}{\partial t} &= \frac{\partial(H_{31}+H_{32})}{\partial x_2} - \sigma_2 E_1\\
\epsilon\frac{\partial E_2}{\partial t} &= -\frac{\partial(H_{31}+H_{32})}{\partial x_1} - \sigma_1 E_2\\
\mu\frac{\partial H_{31}}{\partial t} &= -\frac{\partial E_2}{\partial x_1} - \sigma_1^* H_{31}\\
\mu\frac{\partial H_{32}}{\partial t} &= \frac{\partial E_1}{\partial x_2} - \sigma_2^* H_{32}
\end{aligned}$$

If $\sigma_1 = \sigma_2$ and $\sigma_1^* = \sigma_2^*$, the equations reduce to Maxwell's. However, the PML is more

T.L. Geers (ed.), IUTAM Symposium on Computational Methods for Unbounded Domains, 95–102.

concerned with the choice

$$\frac{\sigma_n}{\epsilon} = \frac{\sigma_n^*}{\mu} \tag{1}$$

for $n = 1, 2$. If σ_n and σ_n^* are positive, waves are attenuated when traveling in the x_n-direction. A sketch of the usual PML situation is given in Figure 1.

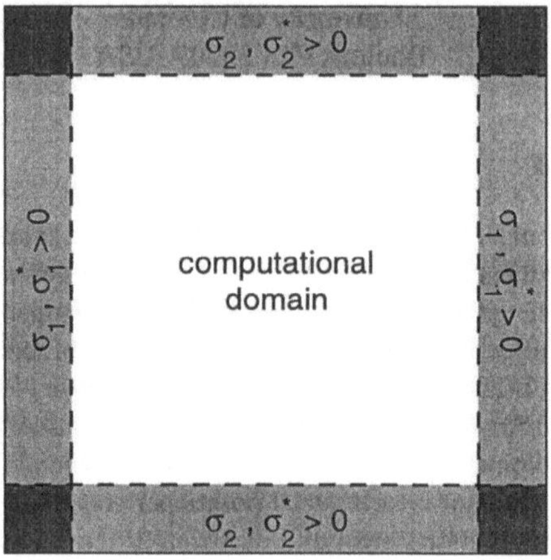

Figure 1: Sketch of typical PML usage.

2 Fourier pseudospectral

Our first example is a rectangular geometry (although the medium in it might be rather complex, as in bioacoustic modeling of human tissue (Hinkelman et al. [submitted, 1996])). For convenience, we nondimensionalize the equations by selecting $\epsilon = \mu = 1$.

While Fourier PS methods show promise for such problems, the physical domain does not exhibit the necessary periodicity for such methods. We propose padding the physical domain with a PML on all sides. Thus, the total computational domain is $[-\pi, \pi] \times [-\pi, \pi]$, while the physically meaningful domain is confined to $[-a\pi, a\pi] \times [-a\pi, a\pi]$ for $a < 1$. Because the imposition of boundary conditions at the outer edges of the PML is rather arbitrary, we replace the conventional perfect conductor with a periodicity condition: $U(x, \pi) = U(x, -\pi)$ and $U(-\pi, y) = U(\pi, y)$, where U is any field component. Hence exiting waves can re-enter the domain, but only after having been damped by a double thickness of the PML.

Berenger (Berenger [1994]) noted for second-order finite differences that it was important not to activate the quantities σ_x, σ_x^*, etc. discontinuously, as that leads to discontinuity of the normal fields and their derivatives. For PS methods the issue of smoothness is even more vital. Moreover, the smoothness is important not just at the domain–PML interface but at the periodic PML–PML interface as well.

Let the normalized depth of a point in a PML layer be denoted by $\delta \in [0,1]$. Then the functional form of the appropriate split σ or σ^* is

$$\frac{S}{8}\Big(3 - 4\cos(\pi\delta) + \cos(2\pi\delta)\Big), \tag{2}$$

which ranges from 0 to S, with three zero derivatives at the ends. In Figure 2 we display the dependence of σ_x on x. In principle, one can accommodate any number of smooth derivatives,

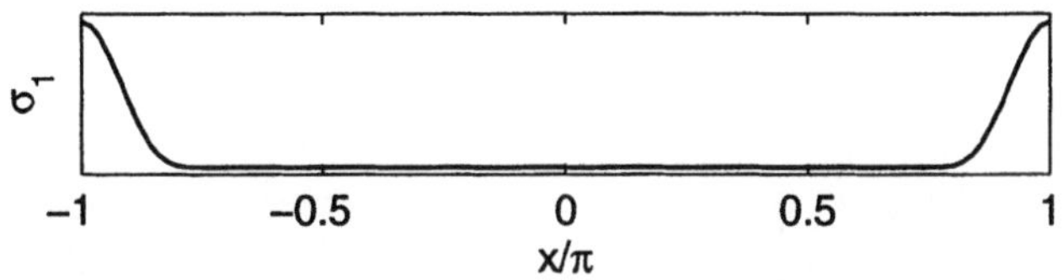

Figure 2: Dependence of σ_1 (or σ_1^*) on x for Fourier PS method.

but we have not seen improvement from going beyond (2). A wave at normal incidence, after having passed through the PML twice, is theoretically damped by a factor

$$e^{-3S(1-a)\pi/4}. \tag{3}$$

In our experiments, typical values are $S = 12$ and $a = 0.75$, which leads to a damping factor of about 8×10^{-4}.

In Figure 3 we show four snapshots of a solution. The H_3 field is shown after being initialized with a fairly smooth pulse centered at $(0.1\pi, 0.2\pi)$; the exact pulse completely leaves the domain after a short time. With just 32 grid points in each direction, the pulse is sharply resolved and travels without dispersion. Figure 4 shows the maximum norm of H_3 against time. The pulse is reflected with an amplitude of about 1%.

3 Pseudospectral elements

For complex geometries, spectral elements retain many of the good qualities of spectral methods while allowing domain decomposition. Some sort of coupling between elements is necessary. In hyperbolic problems, the standard coupling procedure is through outflowing characteristics. This approach has difficulties concerning:

- wasted resolution near interfaces
- higher dimensions, especially corners
- nonlinearities
- material interfaces

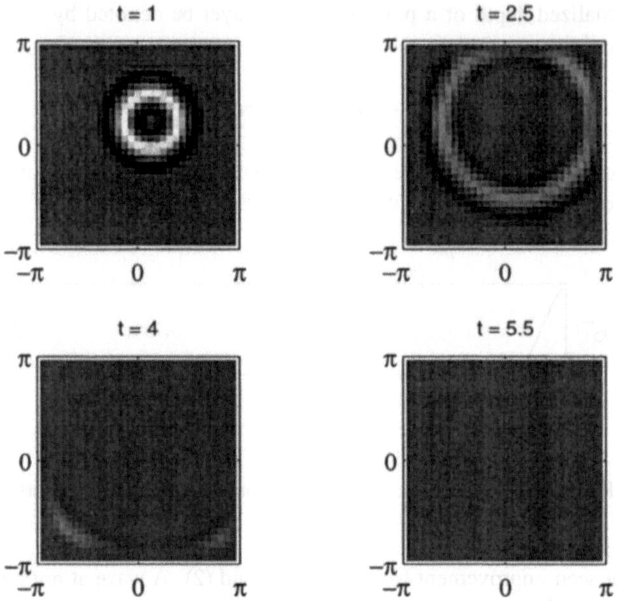

Figure 3: Snapshots of a pulse in H_3 over time, using the Fourier PS method with Berenger PML. The grid is 32×32, with a PML thickness of 4 points.

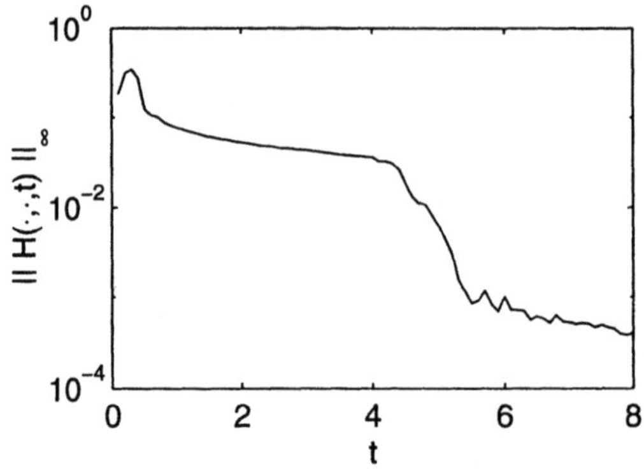

Figure 4: Maximum of H_3 over time for Fourier solution. The ultimate behavior is numerical error.

• slight diffusion

A new method, called the *pseudospectral element method* (PSE), attempts to deal with these difficulties. The method is described in detail for one-dimensional problems in (Driscoll and Fornberg [1997]). A major component is to view PS methods as finite difference methods of maximal order. This allows the use of arbitrary grids, not just those tied to orthogonal polynomials. We choose "open" grids that exclude the endpoints of the interval, so that corners in higher dimensions will pose no problems. To couple domains, we derive relationships for the field components and several derivatives at artificial, material, and conducting interfaces. These interface relationships are directly incorporated into the finite-difference stencils.

A key feature of PSE is that, unlike characteristic-based spectral elements, material interfaces are handled just as accurately and stably as artificial ones. Consider a test problem consisting of two media, one with wave speed 1 and the other with speed 1/2, and periodic boundary conditions. As illustrated by Figure 5, a pulse solution will have multiple reflections and transmissions that, at integer times, coalesce into three pulses of analytically known amplitudes. In Figure 6 we show the max-norm error of the solutions obtained by PSE, standard

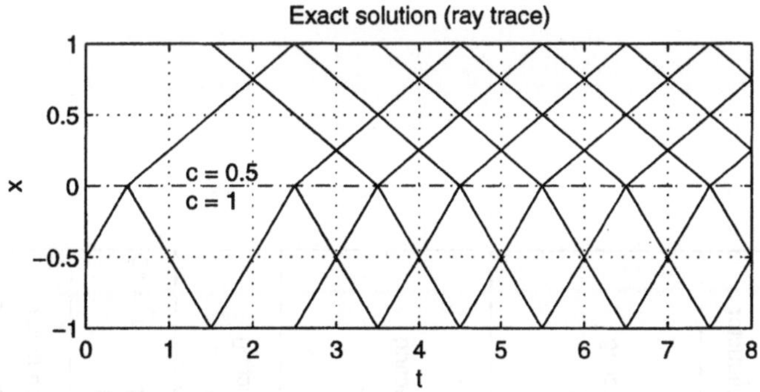

Figure 5: Schematic solution of periodic two-media problem. At integer times, the multiple reflections and transmissions coalesce into just three pulses.

spectral elements, and the Yee and fourth-order finite difference methods on 34 grid points. The low-order methods become inaccurate quickly due to dispersive error, and the spectral element method is actually unstable for this problem. However, PSE remains accurate for very long times. From the standpoint of PSE, the PML medium just presents a certain type of material interface. There should be no need to impose smoothness on σ_1, etc., because the discontinuities that arise in the fields are explicitly accounted for. Figure 7 shows a test problem that includes an absorbing PML. In Figure 8 we show the time history of the solution up to $t = 5$. The changes of medium, including the PML interface, cause no visible spurious reflections.

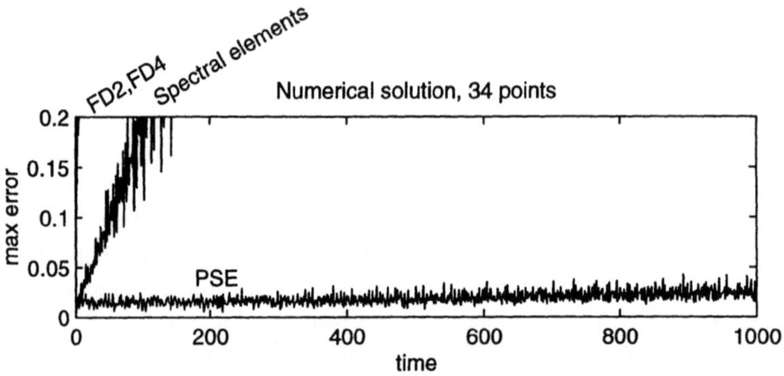

Figure 6: Error in solutions of the two-media problem with 34 grid points. The low-order methods are dominated by dispersion, whereas the spectral element method is unstable. The PSE method is excellent for very long times.

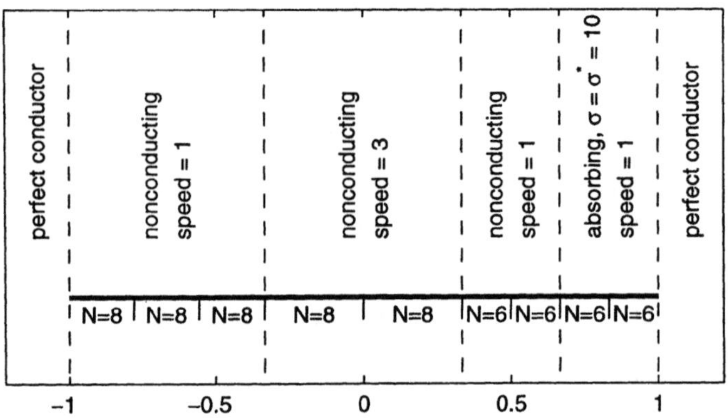

Figure 7: Test problem for PSE with a PML, with 64 total grid points.

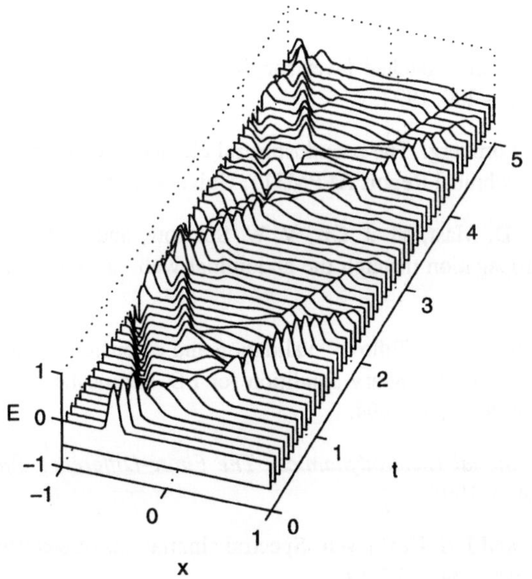

Figure 8: Time history of PSE test problem solution.

4 Conclusion

The PML technique has already been established as useful and important for low-order time-domain simulations of electromagnetic phenomena. Our experiments clearly indicate that the PML is also well-suited to high-order (pseudospectral) methods. We believe that high-order methods offer potentially drastic reductions in computational storage and effort. Work is currently underway in extending our results to practical situations and (in the case of PSE) to higher dimensions. Preliminary work is very encouraging.

References

J.-P. Berenger. A perfectly matched layer for the absorption of electromagnetic waves. *J. Comp. Phys.*, 114:185–200, 1994.

T. A. Driscoll and B. Fornberg. The pseudospectral element method for one-dimensional hyperbolic systems. Technical Report APPM311, University of Colorado at Boulder, 1997.

L. M. Hinkelman, T. D. Mast, M. J. Orr, V. W. Sparrow, and R. C. Waag. Simulation of ultrasonic pulse propagation through the abdominal wall. *J. Acoust. Soc. Am.*, submitted, 1996.

D. S. Katz, E. T. Thiele, and A. Taflove. Validation and extension to three dimensions of the Berenger PML absorbing boundary condition for FD-TD meshes. *IEEE Microwave and Guided Wave Let.*, 4:268–270, 1994.

A. Taflove. *Computational Electrodynamics: The Finite-Difference Time-Domain Method.* Artech House, Boston, 1995.

B. Yang, D. Gottlieb, and J. S. Hesthaven. Spectral simulations of electromagnetic wave scattering. *J. Comp. Phys.*, 134:1–15, 1997.

ETA Expressions for the High Frequency Impedance of a Uniformly Pulsating Submerged Torus

C. A. FELIPPA
Department of Aerospace Engineering Sciences and Center for Aerospace Structures
University of Colorado, Boulder, CO 80309-0429

1. Introduction

The present abstract was motivated by the search for clean yet nontrivial benchmark problems for wave propagation and scattering in an infinite acoustic medium. The submerged sphere, cylinder and disk illustrate important aspects of scattering and radiation while admitting complete analytical solutions. These three classical configurations are thus indispensable part of any set of benchmark problems.

This article begins the study of a fourth benchmark geometry: the circular torus of revolution, called simply *torus* in the sequel. This geometry is generated by rotating a circle of radius a about an axis located at a distance $d \geq 0$ from the circle center. The torus is submerged in an infinite acoustic fluid. This benchmark configuration has two advantages:

1. It illustrates the phenomenon of *fluid entrainment* associated with the boundary concavity in its inner region. The fluid trapped in this volume reacts partly as an internal fluid, acting as a distributed spring stiffness rather than as an inertial medium.
2. It reduces to the sphere and infinite circular cylinder as limit cases when $d = 0$ and $d \to \infty$, respectively. This reduction facilitates checking some derivations.

The key disadvantage of the torus is that exact expressions for its acoustic impedance are not known outside of the two limit cases noted above. It is doubtful that closed form solutions can be obtained for arbitrary frequencies. (Even the calculation of the added mass leads to integrals of hypergeometric functions). In the present work approximate expressions for the high-frequency impedance of a uniformly pulsating torus are investigated using Early-Time Approximations (ETAs). These are derived obtained using high-frequency asymptotics from Kirchhoff's Retarded Potential Formulation (RPF). Although the cylinder and sphere can be processed by hand, the torus involves highly complex and error prone power-series manipulations. These are possible only with the help of computer-algebra programs. For the present derivations *Mathematica 3.0*, a tool not available in 1980, was used to carry out the necessary symbolic computations.

The following three sections outline the ETA derivation method. They are largely extracted from Felippa (1980) and Geers (1991). The former article developed a local source analysis technique to construct ETAs but did not considered spatial variation of field variables. This omission was accounted for by Geers (1991). The impedance analysis of Section 5 is new. Numerical results are presented for a torus with $d = 2a$, a geometry depicted in Figure 1.

2. The Retarded Potential Formulation

We consider an infinite acoustic fluid of density ρ and sound speed c governed by the isotropic, scalar wave equation $c^2\Delta\phi = \ddot{\phi}$, where c is the acoustic wave speed, superposed dots denote temporal differentiation, and the velocity potential ϕ is related to acoustic excess pressure p and fluid-particle velocity $\mathbf{v}$ as

$$p = \rho\dot{\phi}, \qquad \mathbf{v} = \nabla\phi. \tag{1}$$

T.L. Geers (ed.), IUTAM Symposium on Computational Methods for Unbounded Domains, 103–110.

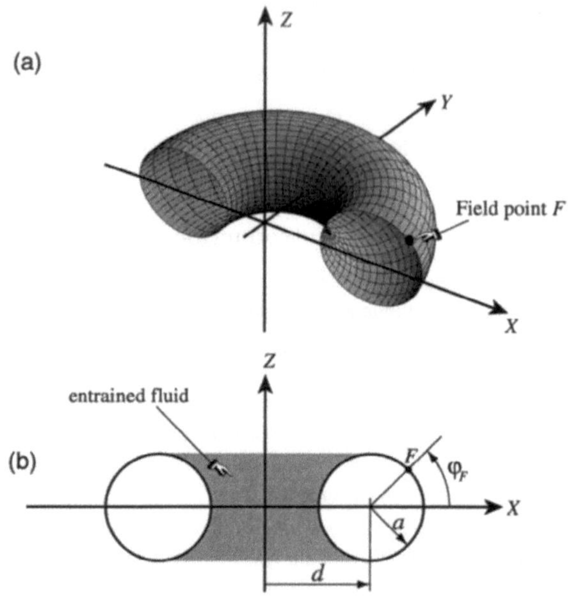

Figure 1. The circular torus, plotted for $d = 2a$. Only the $Y > 0$ half is shown for clarity.

A structure (radiator or scatterer) with "wet surface" S is submerged in the fluid. We assume that the solution of the acoustic problem is expressed by Kirchhoff's Retarded Potential Formulation (RPF) as discussed, for example, e.g., in the monograph of Baker and Copson (1949):

$$4\pi\epsilon\phi = -\int_S \left\{ R^{-1}\left[\frac{\partial\phi}{\partial n}\right] + \beta R^{-2}\left([\phi] + \frac{R}{c}[\dot{\phi}]\right)\right\} dS + 4\pi\epsilon\phi_w, \tag{2}$$

Here $\phi = \phi(\mathbf{X}, t)$ is the velocity potential at a field point F of coordinates $\mathbf{X} = (X, Y, Z)$, $R = ||\mathbf{R}||_2 = |\mathbf{X} - \mathbf{X}_F|$ is the distance from F to a source point P on S of coordinates $\mathbf{X}$, $\mathbf{n}$ is the outward normal to S at the source point (positive going into the fluid) $\beta = \partial R/\partial n = \cos(\mathbf{R}, \mathbf{n})$, $[f] =$ is the *retarded value*, $f(\mathbf{X}, t - R/c)$ of a field function $f(\mathbf{X}, t)$, and $\phi_w =$ is the incident wave potential. Finally, $\epsilon = 1, 1/2$ or 0 depending on whether the field point is within the fluid, on the surface S (assumed to be smooth), or inside S, respectively. (At a sharp corner of S with enclose solid angle α, $\epsilon = \alpha/4\pi$.)

For deriving fluid interaction approximations, the field point F is considered to be on S so that $\epsilon = 1/2$ and the incident wave potential is dropped; that is, only the scattered component of the pressure field will be retained. The resulting expression is

$$2\pi\phi = \int_S \left\{ R^{-1}[u] - \beta R^{-2}\left([\phi] + \frac{R}{c}[\dot{\phi}]\right)\right\} dS, \tag{3}$$

in which $u = -(\partial\phi/\partial n)_S$ denotes the normal fluid-particle velocity at S. Equation (3) is a homogeneous "retarded" integro-differential equation in which all variables pertain to S. This is complemented by the initial rest conditions

$$\phi(t) = u(t) = 0, \quad t \leq 0, \tag{4}$$

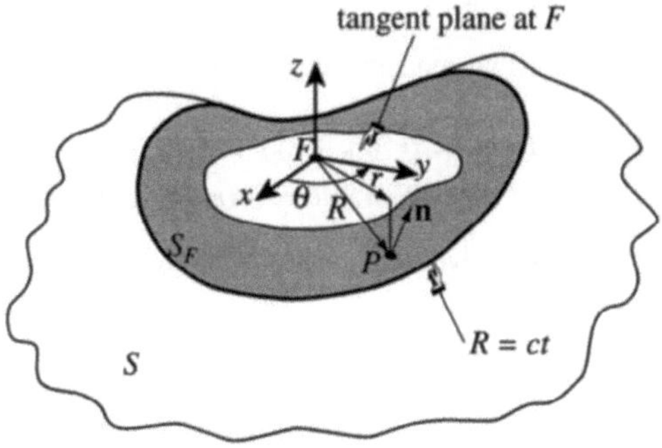

Figure 2. Coordinate systems for local source analysis.

for every field point on S. Note that $\phi(0+)$ may be nonzero, however; that is, occurrence of a step-pressure at $t = 0$, or at other positive values of t, is not precluded.

3. Local Geometry Analysis

The Contributing Surface. Early Time Approximations (ETA) are interaction expressions asymptotically valid for $ct << a$, where a is a characteristic dimension of S. These approximations are spatially *local*, that is, depend only on geometric information collected in the neighborhood of an individual field point, F. The "contributing cap" $S_F(t)$ for $t > 0$ is constituted by surface source points P located at a distance $R \leq ct$ from F, as illustrated in Figure 2.

A local orthogonal Cartesian system (x, y, z) is defined with (x, y) in the tangent plane and z along the external normal at F, respectively. (The approximations will turn out to be invariant with respect to the orientation of x and y, as could be expected.) A cylindrical coordinate system (r, θ, z) is also used as shown in Figure 1. The distance R between P and F is then expressable as $R^2 = r^2 + z^2$. Now the Taylor expansion of z about a regular point F can be written as

$$z = Ar^2 + Br^3 + Cr^4 + Dr^5 + \ldots = AR^2 + BR^3 + (C - A^3)R^4 + (D - \tfrac{7}{2}A^2B)R^5 + \ldots \quad (5)$$

where

$$\begin{aligned} A &= (z_{xx}\cos^2\theta + 2z_{xy}\cos\theta\sin\theta + z_{yy}\sin^2\theta)/2!, \\ B &= (z_{xxx}\cos^3\theta + 3z_{xxy}\cos^2\theta\sin\theta + 3z_{xyy}\cos\theta\sin^2\theta + z_{yyy}\sin^3\theta)/3!, \quad \text{etc.} \end{aligned} \quad (6)$$

in which the derivatives $z_{xx} = \partial^2 z/\partial x^2, \ldots$, etc., are assumed to exist at F. The expansion (5) permits us to express the "cap" S_F in the parametric series form

$$x = r(R, \theta)\cos\theta, \quad y = r(R, \theta)\sin\theta, \quad z = z(R, \theta), \quad (7)$$

in which $r(R, \theta) = +[R^2 - z^2(R, \theta)]^{1/2}$. Hence R and θ will play the role of surface coordinates.

Following Geers (1991) the field variables ϕ and u are also spatially expanded about F:

$$\begin{aligned} \phi &= \phi_F + (\phi_x\cos\theta + \phi_y\sin\theta)r + \tfrac{1}{2}(\phi_{xx}\cos^2\theta + 2\phi_{xy}\sin\theta\cos\theta + \phi_{yy}\sin^2\theta)r^2 + \ldots \\ u &= u_F + (u_x\cos\theta + u_y\sin\theta)r + \tfrac{1}{2}(u_{xx}\cos^2\theta + 2u_{xy}\sin\theta\cos\theta + u_{yy}\sin^2\theta)r^2 + \ldots \end{aligned} \quad (8)$$

in which all derivatives are evaluated at F.

Surface Metric Coefficients. The two geometric quantities needed for (3) are

$$\begin{aligned} dS = W_1\, dR\, d\theta &= \left\{r^2(r_R^2 + z_R^2) + (r_R z_\theta - z_R r_\theta)^2\right\}^{1/2}, \\ \beta R^{-2}\, dS = W_2\, dS\, d\theta &= -(r/R)^3 (z/r)_R\, dR\, d\theta \end{aligned} \tag{9}$$

where subscripts denote partial derivatives with respect to the surface coordinates R and θ. Insertion of (5) into the above yields the series

$$\begin{aligned} W_1/R &= 1 + \tfrac{1}{2} A_\theta^2 R^2 + (AB + A_\theta B_\theta) R^3 + \ldots, \\ -W_2 &= A + 2BR + 3(C - A^3)R^2 + (4D - 144A^2B)R^3 + \ldots \end{aligned} \tag{10}$$

For a spherical surface $W_1 = R$ and $W_2 = -A$ exactly, which provides checks on the coefficients of the series (10). For an infinite circular cylinder of radius a, $W_1/R = [1 - \sin^2\theta \cos^2\theta (R/a)^2]^{-1/2}$ and $W_2 = (1 - W_1/R)/(R^2 \sin^2\theta)$, which can provide further verification.

Retarded Field Integrals. Insertion of the series (7)-(10) into the RPF (4) leads to integrals of the form

$$\bar{g} = \frac{1}{2\pi}\int_0^{2\pi} g(\theta)\, d\theta, \quad J_m = \int_0^{ct} R^m [u]\, dR, \quad K_m = \int_0^{ct} R^m \left([\phi] + (R/c)[\dot{\phi}]\right) dR \tag{11}$$

for $m = 0, 1, \ldots$. The first expression denotes the circumferential average of a function $g(\theta)$. The "retarded" integrals J and K are transformed into ordinary momentum integrals through the change of variable $t' = t - R/c$, $R = c(t - t')$. Repeated integration by parts yields $J_0 = j_0 c u^*$, $J_1 = j_1 c^2 u^{**}$, ..., $K_0 = k_0 c\phi^*$, $K_1 = k_1 c^2 \phi^{**}$, ..., with $j_m = m!$, $k_m = (m+1)! + m!$, and in which each superposed asterisk denotes temporal integration from $t = 0$ through t.

4. Early Time Approximation Family

Derivation. The formal procedure for obtaining the ETA family is as follows: insert (8) through (10) into (3); integrate in θ, from 0 to 2π, and divide through by 2π; integrate in R, from 0 to ct, and insert the values given above for the momentum integrals. Finally time differentiate once and use the first of (1) to pass from velocity potential to pressure variables on the left-hand side. The resulting form is

$$p + P_1 p^* + P_2 p^{**} \ldots = \rho c(u + U_1 u^* + U_2 u^{**} + \ldots) \tag{12}$$

Felippa (1980) gives expressions for P_i and U_i in terms of the circumferential averages of $A, B, \ldots$, if u and p are assumed constant over the contributing surface. Geers (1991) generalized (12) to the case of u and p varying as per the expansions (8). In both cases, it can be shown that at a regular field point, that is, one at which the Taylor expansion (5) holds, all coefficients P_{2j} and U_{2j-1}, for $j = 1, 2, \ldots$, vanish. The remaining ones can be expressed in terms of *geometric* and *field invariants*. For example P_1 is κc, in which $\kappa = \bar{A}/2 = (z_{xx} + z_{yy})/2 = \frac{1}{2}\nabla^2 z$ is the mean wet surface curvature. Higher coefficients rapidly get more complicated. For example P_7 has over 500 terms in the case of a general geometry.

ETA Instances. Approximations called ETA$_k$ ($k = 1, 2, \ldots$) result by truncation of (12) keeping $\lfloor (k+2)/2 \rfloor$ and $\lfloor (k+1)/2 \rfloor$ terms on on the left and right, respectively. [This method of counting agrees with Geers (1991) but differs from that of Felippa (1980) if $k \geq 3$.] ETA$_1$, which is $p = \rho c u$, is the Plane Wave Approximation (PWA) of Mindlin and Bleich (1953). ETA$_2$, which is $p + \kappa c p^* = \rho c u$, agrees with the Curved Wave Approximation (CWA) of Haywood (1958), later generalized by Bedrosian and DiMaggio (1972). ETA$_3$ is $p + \kappa c p^* = \rho c u + \frac{1}{2}\rho c^3(\kappa^2 - \tau + \nabla^2)u^{**}$, where τ is the wet surface total curvature and $\nabla^2 = \partial^2/\partial x^2 + \partial^2/\partial y^2$ is the Laplace operator in the local system at F.

Of interest in some applications is the case of a *plane* wet surface, in which case (12) reduces to

$$p = \rho c(u + \frac{2!}{2^2}c^2\nabla^2 u^{**} + \frac{4!}{8^2}c^4\nabla^4 u^{****} + \frac{6!}{48^2}c^6\nabla^6 u^{******} + \frac{8!}{384^2}c^8\nabla^8 u^{********} + \ldots) \quad (13)$$

This can be recognized as the time-domain version of $\hat{p} = \rho c\hat{u}/\sqrt{1 - c^2\nabla^2 s^{-2}}$, in which $\hat{p}(s)$ and $\hat{u}(s)$ are the one-sided Laplace transforms of $p(t)$ and $u(t)$, respectively.

Acoustic Impedance. Following Junger and Feit (1972) and Geers (1978) the *specific acoustic impedance* (impedance per unit of surface area) of a pressure-velocity relation such as (12) is defined as the ratio

$$\zeta(\omega) = \frac{\hat{p}(j\omega)}{\hat{u}(j\omega)} = \xi(\omega) + j\omega\mu(\omega), \quad (14)$$

where ω is driving circular frequency, $s = j\omega$, and j is the imaginary unit. The *specific acoustic resistance* $\xi = Re(\zeta)$ characterises the radiation damping power and must be always positive. On the other hand, $\mu = -Im(\zeta)/\omega$ is interpreted as either *specific acoustic inertia*, also called *specific accession to inertia*, if positive (pressure in phase with acceleration) or effective *acoustic spring stiffness* if negative (pressure in phase with displacement).

5. Torus Impedance Analysis

The surface S is now specialized to the torus generated by a circle of radius a rotating about the Z axis, which is placed at a distance $d > 0$ from the circle center. The axis X passes through the center of the generating circle, and the plane XY is a symmetry plane. See Figure 1. The parametric equations of the torus are

$$X = (d + a\cos\varphi)\cos\lambda, \quad Y = (d + a\cos\varphi)\sin\lambda, \quad Z = a\sin\varphi \quad (15)$$

where the parameters λ and φ are the longitude and latitude angles, respectively. Because of circumferential symmetry it is sufficient to study field points on the plane $Y = 0$, or $\lambda = 0$. The field point F is located by $\varphi = \varphi_F$, and has coordinates $X_F = d + a\cos\varphi_F$, $Y_F = 0$, $Z_F = a\sin\varphi_F$. The unit normal $\mathbf{n}_F$ at F has direction cosines $\cos\varphi_F$, 0 and $\sin\varphi_F$. A generic point (φ, λ) on the torus has coordinates (15). The distance from the field point to an arbitrary point is

$$\begin{aligned} R^2 &= (X_F - X)^2 + (Y_F - Y)^2 + (Z_F - Z)^2 \\ &= 2(a^2 + (d(\cos\varphi + \cos\varphi_F) - a\sin\varphi\sin\varphi_F)a + d^2 - (d + a\cos\varphi)(d + a\cos\varphi_F)\cos\lambda) \end{aligned} \quad (16)$$

Uniformly Pulsating Torus. To reduce the complexity of the benchmark, it is assumed that the torus is *uniformly pulsating* at a frequency ω, with a driven wet surface velocity $u(t) = u_0 e^{j\omega t}$ which is the same at all points. Because of axisymmetry the field point pressure p must depend spatially only on φ_F and be an even function of it. Hence $p(\varphi_F, t) = p(\varphi_F)e^{j\omega t}$, where $p(\varphi_F)$ is conveniently expanded in the toroidal zonal-harmonic series

$$p(\varphi_F) = p_0\psi_0 + p_1\psi_1(\varphi_F) + p_2\psi_2(\varphi_F) + \ldots \qquad \text{with, if } d > a,$$

$$\psi_0 = 1, \quad \psi_1 = \cos\varphi_F - \frac{a}{2d}, \quad \psi_2 = \cos 2\varphi_F - \frac{ad}{2d^2 - a^2}\cos\varphi_F + \frac{a^2}{2(2d^2 - a^2)}, \quad \ldots \quad (17)$$

The zonal harmonics $\psi_0, \psi_1, \psi_2, \ldots$ form a system of orthogonal functions over the torus wet surface. The mean value of $\psi_i, i \geq 1$, over that surface is zero. The necessary partial derivatives of p are easily obtained since $p_x = p_\varphi/a$, $p_{xx} = p_{\varphi\varphi}/a^2 \ldots$, and all partials containing y are zero.

ETAs for Torus Geometry. For a preliminary study only the first two pressure harmonics in (17) were retained. The resulting Laplace transformed ETA expression in terms of u_0, p_0 and $p1$ is

$$(1 + P_1^0 s^{-1} + P_3^0 s^{-3} + \ldots)\hat{p}_0 + (1 + P_1^1 s^{-1} + P_3^1 s^{-3} + \ldots)\hat{p}_1 = \rho c(1 + U_2^0 s^{-2} + \ldots)\hat{u}_0 \quad (18)$$

Coefficients indexed 1 through 8 were recursively calculated by *Mathematica* for a torus with $d > a$ (a torus with a "hole"). Only the U coefficients are listed below to give an idea of their increasingly complex structure:

$$\begin{aligned}
U_2^0 &= c^2 d^2/[16a^2(d + a\cos\varphi_F)^2] \\
U_4^0 &= 6c^4[d(8a^2 d + 9d^3 - 16a^3\cos\varphi_F + 24ad^2\cos\varphi_F)]/[1024a^4(d + a\cos\varphi_F)^4] \\
U_6^0 &= 120c^6[d(-24a^4 d + 108a^2 d^3 + 25d^5 - 128a^5\cos\varphi_F + 96a^3 d^2\cos\varphi_F + 120ad^4\cos\varphi_F \\
&\quad - 96a^4 d\cos 2\varphi_F + 100a^2 d^3\cos 2\varphi_F)]/[16384a^6(d + a\cos\varphi_F)^6] \\
U_8^0 &= 5040c^8[d(-7424a^6 d + 8544a^4 d^3 + 12000a^2 d^5 + 1225d^7 - 14592a^7\cos\varphi_F - 4608a^5 d^2\cos\varphi_F \\
&\quad + 25920a^3 d^4\cos\varphi_F + 8400ad^6\cos\varphi_F - 19584a^6 d\cos 2\varphi_F + 9600a^4 d^3\cos 2\varphi_F \\
&\quad + 11760a^2 d^5\cos 2\varphi_F + 1536a^7\cos 3\varphi_F - 9600a^5 d^2\cos 3\varphi_F + 7840a^3 d^4\cos 3\varphi_F)] \\
&\quad /[4194304a^8(d + a\cos\varphi_F)^8]
\end{aligned} \quad (19)$$

To determine $\hat{p}_0$ and $\hat{p}_1$, (18) was integrated over the torus surface with weights 1 and $\cos\varphi_F$. This provides two surface-averaged zonal-harmonic equations, which are weakly coupled because the P and U coefficients depend on ϕ_F. From these the two averaged impedances $\zeta_0 = \hat{u}_0/\hat{p}_0$ and $\zeta_1 = \hat{u}_0/\hat{p}_1$ were obtained. The ETA impedances for the infinite cylinder result by making $d \to \infty$. Since those coefficients are well known a valuable check is obtained.

The impedances of ETA_2 through ETA_8 for zonal harmonics 0 and 1 are displayed in Figures 3 and 4 for the $d = 2a$ torus. These show dimensionless ratios $\xi/(\rho c)$ and $\mu/(\rho a)$ as function of the dimensionless frequency $\omega a/c = ka$, where k is the acoustic wave number. Good agreement is obtained only for $ka > 2$ for harmonic 0, which is the most important mean response. It should also be noted that ETA_3 and ETA_7 give negative radiation damping over certain frequency bands for that harmonic, while ETA_5 give grossly overdamped behavior. The inertial curves of ETA_3, ETA_5 and ETA_7 display significant stiffness-like behavior from the others for $ka < 1$.

6. Discussion

Preliminary results for this benchmark problem suggest that the direct use of ETA_k beyond $k = 2$ may not necessarily attain higher accuracy when paired with Virtual Mass Approximations (VMAs) to build DAAs. The nonphysical negative damping displayed by ETA_3 and ETA_7, as well as the anomalous damping behavior of ETA_5 suggests that even numbered ETAs, which maintain a consistent number of terms from both sides of (12), are preferable. For the uniformly pulsating circular cylinder Felippa (1980) found that the range of application of the ETAs can be extended to $1/2 \leq ka \leq 2$ by using Padé approximants, but the application of this technique to the torus has not been completed.

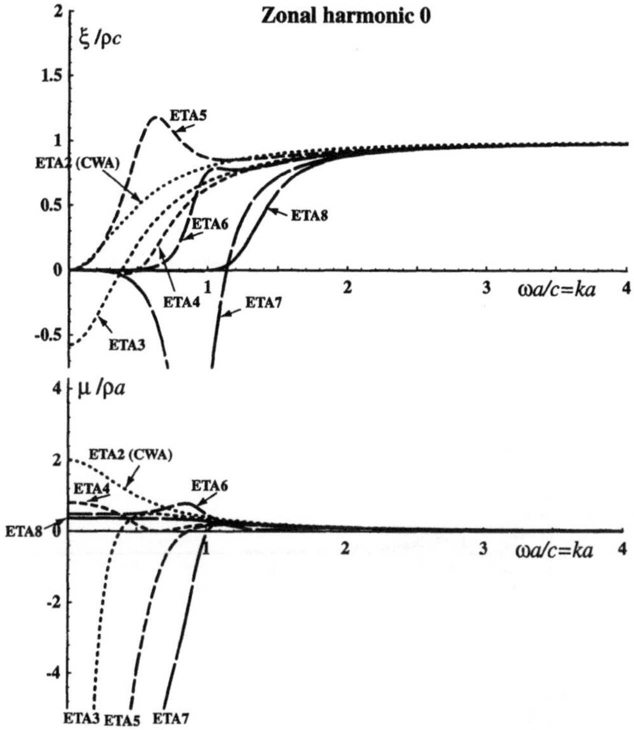

Figure 3. Surface-averaged specific acoustic impedance and inertia for the zero pressure zonal harmonic ψ_0 of a torus with $d = 2a$ surrounded by a ρc infinite acoustic medium. The torus is uniformly pulsating, velocity-driven at frequency ω. Results are plotted in dimensionless form for ETA_k ($k = 2$ to 8).

To continue the study of the torus as benchmark it will be necessary to derive a VMA by an accurate BEM added-mass analysis, as discussed by DeRuntz and Geers (1977) and O'Brien and Geers (1995). A DAA family can be then constructed by pairing VMA and selected Padé reformulated ETAs. Finally, an exact RPF analysis at specific intermediate frequencies, again using a BEM discretization, should be done to calibrate the accuracy of such DAAs for this configuration.

References

Baker, B. B. and Copson, E. T. (1949), *The Mathematical Theory of Huygens' Principle*, Clarendon Press, Oxford.

Bedrosian, B. and DiMaggio, F. L. (1972), Acoustic Approximations to Fluid-Shell Interaction, *J. Engrg. Mech. Div. ASCE*, **98**, pp. 731-742.

DeRuntz, J. A. and Geers, T. L. (1977), Added Mass Computation by the Boundary Integral Method, *Int. J. Numer. Meth. Engrg.*, **12**, pp. 531-549.

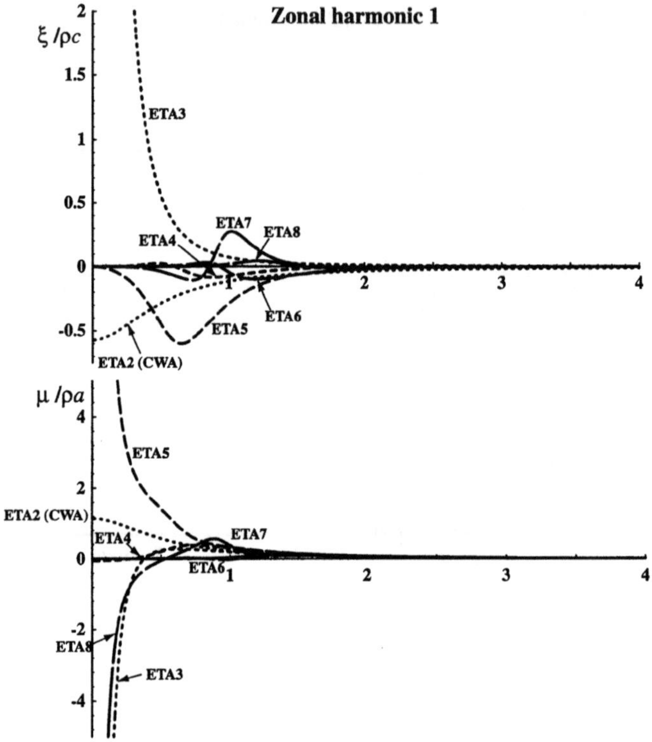

Figure 4. Similar to Figure 3, but for the pressure harmonic ψ_1.

Felippa, C. A. (1980), A Family of Early Time Approximations for Fluid Structure Interaction, *J. Appl. Mech.*, **47**, No. 4, pp. 703–708.

Geers, T. L. (1975), Transient Response Analysis of Submerged Structures, in *Finite Element Analysis of Transient Nonlinear Structural Behavior*, eds. Belytschko T., et al., AMD Vol. 26, ASME, New York, pp. 59–84.

Geers, T. L., A Fully Consistent Formulation of Early Time Approximations for Acoustic Media, in *The Finite Element Method in the 1990's*, eds. E. Oñate et al., Springer-verlag, 1991, pp. 521–528

Haywood, J. H. (1958), Response of an Elastic Cylindrical Shell to a Transverse Step Shock Wave, *Quart. J. Mech. Appl. Math..*, **11**, pp. 129–141.

Junger, M. C. and Feit, D. (1972), *Sound, Structures and their Interaction*, MIT Press, Cambridge.

Mindlin, R. D. and Bleich, H. H. (1953), Response of an Elastic Cylindrical Shell to a Transverse Step Shock Wave, *J. Appl. Mech.*, **20**, pp. 189–195.

O'Brien, J. L. and Geers, T. L. (1995), Variational Boundary-Element Analysis of Small-Amplitude Fluid Oscillation in a Low-Gravity Environment, *J. Fluids Struct.*, **9**, pp. 279-301.

WAVE VS. GEOMETRIC BASED MODELING OF BARRIERS WITH ATMOSPHERIC EFFECTS

Ken R. Fyfe & Amir Muradali

4-9 Mechanical Engineering
University of Alberta,
Edmonton, Alberta T6G 2G8
Canada

Abstract

Wave based solutions of noise barrier geometries accurately model the complex direct, reflected and diffracted sound field interactions. However, typical boundary element and finite element approaches are limited to small geometries and lower frequency ranges because the solutions are very computer intensive and are thus not practical as a design tool. A new class of diffraction based methods, that include phase, are being applied that yield wave-like accuracy with trivial calculation times.

These results, however accurate, typically over-predict the actual performance of noise barriers, because non-homogeneous atmospheric effects such as wind, temperature gradients and turbulence have not been considered. To overcome this limitation, a combined diffraction based barrier model and a heuristic atmospheric model have been developed. Comparisons with the FFT and PE approach show close agreement. Application of this model yield the expected sound barrier performance degradation due to the acoustic medium non-homogeneity.

1. Introduction

There are numerous techniques for determining the performance of noise barriers. The earliest methods in use were simple diffraction based models. The most popular was that due to Maekawa [Maekawa, 1968]. It was based upon empirical data and gave designers a simple tool to predict insertion loss. More exact approaches of estimating barrier insertion loss are with wave based methods such as the Boundary Element Method (BEM) [Hothersall et al, 1991; Fyfe & Harrison, 1995], and the Finite Element Method (FEM) and in particular the Infinite Wave Envelope Method (IFEM) [Astley 1983; Cremers & Fyfe, 1995]. Wave based methods solve the governing wave equation and model exactly the reflection, diffraction and the phase interferences in the sound field around the barrier. However, wave based methods require excessive computational times as the frequency of analysis gets higher, and the geometry more complicated.

Muradali and Fyfe [Muradali and Fyfe, 1997] recently demonstrated how an improved diffraction based model (extensions to the method of Lam [Lam 1994]) could yield good agreements between wave based (BEM and FEM) and the new diffraction based methods, with

T.L. Geers (ed.), IUTAM Symposium on Computational Methods for Unbounded Domains, 111–123.

the latter taking only a fraction of the calculation times. Thus the diffraction based methods make excellent design tools providing results with wave based accuracy.

Although these techniques have received wide acceptance in the scientific community, it has been commonly accepted that the actual acoustic performance of barriers is different from what is predicted by these techniques. This is because all the above mentioned techniques assume a homogeneous, stationary atmosphere. In actuality, the atmosphere has wind and temperature gradients, turbulence, and viscous dissipation (or absorption of the sound energy). These inhomogeneities greatly affect the way in which the sound energy is transmitted.

In this paper we briefly discuss the older diffraction based models and the more recent wave based models. We then highlight the preliminary work of Muradali and Fyfe [Muradali and Fyfe, 1997] showing the excellent agreement with wave based calculations. Finally we present a simple model for including atmospheric considerations with barrier models.

2. Diffraction Based Models

Current literature contains a wide variety of methods for evaluating the attenuation due to a barrier of finite length [Maekawa, 1968; Kurze & Anderson, 1971; Pierce, 1974]. These diffraction methods extend the optical geometric theory of diffraction to that of acoustic waves.

One of the simplest and most widely used in the engineering community, is that of Maekawa [Maekawa, 1968]. He introduced an empirically based diffraction model that provides the insertion loss due to a thin-walled barrier in terms of the Fresnel number. Maekawa then suggested that the insertion loss for a finite-length barrier could be determined by multiple applications of this curve to the diffraction paths around the barrier and then summing the energy contributions of these paths.

There exist more sophisticated mathematical models for determining barrier diffraction coefficients. One model is Pierce's work for diffraction due to a semi-infinite wedge of arbitrary angle with ideal (rigid) boundary conditions [Pierce, 1974]. This model can determine the diffraction field for both 2D and 3D geometries, and involves the evaluation of Airy functions and Fresnel integral functions. Another model is that of Kurze and Anderson's for sound attenuation by semi-infinite barriers [Kurze & Anderson, 1971]. This model is semi-empirical and also addresses both 2D and 3D geometries. In a paper by Hayek [Hayek, 1990], other diffraction models are discussed. These include the diffraction due to a half plane and a wedge, with either ideal boundary conditions or impedance boundary conditions on either face of the plane/wedge.

3. Boundary Element and Wave Envelope Methods

The more exact approach to estimating the insertion loss of a barrier is with wave-based methods such as the Boundary Element Method (BEM) [Hothersall et al,; Fyfe & Harrison, 1995] and Finite Element Methods (FEM) [Astley; Cremers & Fyfe, 1995]. These methods solve the wave equation and thus *exactly* model the phase interaction and the diffraction field around the barrier instead of approximating this effect with diffraction coefficients. For two dimensional and axisymmetric geometries, this approach yields excellent results in very reasonable calculation times. However, when considering a real-life geometry of a finite length barrier (i.e. an exterior three-dimensional problem with large geometry), the BEM/FEM methods quickly approach their limitations. Typical barrier geometries at road noise frequencies prove to be computational intensive and take in the order of hours, and for some cases days, to solve for one frequency on a engineering workstation.

Consider the simple case of a 15m long by 3m tall barrier (this is a very short barrier!). Using the boundary element method with an analysis frequency of 250 Hz (this is a low frequency by barrier standards) requires a mesh of 75 by 15 nodes (using the 6 node per wavelength rule). This totals 1125 nodes for this 3D model. Utilizing the variational method and double precision variables, the coefficient matrix alone requires about 10 megabytes of storage. This size is reasonable by today's standards. However, if the analysis frequency was simply doubled to 500 Hz, a total of 160 megabytes would be required. This requires a large workstation with a large amount of core memory to enable efficient calculations (i.e. minimal swapping). Remember that this says nothing about what would be required to bring the analysis up to a reasonable length of 500m or so. Obviously this makes it impossible to conduct typical barrier design tests such as frequency sweeps to high frequencies, broad-band tests, and parametric studies on barrier geometries with the BEM/FEM methods.

Infinite element strategies have been suggested and developed to address the above mentioned memory difficulties. This includes work by Astley, Coyette and Cremers/Fyfe. The advantages of this approach are that the domain is sub-divided into smaller domains and this yields banded coefficient matrices which are smaller and faster to solve. However, extensive testing by Cremers and Fyfe have shown many real limitations of this approach in that only low *ka* geometries can be modeled (i.e. low frequencies and/or geometries). This work found that to obtain accurate modeling, $ka < 20$ was required. If a 3m high barrier was considered ($a = 20$), this would mean $k = 6.67$ or $f = 365$. Clearly, this renders this approach useless for the type of frequency ranges required (up to 2000 Hz) and geometries used (500m to 2000m) in road noise barrier design.

4. Improved Diffraction Method

Older diffraction based methods relied on summing the acoustic energies of each diffracted path. Lam proposed an improvement to this by summing the complex acoustic pressure of each path [Lam, 1994]. The pressure due to the diffracted path *i* is

$$p_i = A_i e^{j\psi_i} G(kd_i)$$

where A_i, ψ_i and d_i are the amplitude change at diffraction, phase shift at diffraction and the path length of the diffracted wave, respectively. The function, G, is the geometric spreading for the source being considered. The total pressure at the receiver is the sum of the individual paths,

$$p_R = \sum_{i=1}^{n} p_i = \sum_{i=1}^{n} A_i e^{j\psi_i} G(kd_i)$$

where n is the total number of paths being considered. For 2D geometries, $n = 4$ and $G = H_0^{(1)}(kd_i)$, which is a Hankel function of the first kind of order zero. It is used to determine the scattering due to a cylindrical source (see Figure 1). For the 3D case, $n = 8$ and $G = e^{-jkd_i}/d_i$.

4.1 Diffraction Models

The amplitude and phase changes at the diffraction edges can be computed using different diffraction models. In this paper, we will compare two fairly simple methods (Maekawa's and Kurze and Anderson's), which only adjust the amplitude at diffraction, with the Pierce's method, which alters both the magnitude and phase of the acoustic ray at the diffraction edge.

Maekawa's method is derived from empirical data and thus the acoustic pressure due to any path is derived from a curve-fit of experimental data. His expressions are give in terms of the Fresnel number, which is a function of the path length difference (from source to receiver with and without the barrier) and the wavelength. This is appropriate for simple 3D geometries due to the nature of the experimental tests from which the results were derived.

The method of Kurze and Anderson yields both 2D line source and 3D point source expressions for the diffraction edge changes. Its parameters are determined from purely geometrical factors.

Finally, Pierce formulated an approximate solution to the wave equation for the single-edge diffraction by a semi-infinite wedge. His solution considers cases where the source and receiver are at large distances from the barrier in terms of wavelength. The diffraction due to a semi-infinite thin screen is a special case where the wedge angle is zero. When considering a point source the pressure due to an individual path p_i is found from [Pierce, 1974],

$$p_i = [A_D(X_+)+A_D(X_-)]\,\frac{e^{j\frac{\pi}{4}}}{\sqrt{2}}\,\frac{e^{-jkd_i}}{d_i}$$

where

$$\begin{aligned} A_D(X) &= sign(X)[f(|X|) - jg(|X|)] \\ X_+ &= X(\theta + \theta_o) \\ X_- &= X(\theta - \theta_o) \\ X(\theta) &= \left[\frac{2r_o r}{\lambda d_i}\right]^{\frac{1}{2}} [-2\cos\left(\frac{\theta}{2}\right)] \\ d_i &= r_o + r \end{aligned}$$

(see Figure 2 for a visualization of the various terms). In the above equations, the functions f and g are series representation of Fresnel integrals and A_D is the Airy function.

For a 2D geometry, where the source is now cylindrical, the only change introduced is the term for cylindrical spreading,

$$p_i = [A_D(X_+)+A_D(X_-)]\left(\frac{e^{j\frac{\pi}{4}}}{\sqrt{2}}\right)(-jH_0^{(1)}(kd_i))$$

4.2 Modeling Examples

To demonstrate the modeling ability of Lam's method, several examples will be presented. The first is a 2D modeling case, where a line source is in front of an infinite barrier. Figure 3 shows the insertion loss (IL) for a line of receivers as depicted in the figure. The insertion loss is the difference in acoustic pressure with and without the barrier. It is observed that both the Lam/Pierce and the Lam/K&A (Kurze and Anderson) models both closely follow the boundary element solutions. In Figure 4, the same geometry is used, except that the line of receivers is replaced by a single receiver point. Now a frequency sweep is performed up to 2000 Hz. It is observed that all methods agree well, except for small deviations in the low frequency range. This result is not surprising as diffraction based techniques are more accurate for high frequencies.

Lam's method is also easily extended to parallel barriers. In this situation, a single barrier is used in the model, while the reflections between the barriers are modeled by multiple image sources. Figure 5 shows the insertion loss for a line of receivers for the geometry depicted in the figure. It is seen that even for this more complex geometry, the diffraction based models do an admirable job for predicting the acoustic pressures/insertion losses.

Figure 6 shows the insertion loss for a 3D finite length barrier. Again it is shown that the diffraction models follow the accepted results of the boundary element solution. Over most of the line of receivers, the differences in the solutions are less that 2 dB. It is interesting to note that the differences in the calculation times for these 3D simulations was quite dramatic. For the diffraction models, the calculations were complete in a few seconds, while the 3D boundary element solution required over 30 minutes!

5. Atmospheric Modeling

Although the various noise barrier performance prediction techniques have received wide acceptance in the scientific community, it is also known that they consistently over-predict the barrier insertion loss compared to full scale measurements. This is because these techniques assume a homogeneous, stationary atmosphere. In actuality, the atmosphere has wind and temperature gradients, turbulence, and viscous dissipation (or absorption of the sound energy). These inhomogeneities greatly affect the way in which the sound energy is transmitted.

More recently, techniques have been developed to account for non-homogeneous medium properties. These include the Fast Field Program (FFP) [Lee, 1986] and the Parabolic Equation (PE) [White, 1989], to name a few. These techniques can take into account various types of sound speed profiles, and the PE formulations can also model the effects of turbulence [Gilbert, 1990]. However, these techniques are generally restricted to flat ground propagation, and all require excessive computational times and a detailed understanding of the controlling parameters to provide results that are representative of the atmospheric conditions being considered.

5.1 The Heuristic Model

To counter the problems associated with excessive computational time, L'Esperance recently developed a model based on the geometrical ray theory to predict outdoor sound propagation [L'Esperance et al., 1992]. This model, hereby referred to as the *heuristic model*, assumes a linearly varying sound speed profile (of slope *a*) which permits analytical determination of all possible rays between a source and receiver, and the associated ray parameters. The curvature of the rays is based on Snell's law which states that the direction of the sound ray changes in the presence of a varying sound speed. Wind and temperature gradients yield an effective sound speed profile. For a linear profile, the resulting sound rays are circular.

As the sound speed increases with height (downwind or nighttime conditions), the rays tend to curve back towards the ground, and vice-versa, the sound rays curve towards the sky as the sound speed decreases with height (upwind or daytime conditions). The downwind conditions and associated ray paths are depicted in Figure 7. The heuristic model handles the two sound speed gradient cases separately. For the downwind conditions (positive sound speed profile), many rays can appear between a source and a receiver with multiple reflections on the ground, as is seen in Figure 7 [Embleton, 1976]. For the upwind conditions (negative sound speed profile), a shadow zone emerges past which the sound pressures are determined using a diffraction theory based on a residue series solution. The heuristic model can also include the effects of ground, atmospheric absorption and turbulence.

When a barrier is introduced into a non-uniform sound field, it also affects the propagation path. Figure 8 shows the effect of placing a barrier in a fairly strong field. It is observed that the gradient field not only effects the rays once they pass over the barrier (due do the multiple reflections), but also extra rays reach the receiver by *walking over* directly from the source (as depicted by the long curved rays). This model is referred to as the combined heuristic model.

5.2 Modeling Examples

The first example of modeling with the combined heuristic method is a verification with the parabolic equation (PE) solution. Figure 9 shows the relative pressure (in dB) relative to free field propagation against the receiver position. Plotted are the homogeneous case and the PE and heuristic model for the depicted geometry. It is observed how the heuristic modeling follows the PE solution in a series of *steps*. This is due to the fact that as the receiver moves away from the source, additional rays are added which contribute to the sound field at that point.

Figure 10 shows the insertion loss vs. receiver position for two gradients and also that for an averaged gradient field and the homogeneous case. It is observed that any particular gradient field can yield wildly varying results. In the actual environment, however, the sound field will not identical gradients throughout the propagation range due to local deviations and turbulence, therefore an averaged case better depicts reality [Daigle, 1982].

Figure 11 shows the same test case as the previous example, except that now the averaged insertion loss is compared for both hard and soft ground conditions at 500 Hz. This averaged case is designated as the turbulent condition. First of all it is observed that the insertion loss is lower for the soft ground case as the soft ground itself already contributes a great deal to the sound absorption. Secondly it is seen that for either case, the barrier insertion loss greatly deteriorates after about 75m. After this distance, the rays that walk over the barrier, negate most of the barrier shielding. Figure 12 again shows this gradient averaged case, except that now a weighted frequency average is performed between 100 and 2000 Hz instead of just a single frequency. Similar responses are observed.

6. Summary & Conclusions

This paper has outlined the difficulties in using wave-based methods (BEM and IFEM) to perform realistic road noise barrier modeling. To overcome these limitations, an improved diffraction based model is introduced which yields close to wave-based accuracy with diffraction based solution speed. This model is then enhanced to included the effects of linear sound speed gradients in the calculation of barrier insertion loss. Results from this combined model better agree with trends observed in real world measurements. Experimental validation and comparisons are required to fully assess the validity of this new model.

7. References

Astley, R.J (1983) Wave envelope and infinite element elements for acoustical radiation, Int. J. Numerical Methods Fluids, **3**, 507-526.

Cremers, L. & Fyfe, K.R., (1995) On the use of variable order infinite wave envelope elements for acoustic radiation and scattering, J. Acoust. Soc. Am., **97** (4) 2028-2040.

Embleton, T. F. W., Thiessen, G. J., & Piercy, J. E., (1976) Propagation in an inversion and reflections at the ground, J. Acoust. Soc. Am., **59** 278-282.

Daigle, G. A., (1982) Diffraction of sound by a noise barrier in the presence of atmospheric turbulence, J. Acoust. Soc. Am., **71** 847-854.

Fyfe, K.R. & Harrison, C.C., (1995) Modeling of road noise and optimal barrier design,\ CMHC report #6585/F039.

Gilbert, K. E., Raspet, R., & Di, X., (1990) Calculation of turbulence effects in an upward-refracting atmosphere, J. Acoust. Soc. Am., **87** 2428-37.

Hayek, S., (1990) Mathematical modeling of absorbent highway noise barriers. Appl. Acoust., **31** 77-100.

Hothersall, D.C. Chandler-Wilde, S.N. & Hajmirzae, M.N., (1991) Efficiency of single noise barriers. J. Sound Vib., **146** (2) 303-322.

Kurze, U. J. & Anderson, G. S., (1971) Sound attenuation by barriers, Appl. Acoust., **4** 35-53.

Lam, Y. W., (1994) Using Maekawa's chart to calculate finite length barrier insertion loss, Appl. Acoust., **42** 29-40.

Lee, S. W., Bong, N., Richards, W. F. & Raspet, R., (1986) Impedance formulation of the fast-field program for acoustic wave propagation in the atmosphere. J. Acoust. Soc. Am., **79** 628-34.

L'Esperance, A., Herzog, P., Daigle, G. A. & Nicolas, J. R., (1992) Heuristic model for outdoor sound propagation based on an extension of the geometrical ray theory in the case of a linear sound speed profile, Applied Acoustics, **37** 111-139.

Maekawa, Z., (1968) Noise reduction by screens, Appl. Acoust., **1** 157-173.

Muradali, A. & Fyfe, K.R., (1997) A Study of 2D and 3D Barrier Insertion Loss Using Improved Diffraction Based Techniques, Accepted for publication in Applied Acoustics.

Pierce, A. D., (1974) Diffraction of sound around corners and over wide barriers. J. Acoust. Soc. Am., **55** 941-955.

White, M. J. & Gilbert, K. E., (1989) Application of the parabolic equation to the outdoor propagation of sound. Applied Acoustics, **27** 227-38.

2D

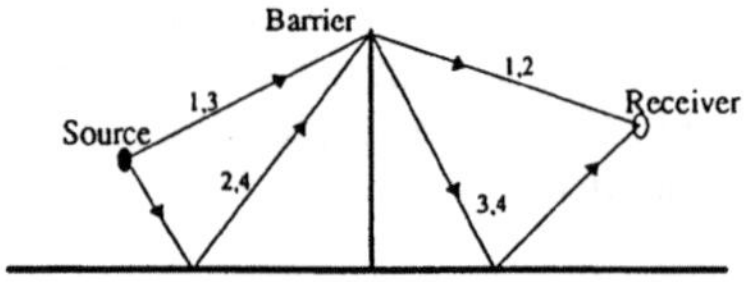

$$p_i = A_i e^{j\psi_i}(-jH_o^{(1)}(kr_i))$$

$$p_R = \sum_{i=1}^{4} p_i = \sum_{i=1}^{4} A_i e^{j\psi_i}(-jH_o^{(1)}(kr_i))$$

diffraction models:

- Kurtz and Anderson
- Maekawa
- Pierce

3D

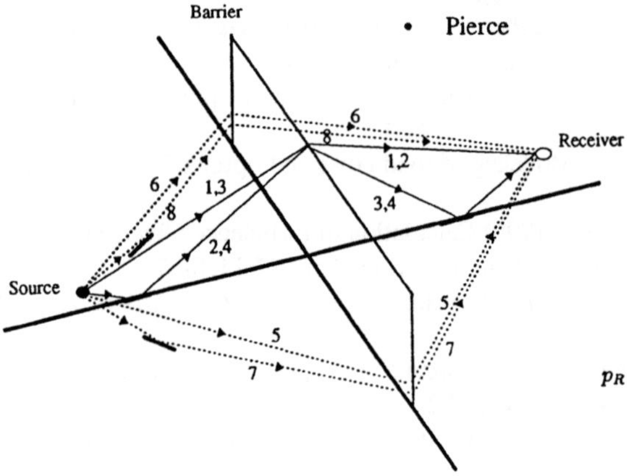

$$p_i = A_i e^{j\psi_i}\frac{e^{-jkr_i}}{r_i}$$

$$p_R = \sum_{i=1}^{8} p_i = \sum_{i=1}^{8} A_i e^{j\psi_i}\frac{e^{-jkr_i}}{r_i}$$

Figure 1 – 2D and 3D Diffraction Paths

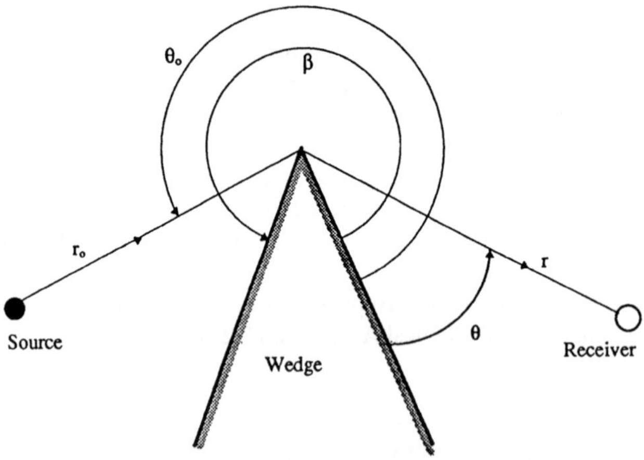

Figure 2 – Terms in Pierce's Diffraction Model

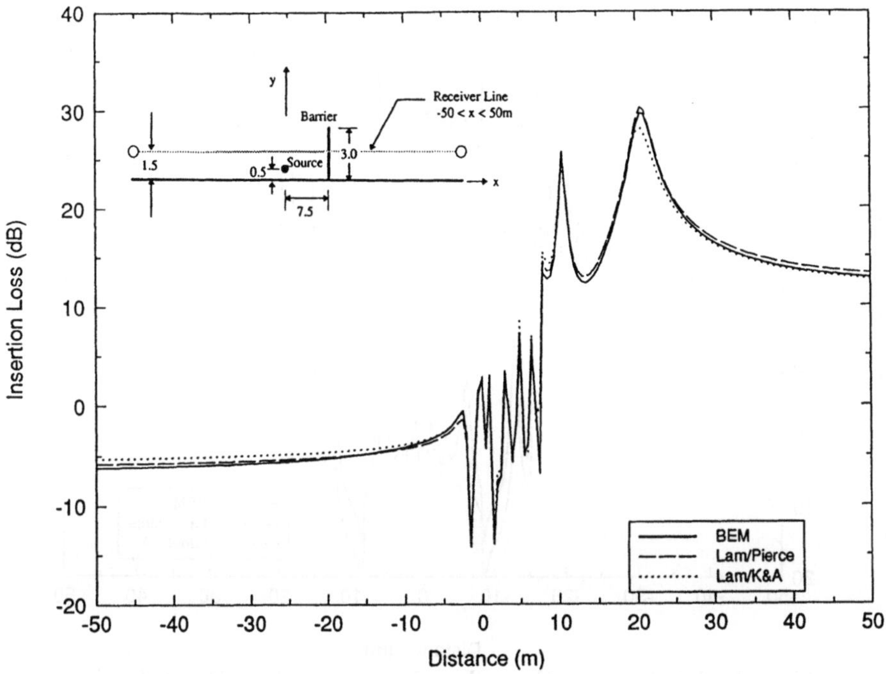

Figure 3 – Insertion Loss Along a Line Through an Infinite Single Barrier

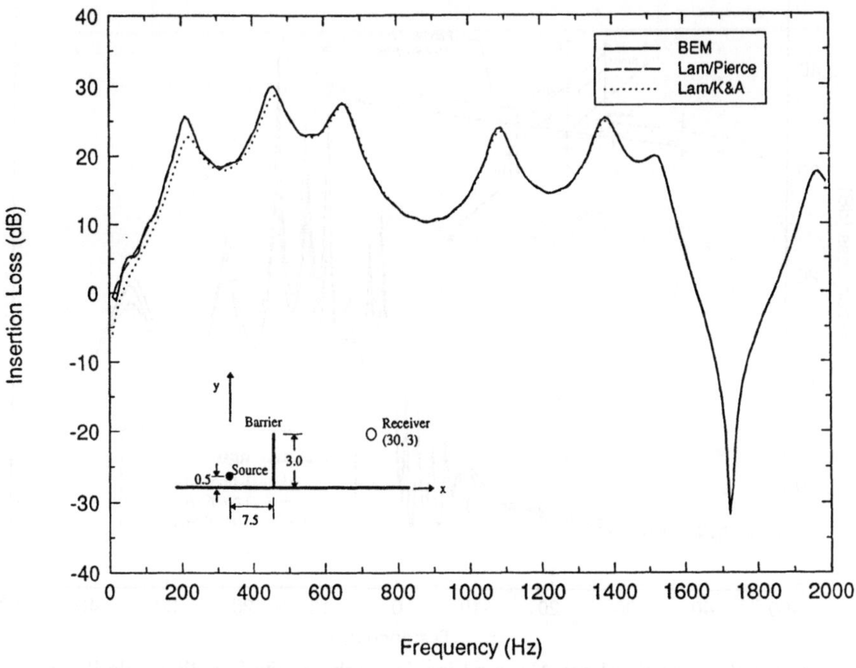

Figure 4 – Insertion Loss Frequency Response of an Infinite Single Barrier

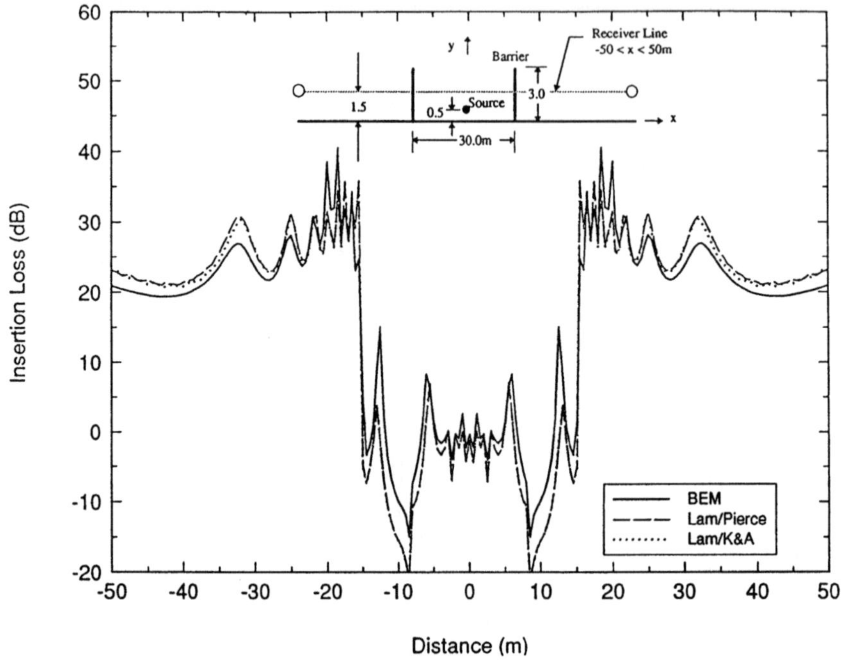

Figure 5 – Insertion Loss Along a Line Through an Infinite Double Barrier

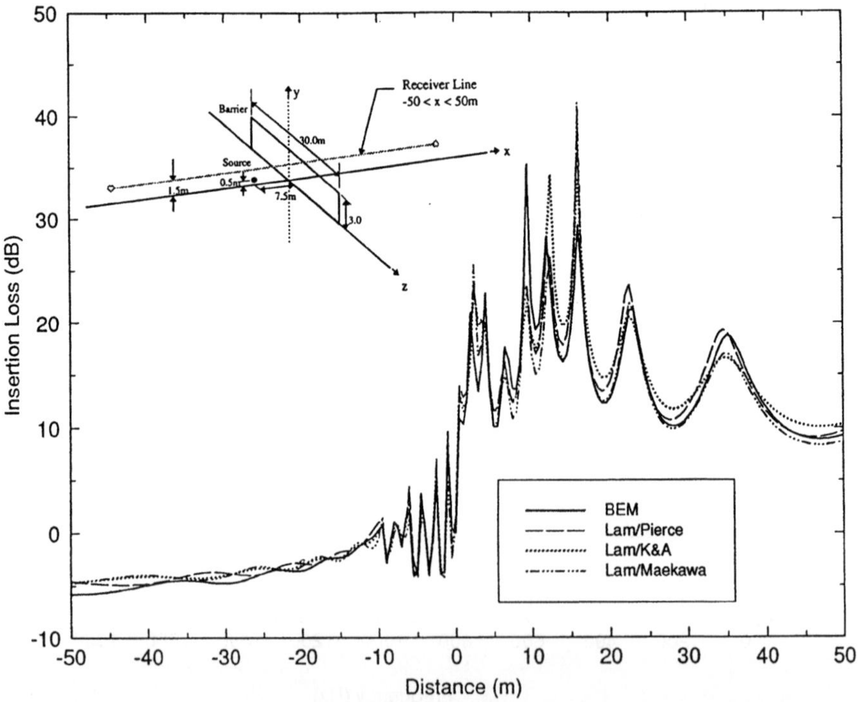

Figure 6 – Insertion Loss Along a Line Through a Finite Length Single Barrier

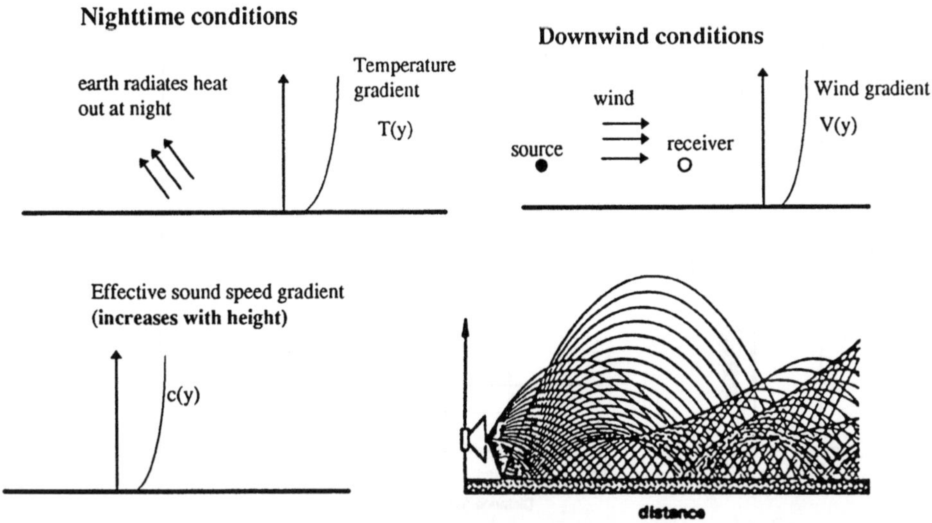

Figure 7 – Downwind Atmospheric Conditions

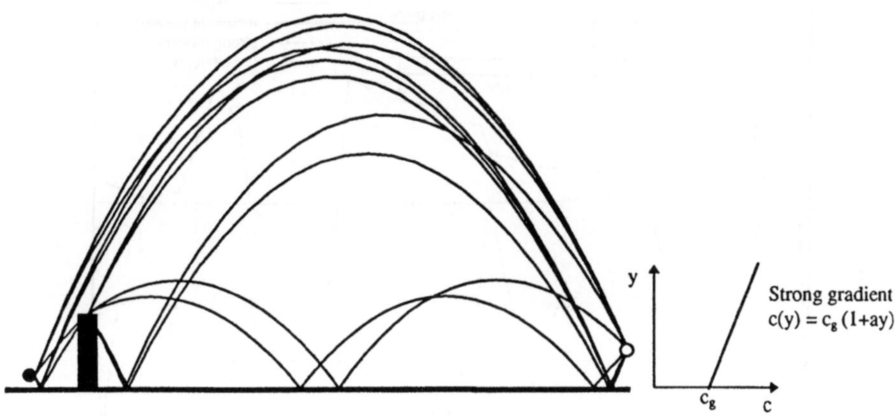

Figure 8 – Propagation Paths Over a Barrier in a Strong Downwind Field

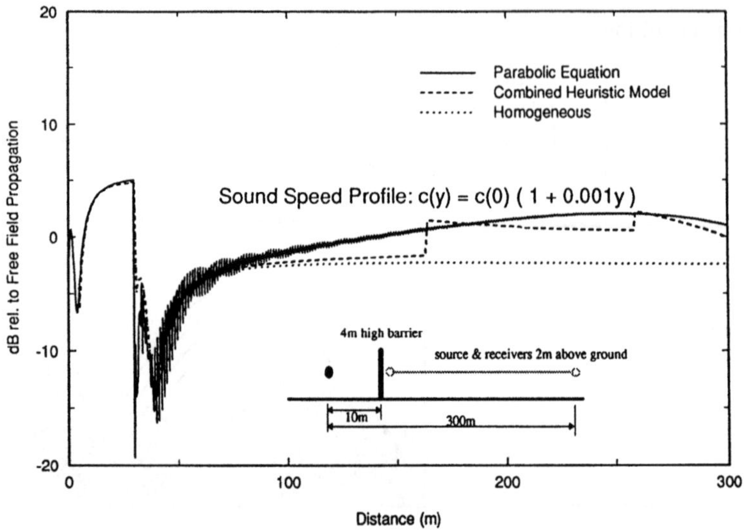

Figure 9 – Model Validation for a Barrier on Soft Ground

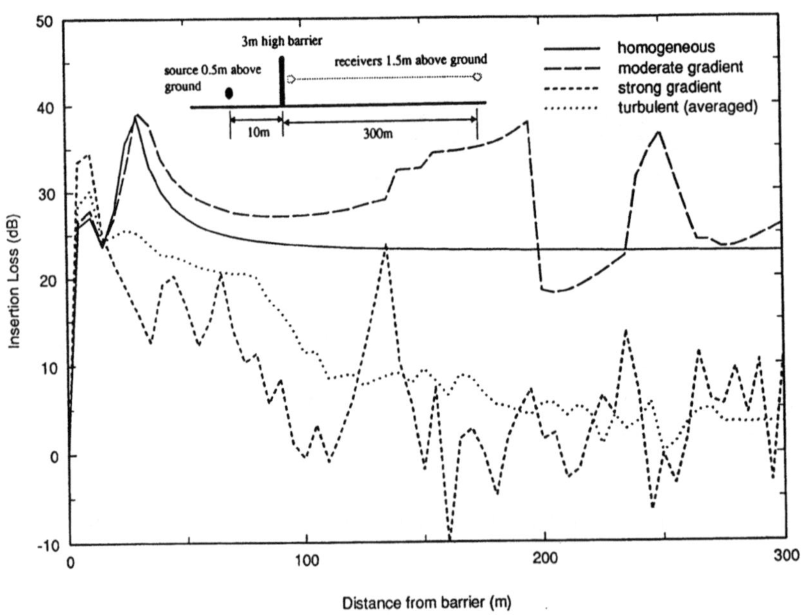

Figure 10 – Barrier Insertion Loss for Different Sound Speed Gradients

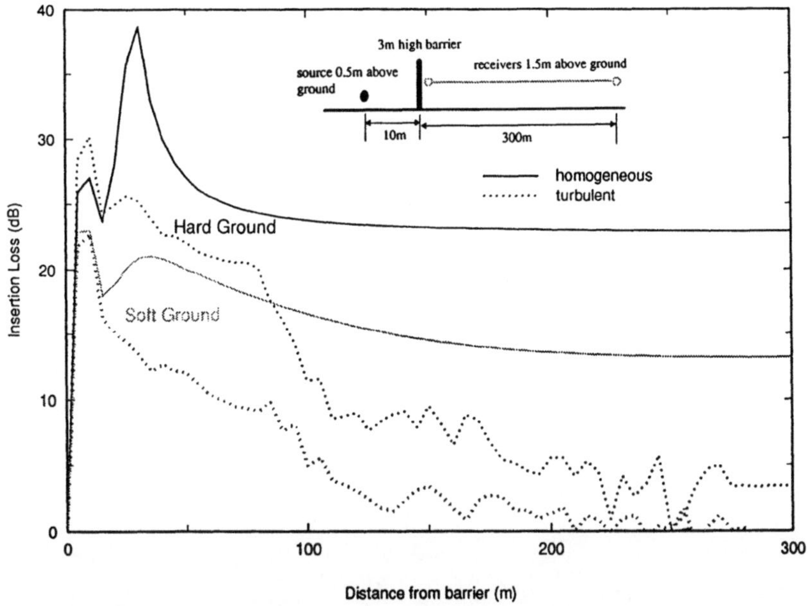

Figure 11 – Insertion Loss vs. Distance for 3m High Barrier at 500Hz

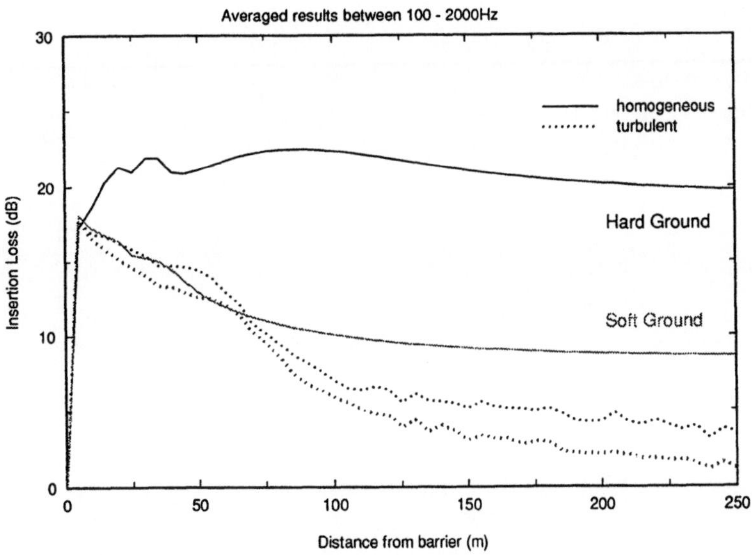

Figure 12 – Frequency Averaged Insertion Loss vs. Distance for 3m High Barrier

CALCULATION OF TRANSIENT RESPONSE OF VISCOELASTIC UNBOUNDED DOMAINS BY DIRECT BOUNDARY ELEMENT METHOD

Lothar Gaul
Institute A of Mechanics
University of Stuttgart
Stuttgart, Germany

Martin Schanz
Institute of Applied Mechanics
Technical University of Braunschweig
Braunschweig, Germany

Abstract

Mixed boundary value problems of solid mechanics with bounded as well as unbounded domains are treated efficiently by numerical solutions of Boundary Integral Equations (BIE) in time domain with the Boundary Element Method (BEM). Thus, the spatial problem dimension is reduced by one. Viscoelastic constitutive behaviour is implemented by means of a Laplace transform technique based on an elastic–viscoelastic correspondence principle. The concept of fractional differintegration generalizes conventional constitutive equations and provides improved curve fitting of measured material response with less parameters. As the implementation of viscoelasticity is provided in each time step in the Laplace domain, efficient algorithms for the inverse transformation in time domain are needed. This is why the performance of adapted algorithms by TALBOT, DURBIN and CRUMP are compared. The impact response of a base plate bonded on a viscoelastic soil halfspace is discussed as numerical example. Viscous forces increase the velocities of surface wave propagation and cause attenuation in addition to the so called geometrical damping by radiation.

1. Introduction

Mixed boundary value problems of solid mechanics with bounded as well as unbounded domains can be treated efficiently by the Boundary Element Method (BEM). In linear elastodynamics the Lamé-Navier equations of motion, along with the boundary conditions and initial conditions are cast into Boundary Integral Equations (BIE). The calculation of transient response requires besides time discretization spatial discretization on the boundary only, thus reducing the problem dimension by one. Infinite or semi-infinite domains such as a half-space in geomechanics are effectively treated by boundary discretization because the energy radiation (sometimes referred to as geometrical damping) is accounted for by the Sommerfeld radiation condition fulfilled by the boundary integral equation at infinity [Eringen and Suhubi, 1975]. For problems of soil-structure-interaction the boundary data are often of primary interest because they govern the propagation of surface waves, the efficiency of isolation measures and the interaction with structures. BEM formulations in frequency domain based on the complex moduli concept account for viscoelastic constitutive equations. In the paper at hand viscoelastic behaviour is implemented in a time domain approach by means of a transform technique [Schanz,

T.L. Geers (ed.), IUTAM Symposium on Computational Methods for Unbounded Domains, 125–133.

1994; Gaul and Schanz, 1994]. Conventional viscoelasticity can be described by integer time derivatives in constitutive equations or by exponential kernels in hereditary integrals. Improved curve-fitting of measured material properties by constitutive equations with less parameters is achieved with the concept of fractional differintegration [Gaul, Klein, and Kempfle, 1991]. Implementation of such constitutive equations is performed on the time step level. As the BIE can be analytically integrated with an elastic fundamental solution in each time step [Steinfeld, 1993], this forms a basis for an application of the elastic-viscoelastic correspondence principle. The elastic constitutive description is replaced by the viscoelastic one after Laplace transformation of the time discretized BIE. Numerical inverse Laplace transformation leads to a viscoelastic BEM in time domain. The solution by point collocation and recursion is shown to be equivalent to the elastic problem [Gaul and Schanz, 1994]. Only few viscoelastic models lead to analytical inverse Laplace transformation. This is why the performance of numerical inverse transformations is crucial. The algorithms of TALBOT, DURBIN and CRUMP are compared with respect to accuracy and CPU time for transforms of decaying waves including those with fractional derivatives [Schanz, 1994]. The numerical example of elastic compared to viscoelastic wave-propagation in a 3-d soil interacting with a bonded foundation slab is discussed. The impact of viscoelasticity increases wave propagation velocities and causes wave attenuation. Because of time discretization, attenuation as well as instability may occur. The dependence of these phenomena on time step size have been discussed in a previous paper [Gaul and Schanz, 1994] and were demonstrated for waves in a 3-d waveguide.

2. Constitutive equation

Two independent sets of constitutive equations for viscoelastic materials exist after decomposing stress and strain tensors $\sigma_{ij}, \varepsilon_{ij}$ into the hydrostatic parts $\delta_{ij}\sigma_{kk}/3, \delta_{ij}\varepsilon_{kk}/3$ and the deviatoric parts s_{ij}, e_{ij}

$$\sum_{k=0}^{N} p'_k \frac{d^k}{dt^k} s_{ij} = \sum_{k=0}^{M} q'_k \frac{d^k}{dt^k} e_{ij} \qquad \sum_{k=0}^{N} p''_k \frac{d^k}{dt^k} \sigma_{ii} = \sum_{k=0}^{M} q''_k \frac{d^k}{dt^k} \varepsilon_{ii}. \tag{1}$$

More flexibility in fitting measured data in a large frequency range is obtained by replacing the integer order time derivatives by fractional order time derivatives [Gaul, Klein, and Kempfle, 1991].

The derivative of fractional order α is defined by

$$\frac{d^\alpha x(t)}{dt^\alpha} = \frac{1}{\Gamma(1-\alpha)} \frac{d}{dt} \int_0^t \frac{x(t-\tau)}{\tau^\alpha} d\tau \qquad 0 \le \alpha < 1 \tag{2}$$

with the Gamma function $\Gamma(1-\alpha) = \int_0^\infty e^{-x} x^{-\alpha} dx$, as the inverse operation of fractional integration attributed to Riemann and Liouville [Oldham and Spanier, 1974].

The fractional derivative in equation (2) appears complicated in time domain. However, Laplace transform reveals the useful result

$$\mathcal{L}\left\{\frac{d^\alpha x(t)}{dt^\alpha}\right\} = s^\alpha \mathcal{L}\{x(t)\} - \sum_{k=0}^{n-1} s_k \frac{d^{\alpha-1-k}}{dt^{\alpha-1-k}} x(0), \quad n-1 < \alpha \le n, \tag{3}$$

where s is the Laplace variable.

For $N = M = 1$ and fractional time derivatives of order α and β a possible representation of a uniaxial stress strain equation with 5 parameters is given by

$$p\frac{d^\alpha}{dt^\alpha}\sigma + \sigma = E\left(\varepsilon + q\frac{d^\beta}{dt^\beta}\varepsilon\right). \tag{4}$$

The restrictions of nonnegative internal work and nonnegative rate of energy dissipation [Bagley and Torvik, 1986] hold, if the parameters of (4) fulfil the constraints [Schanz, 1994]

$$0 < \alpha = \beta < 2 \quad E \geq 0 \quad q > p > 0. \tag{5}$$

If the same damping mechanisms are assumed in hydrostatic and deviatoric stress strain state, the corresponding 3–d constitutive equations are gained by replacing the uniaxial stress and strain by the hydrostatic and the deviatoric states.

A powerful tool for calculating viscoelastic behaviour from a known elastic response is the elastic–viscoelastic correspondence principle. According to this principle [Flügge, 1975] the viscoelastic solution is calculated from the analytical elastic solution by replacing the elastic moduli in the Laplace transformed domain by the transformed impact response functions of the viscoelastic material model. The viscoelastic solution is then obtained by inverse transformation.

For the above mentioned generalized three parameter model the elastic–viscoelastic correspondence is given by

$$3K \to 3K\frac{1+qs^\alpha}{1+ps^\alpha} \qquad 2G \to 2G\frac{1+qs^\alpha}{1+ps^\alpha} \tag{6}$$

where K is the elastic bulk modulus and G the elastic shear modulus. In equation (6) the transformation property (3) is used.

3. Elastic BE–formulation

The Lamé field equations of a homogeneous isotropic elastic domain Ω with boundary Γ are given by

$$\left(c_1^2 - c_2^2\right) u_{i,ij} + c_2^2\, u_{j,ii} + b_j = \ddot{u}_j \quad \mathbf{x} \in \Omega \tag{7}$$

with displacement coordinates $u_j(\mathbf{x}, t)$ and wave speeds

$$c_1^2 = \frac{K + \frac{4}{3}G}{\varrho}, \qquad c_2^2 = \frac{G}{\varrho}. \tag{8}$$

The corresponding traction and displacement boundary conditions are

$$\begin{aligned} t_i(\mathbf{x},t) = \sigma_{ik} n_k &= p_i(\mathbf{x},t) \qquad \mathbf{x} \in \Gamma_t, \\ u_i(\mathbf{x},t) &= q_i(\mathbf{x},t) \qquad \mathbf{x} \in \Gamma_u \end{aligned} \tag{9}$$

and the initial conditions are

$$u_i(\mathbf{x},0) = u_{0i}(\mathbf{x}), \qquad \dot{u}_i(\mathbf{x},0) = v_{0i}(\mathbf{x}) \qquad \mathbf{x} \in \Omega. \tag{10}$$

The 3-d Stokes fundamental displacement tensor of the Lamé equation (7) in an unbounded space, excited by the volume force $b_j(\mathbf{x}, \boldsymbol{\xi}, t, \tau) = \delta(t-\tau)\,\delta(\mathbf{x}-\boldsymbol{\xi})\,e_j$ is given by (e.g. [Eringen and Suhubi, 1975])

$$\tilde{u}_{ij}(\mathbf{x}, \boldsymbol{\xi}, t, \tau) = \frac{1}{4\pi\varrho}\left\{\frac{t-\tau}{r^2}\left(\frac{3r_i r_j}{r^3}-\frac{\delta_{ij}}{r}\right)\left[H\left(t-\tau-\frac{r}{c_1}\right)-H\left(t-\tau-\frac{r}{c_2}\right)\right]+\right.$$
$$\left.\frac{r_i r_j}{r^3}\left[\frac{1}{c_1^2}\delta\left(t-\tau-\frac{r}{c_1}\right)-\frac{1}{c_2^2}\delta\left(t-\tau-\frac{r}{c_2}\right)\right]+\frac{\delta_{ij}}{r c_2^2}\delta\left(t-\tau-\frac{r}{c_2}\right)\right\} \tag{11}$$

where $r = \sqrt{r_i r_i}$, $r_i = x_i - \xi_i$ is the Euclidian distance between the field point $\mathbf{x}$ and the load point $\boldsymbol{\xi}$. The capital H denotes the unit step function, and δ the Dirac distribution. The corresponding fundamental stress vector components are obtained by substituting (11) into the constitutive equation and adopting Cauchy's stress formula.

The dynamic extension of Betti's reciprocal work theorem combines two states of displacements and tractions, $\left(\tilde{u}_{ij}, \tilde{t}_{ij}\right)$ and (u_i, t_i) respectively, and leads to the boundary integral equation

$$c_{ij}u_j = \int_\Gamma \left[\tilde{u}_{ij} * t_j - \tilde{t}_{ij} * u_j\right] d\Gamma + \int_\Omega \varrho\left[\tilde{u}_{ij} * b_j + \tilde{u}_{ij} v_{0j} + \dot{\tilde{u}}_{ij} u_{0j}\right] d\Omega \tag{12}$$

where $c_{ij} = \frac{\delta_{ij}}{2}$ if $\boldsymbol{\xi}$ is located on a smooth boundary. The $*$ denotes the convolution with respect to time.

The integral equation (12) contains boundary integrals only if the volume forces b_j and the initial conditions vanish. Discretization of the boundary integral equation in space and time leads to the boundary element formulation. Only the time discretization by n equal steps Δt is discussed here. The simplest nontrivial choice ensuring that no terms drop out in the boundary integral equation are linear shape functions for the displacements u_i and constant shape functions for the tractions t_i in time domain

$$u_i(\mathbf{x}, \tau) = \left(U_{il}^{m-1}\frac{t_m-\tau}{\Delta t} + U_{il}^{m}\frac{\tau - t_{m-1}}{\Delta t}\right)\eta_l(\mathbf{x}) \tag{13}$$
$$t_i(\mathbf{x}, \tau) = T_{il}^{m}\cdot \mu_l(\mathbf{x}). \tag{14}$$

The actual time step is m. The nodal values are U_{il}^m, T_{il}^m for the corresponding boundary element Γ_l at time $t_m = m\Delta t$. After substituting (13,14) the boundary integral (12) reads

$$\int_0^t\int_\Gamma \left[t_i(\mathbf{x}, \tau)\cdot\tilde{u}_{ij}(\mathbf{x}, \boldsymbol{\xi}, t-\tau) - \tilde{t}_{ij}(\mathbf{x}, \boldsymbol{\xi}, t-\tau)\cdot u_i(\mathbf{x}, \tau)\right] d\Gamma d\tau =$$
$$\sum_l\sum_{m=1}^{n}\int_{\Gamma_l}\int_{t_{m-1}}^{t_m}\left[\tilde{u}_{ij}(\mathbf{x}, \boldsymbol{\xi}, t-\tau)\,\mu_l(\mathbf{x})\cdot T_{il}^m\right.$$
$$\left.-\tilde{t}_{ij}(\mathbf{x}, \boldsymbol{\xi}, t-\tau)\cdot\eta_l(\mathbf{x})\cdot\left(U_{il}^{m-1}\frac{t_m-\tau}{\Delta t} + U_{il}^{m}\frac{\tau-t_{m-1}}{\Delta t}\right)\right] d\tau d\Gamma. \tag{15}$$

Analytical time integration of equation (15) has been carried out by [Steinfeld, 1993].

After substituting $t_m = m\Delta t, t_{m-1} = (m-1)\,\Delta t$ and $t = n\Delta t$ it can be seen, that the functions depend on the difference $(n-m)$, between the observation and the excitation time only.

After time and space discretization a system of algebraic equations [Steinfeld, Takemiya, and Antes, 1994] is obtained

$$\left[\frac{1}{2}\mathbf{I}+\overset{(1)}{\mathbf{T}}\right]\overset{(n)}{\mathbf{u}}+\sum_{m=1}^{n-1}\overset{(n-m+1)}{\mathbf{T}}\overset{(m)}{\mathbf{u}}=\sum_{m=1}^{n}\overset{(n-m+1)}{\mathbf{U}}\overset{(m)}{\mathbf{t}}, \tag{16}$$

where $\mathbf{I}$ is the identity matrix, $\overset{(m)}{\mathbf{T}}$ and $\overset{(m)}{\mathbf{U}}$ are the influence matrices of stresses and displacements at the time step m. The vectors $\overset{(m)}{\mathbf{u}}$ and $\overset{(m)}{\mathbf{t}}$ contain all nodal displacements and tractions of the time step m.

4. Viscoelastic BE–formulation

In order to obtain a viscoelastic boundary integral formulation from the elastic formulation (16) the elastic–viscoelastic correspondence principle (6) is applied. This requires the Laplace transformation of (16) first. The kernels of the matrices consist of the fundamental solutions of the displacements and tractions after time integration. For the sake of brevity the procedure of deducing a viscoelastic formulation is explained for the first term on the right side of equation (15) only.

The one sided Laplace transformation of the first term in (15) leads to

$$\int_0^\infty \sum_{m=1}^{n} \int_{t_{m-1}}^{t_m} \tilde{u}_{ij}\left(\mathbf{x},\boldsymbol{\xi},t-\tau\right) d\tau e^{-st} dt =$$

$$\frac{1}{4\pi\varrho}\sum_{m=1}^{n}\Bigg\{\int_{t_{m-1}+\frac{r}{c_1}}^{t_m+\frac{r}{c_1}}\left[f_0\left(r\right)\frac{1}{2}\left(\left(t-t_{m-1}\right)^2-\left(\frac{r}{c_1}\right)^2\right)+f_1\left(r\right)\right]e^{-st}dt$$

$$+\int_{t_m+\frac{r}{c_1}}^{t_{m-1}+\frac{r}{c_2}} f_0\left(r\right)\left(t\,t_m-t\,t_{m-1}-\frac{t_m^2}{2}+\frac{t_{m-1}^2}{2}\right)e^{-st}dt$$

$$+\int_{t_{m-1}+\frac{r}{c_2}}^{t_m+\frac{r}{c_2}}\left[f_0\left(r\right)\left(t\,t_m-\frac{t_m^2}{2}-\frac{t^2}{2}+\frac{1}{2}\left(\frac{r}{c_2}\right)^2\right)+f_2\left(r\right)\right]e^{-st}dt\Bigg\}, \tag{17}$$

with f_0, f_1 and f_2 depending on spatial coordinates only. After integrating one time interval of (17), the elastic constants are replaced by the viscoelastic impact response functions. The elastic bulk and shear moduli appear only in the compression wave speed c_1 and the shear wave speed c_2. Thus the corresponding viscoelastic expressions

$$c_{1v}^2=\frac{1}{\varrho}\left(K\frac{1+qs^\alpha}{1+ps^\alpha}+\frac{4}{3}G\frac{1+qs^\alpha}{1+ps^\alpha}\right) \quad c_{2v}^2=\frac{1}{\varrho}G\frac{1+qs^\alpha}{1+ps^\alpha} \tag{18}$$

	time	TALBOT	exact	DURBIN	CRUMP
function 1	0.7 s	-3.91 10^{50}	0.000000	7.88 10^{-4}	1.60 10^{-9}
	0.9 s	-5.04 10^{12}	0.000000	9.67 10^{-4}	1.95 10^{-8}
	1.1 s	1.000000	1.000000	0.998886	0.999999
	1.3 s	1.000000	1.000000	0.998565	1.000000
	1.5 s	1.000000	1.000000	1.004410	1.000000
function 2	0.5 s	0.394691	0.389400	0.389363	——
	0.6 s	0.708395	0.745920	0.746377	0.745920
	0.7 s	0.715297	0.715298	0.716224	0.715298
	0.8 s	0.686752	0.686759	0.690466	0.686759
	0.9 s	0.660142	0.660141	0.663267	0.660141
function 3	0.5 s	0.906345	——	0.906504	——
	0.6 s	0.926606	——	0.925405	0.926606
	0.7 s	0.938730	——	0.942046	0.938730
	0.8 s	0.947151	——	0.943823	0.947151
	0.9 s	0.953470	——	0.949619	0.953470

Table 1: Comparison of inverse transformation algorithms

are inserted into equation (17) after integration.

The method of TALBOT [Talbot, 1979] has been selected for numerical inversion.

5. Comparison of numerical inverse transformation algorithms

As only few viscoelastic constitutive equations of minor practical importance allow analytical inverse Laplace transformation, the performance of numerical inverse transformation methods is analyzed below [Schanz, 1994].

Comparison of the algorithms by TALBOT [Talbot, 1979], DURBIN [Durbin, 1974] and CRUMP [Crump, 1976] shows advantages of the TALBOT algorithm for BEM application. In this algorithm complex integration along the Bromwich contour with infinite limits is mapped into a real integral with finite integration limits [Schanz, 1994].

Table 1 compares the results of inverse transformation for the following test functions at discrete times

1. $\frac{e^{-sa}}{s} \bullet\!\!-\!\!\circ\; H\left(t-a\right)$

2. $\frac{1}{\sqrt{s(s+a)}}\, e^{-\frac{r}{c}\sqrt{s(s+\alpha)}} \bullet\!\!-\!\!\circ\; I_0\left(\frac{a}{2}\sqrt{t^2-\left(\frac{r}{c}\right)^2}\right) e^{-\frac{a}{2}t} H\left(t-\frac{r}{c}\right)$

3. $\frac{1}{s} e^{-s\frac{r}{c}\sqrt{\frac{1+ps^{0.7}}{1+qs^{0.7}}}} \bullet\!\!-\!\!\circ$ no analytical solution

where $a = 1, \frac{r}{c} = 0.5, p = 1$ and $q = 2$. The test functions were selected because of the following reasons:

- shifting of the Heaviside function in function 1 occurs in the fundamental solution (17)

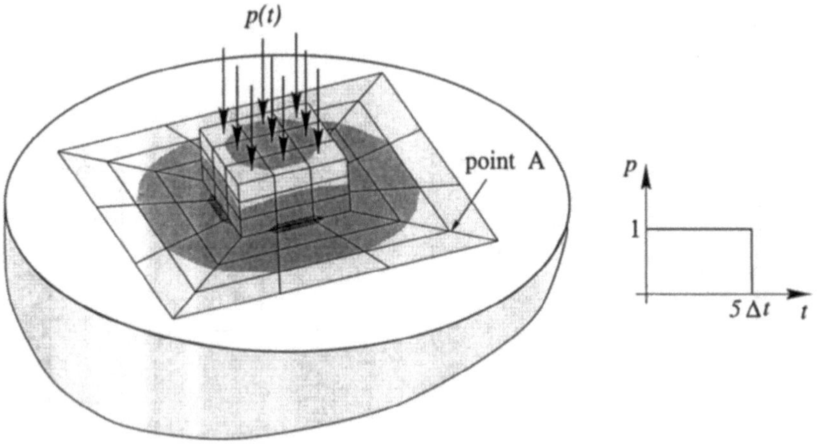

Figure 1: Elastic concrete slab on viscoelastic half-space: boundary element discretization, surface displacement wave fronts, loading function

- function 2 is part of the transformed fundamental solution with viscoelastic Maxwell model [Schanz and Gaul, 1993]
- function 3 is part of (17) after carrying out Laplace transformation with viscoelastic wave speeds (18) which are based on the fractional derivative concept

The results in table 1 for function 1 correspond to times before and after the Heaviside jump at $t = a$. The method of TALBOT fails for $t < a$. Only the results of the method by CRUMP can be considered zero numerically in this range.

Based on more detailed comparison [Schanz, 1994] it can be concluded, that the methods by TALBOT and CRUMP lead to better accuracy than the method by DURBIN, which is as well more time consuming for complicated functions. The choice of the TALBOT method for the BEM in the present work is justified by minimal computer time. Difficulties by jumps and singularities are avoided.

6. Numerical example

The propagation of waves in elastic concrete ($E = 3 \cdot 10^8 \frac{N}{m^2}$, $\varrho = 2000 \frac{kg}{m^3}$, $\nu = 0.2$) foundation slab bonded on a viscoelastic soil half-space ($\nu = 0.35$, $E = 1.38 \cdot 10^8 \frac{N}{m^2}$, $\varrho = 1966 \frac{kg}{m^3}$, $\alpha = 1.3$, $p = 1\mathrm{s}^{-1.3}$) has been calculated by the presented BEM in time domain. Both domains are coupled by a substructure technique based on displacement- and traction-continuity at the interface. The assumption of welded contact does not allow the nonlinear effect of partial uplifting.

The problem geometry and the associated boundary discretization are shown in Fig. 1 The soil discretization is truncated after a distance of the foundation length. The surface of the foundation slab is excited by a positive and negative pressure jump according to Fig. 1. Linear spatial shape functions have been used. Similar to the Courant criteria the time step size Δt has

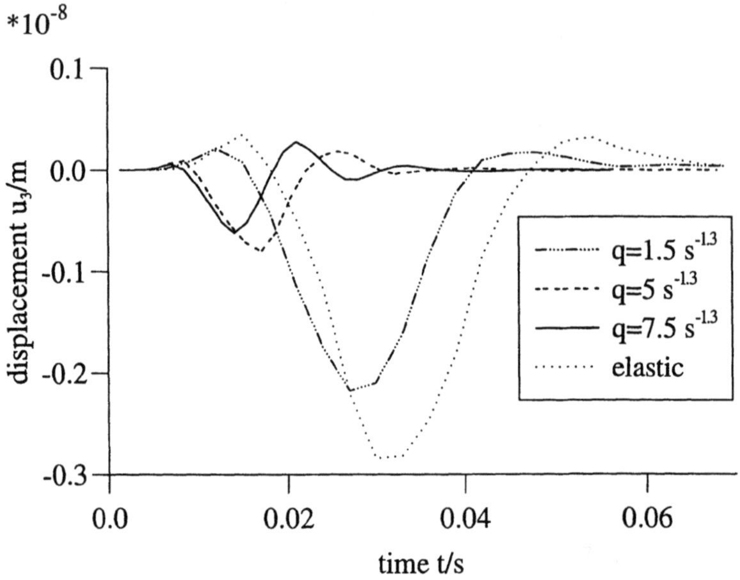

Figure 2: Displacement u_3 perpendicular to the surface on point A for different values q

been chosen close to the time needed by the viscoelastic compression wave to travel across the largest element.

In Fig 2 vertical surface displacement at point A is plotted versus time for different values of the constitutive parameter q. Obviously the wave speeds increase for higher values of q, because of a stiffening of the material with growing influence of viscoelasticity. This is associated with a significant displacement reduction.

7. Conclusions

Current BEM formulations in elastodynamics treat viscoelastic solids in Laplace or Fourier transform domain. Herein the elastic moduli are replaced by the corresponding viscoelastic moduli. The calculation of transient response in time domain requires inverse transformation [Klein, 1990] which is associated with truncation effects when the solution e.g. in frequency domain is known in a limited frequency range only. To avoid this problem the present paper uses the BEM formulation in time domain and converts the elastic formulation to a viscoelastic one. It turned out to be advantageous to convert after integration over a time step Δt rather than converting the fundamental solution itself. The conversion is based on the elastic–viscoelastic correspondence principle in Laplace domain. The inverse transformation in each time step is performed numerically and thus allows to take general constitutive equations into account including those with fractional time derivatives. The wave propagation in an inhomogeneous 3-d continuum is calculated by BEM in time domain. The material behaviour is described by a rheological three parameter model, generalized by two fractional time derivatives.

8. References

Bagley, R., and Torvik, P., 1986, On the Fractional Calculus Model of Viscoelastic Behaviour, *Journal of Rheology*, Vol. 30, pp. 133–155.

Crump, K. S., 1976, Numerical Inversion of Laplace Transforms Using a Fourier Series Approximation, *Journal of the Association for Computing Machinery*, Vol. 23, pp. 89–96.

Durbin, F., 1974, Numerical inversion of Laplace transforms: an efficient improvement to Dubner and Abate's method, *The Computer Journal*, Vol. 17, pp. 371–376.

Eringen, A. C., and Suhubi, E. S., 1975, *Elastodynamics*, Volume II, Academic Press, New York, San Francisco, London.

Flügge, W., 1975, *Viscoelasticity*, Springer–Verlag, New York, Heidelberg, Berlin.

Gaul, L., Klein, P., and Kempfle, S., 1991, Damping Description Involving Fractional Operators, *Mechanical Systems and Signal Processing*, Vol. 5, pp. 81–88.

Gaul, L., and Schanz, M., 1994, Dynamics of viscoelastic solids treated by boundary element approaches in time domain, *European Journal of Mechanics A/Solids*, Vol. 13, pp. 43–59.

Klein, P., 1990, Zur Beschreibung der dynamischen Wechselwirkung von Fundamentstrukturen mit dem viskoelastischen Baugrund durch dreidimensionale Randelementformulierungen, Bericht aus dem Institut für Mechanik, Universität der Bundeswehr Hamburg.

Oldham, K. B., and Spanier, J., 1974, *The Fractional Calculus*, Academic Press, New York, London.

Schanz, M., 1994, Eine Randelementformulierung im Zeitbereich mit verallgemeinerten viskoelastischen Stoffgesetzen, Bericht aus dem Institut A für Mechanik Heft 1, Universität Stuttgart.

Schanz, M., and Gaul, L., 1993, Implementation of viscoelastic behaviour in a time domain boundary element formulation, *Applied Mechanics Reviews*, Vol. 46, pp. S41–S46.

Steinfeld, B., 1993, Numerische Berechnung dreidimensionaler Kontaktprobleme Bauwerk–Boden mittels zeitabhängiger Randintegralgleichungen der Elastodynamik, Technisch–wissenschaftliche Mitteilungen Nr. 93–1, Ruhr–Universität Bochum.

Steinfeld, B., Takemiya, H., and Antes, H., 1994, Analysis of transient 3d wave propagation in an elastic halfspace: The classical approach and the direct boundary element method, Zeitschrift für angewandte Mathematik und Mechanik, GAMM–Conference, Braunschweig.

Talbot, A., 1979, The Accurate Numerical Inversion of Laplace Transforms, *Journal of the Institute of Mathematics and its Applications*, Vol. 23, pp. 97–120.

SINGLY AND DOUBLY ASYMPTOTIC COMPUTATIONAL BOUNDARIES

Thomas L. Geers

Center for Acoustics, Mechanics and Materials
Department of Mechanical Engineering
University of Colorado
Boulder, CO, USA

1. Introduction

The class of problems addressed in this symposium involves the numerical computation of wave fields in and/or around a body residing within an unbounded domain. Each of these problems may be regarded as one of the following types:

Type 1: The analyst is interested in the fields over a large region surrounding the body,

Type 2: The analyst is (a) interested only in the fields in the vicinity of the body, or (b) not interested in the fields outside the body.

Spatial discretization is done, of course, over a subdomain surrounding the body, which, in turn, is enveloped by a computational boundary that portrays the rest of the domain. For a Type 2b problem, the subdomain may not exist, thereby making the computational boundary coincident with the surface of the body.

Let us focus on 3-D problems. For a Type 1 problem, the computational boundary is typically placed out in the *far field*, characterized by $(k_m r)^2 >> 1$, where k_m is the magnitude of the smallest characteristic wavenumber for the wave fields and r is the distance out from the centroid of the body. There, the fields are *radiative*, i.e., virtually all of the field energy is propagating out to infinity. Hence, an appropriate computational boundary for Type I problems is an *absorbing* (or nonreflecting) *boundary* [Givoli, 1991].

For Type 2 problems, however, it is often desirable to place the computational boundary near the body in order to minimize the size of the subdomain to be discretized. There, the fields may be non-radiative, i.e., much or most of the field energy may be trapped near the body. Hence, an absorbing boundary may not be appropriate for a Type 2 problem.

The appropriate boundary for a Type 2 problem is usually an *impedance boundary*. Such a boundary is designed, not merely to absorb field energy, but to portray, with satisfactory accuracy, the infinite domain surrounding the subdomain, usually over some frequency band. An absorbing boundary can do this if the band is such that $(k_m a)^2 >> 1$, where a is a characteristic radius of the body. Thus, an absorbing boundary is *singly asymptotic* in the high-frequency limit, inasmuch as k_m increases with frequency.

Actually, there exists another situation for this class of problems, that in which $(k_M a)^2 << 1$, where k_M is the magnitude of the largest characteristic wavenumber for the wave fields. In this situation, the far field is defined by $(r / a)^2 >> 1$, and the computational boundary for a Type 1 problem can be almost anything, as there is essentially no energy

T.L. Geers (ed.), IUTAM Symposium on Computational Methods for Unbounded Domains, 135–141.

radiation and virtually all of the non-radiative field energy is contained within the subdomain. For a Type 2 problem with $(k_M a)^2 << 1$, the appropriate boundary is a near-field *reactive boundary* that is singly asymptotic in the low-frequency limit; examples include fluid added mass and static soil stiffness.

Now a broadband Type 1 problem is well served by an absorbing boundary because the non-radiative energy is well contained within the large subdomain and the radiative energy is properly absorbed. A broadband Type 2 problem, however, is better served by an impedance boundary that is *doubly asymptotic*, i.e., it approaches exactness in both the high- and low-frequency limits. Doubly asymptotic boundaries have been developed by several researchers [Geers, 1971 & 1978; Felippa, 1980; Underwood and Geers, 1981; Geers and Zhang, 1988 & 1994, Nicolas-Vullierme, 1991; Astley, 1995; Wolf, 1995; Qi and Geers, 1997].

2. The Dilating Sphere in an Acoustic Medium

Perhaps the simplest example problem that illustrates the points made above involves the time-harmonic, uniform vibration of a spherical surface in an infinite acoustic medium. The pressure and radial velocity fields generated in the medium may be expressed [Kinsler, et al., 1982]

$$\begin{aligned} p(r,t) &= \rho c U(r) \cos\zeta(r) \cos(\omega t - kr), \\ u(r,t) &= U(r) \cos[\omega t - kr - \zeta(r)], \end{aligned} \tag{1}$$

in which ρ and c are the fluid's density and speed of sound, and $U(r) = U_a \cdot a \cos\zeta(a) / r \cos\zeta(r)$, where U_a is the magnitude of the surface velocity and $\zeta(r)$ is obtained from $\cot\zeta(r) = kr$.

One way to assess the degree to which the fields at a certain point are radiative is to determine the relation between time-averaged acoustic intensity (referenced to the outward normal) and energy density; this yields for the present problem

$$< I(r) > = \frac{\cos^2\zeta(r)}{1 - \frac{1}{2}[1 - \cos^2\zeta(r)]} \cdot c < E(r) >. \tag{2}$$

For $(kr)^2 >> 1$, $\zeta(r) \approx 0$, which yields $< I(r) > \approx c < E(r) >$, approximately the relation for plane-wave fields, which are purely radiative. For $(kr)^2 << 1$, $\zeta(r) \approx \pi / 2$ and $< I(r) >$ is much smaller than $c < E(r) >$, which means virtually no radiation.

Another way is with specific impedance; for the present problem, the specific acoustic impedance $z_r(\omega) \equiv p(r,\omega) / u(r,\omega)$ is given by

$$z_r(\omega) = \rho c \, \eta_r(\omega) + i\omega\rho r \, v_r(\omega), \tag{3}$$

where the specific acoustic resistance $\eta_r(\omega) = (kr)^2 / [1 + (kr)^2]$ and the specific acoustic intertia $v_r(\omega) = 1 / [1 + (kr)^2]$. For $(kr)^2 >> 1$, $\eta_r(\omega) \approx 1$ and $v_r(\omega) \approx 0$, so the impedance is primarily resistive and the fields are essentially radiative; for $(kr)^2 << 1$, $\eta_r(\omega) \approx 0$ and $v_r(\omega) \approx 1$, so the impedance is primarily inertial and the fields are essentially non-radiative.

3. More Complicated Acoustic Cases

Impedance behavior for higher-order modes of motion of the spherical surface is similar to that for dilatation. Figure 1 from Toothaker, 1997 shows the specific acoustic impedance curves $\eta_r(\omega)$ and $v_r(\omega)/(n+1)$ for dilatation $(n=0)$, translation $(n=1)$ and ovalization $(n=2)$. All of the $\eta_r(\omega)$-curves approach zero as $\omega \to 0$ and unity as $\omega \to \infty$; all of the $v_r(\omega)/(n+1)$-curves approach unity as $\omega \to 0$ and zero as $\omega \to \infty$. The quantity $\rho a/(n+1)$ is the modal specific added mass for incompressible flow. In the figure, the pair of solid curves that never rise above unity pertain to $n = 0$, the pair of solid curves that are just above those pertain to $n = 1$, and the highest pair of solid curves pertain to $n = 2$. The dashed curves and the legend in the figure are discussed below.

The doubly asymptotic behavior illustrated in Figure 1 characterizes the specific acoustic impedance for 3-D motion around all surfaces. For example, Figure 2, also from Toothaker, 1997, shows modal specific acoustic impedance curves for the infinite, circular cylindrical surface vibrating steadily as $u(a,t) = U_a \cos \mu z \cos \omega t$. The specific acoustic impedance curves $\eta_r(\omega)$ and $v_r(\omega)/[\mu K_1(\mu)/K_0(\mu)]$ are shown for $\mu = \pi/10$, $\mu = 1$ and $\mu = \pi$. Note that here $\rho a/[\mu K_1(\mu)/K_0(\mu)]$ is the modal specific added mass for incompressible flow. The exact (solid) impedance curves in Figure 2 exhibit the jump and extreme resonance behavior characteristic of unbounded surfaces [Junger and Feit, 1986]. The other curves in the figure are discussed below.

Modal specific impedance curves in electromagnetics, elastodynamics and poro-elastodynamics exhibit behavior that is certainly different from that seen in Figures 1 and 2. All such curves, however, are doubly asymptotic.

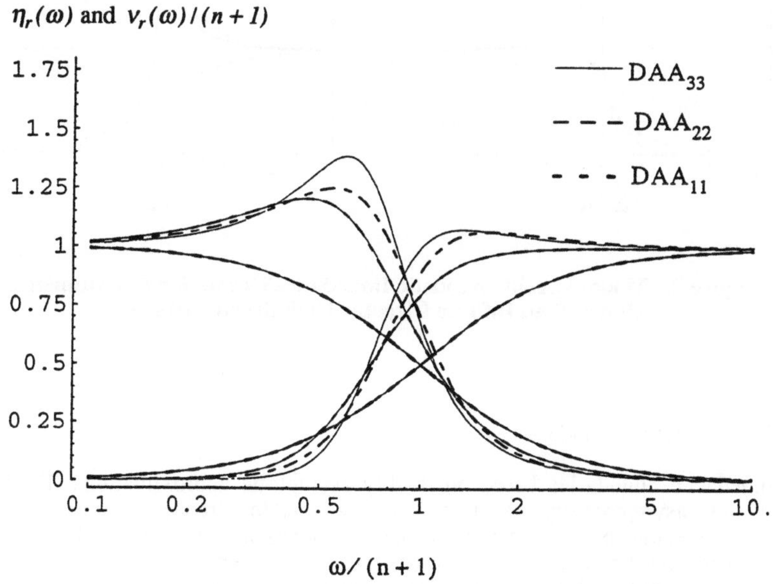

Figure 1. Modal Specific Acoustic Impedance Curves for the Lower Modes of a Spherical Surface

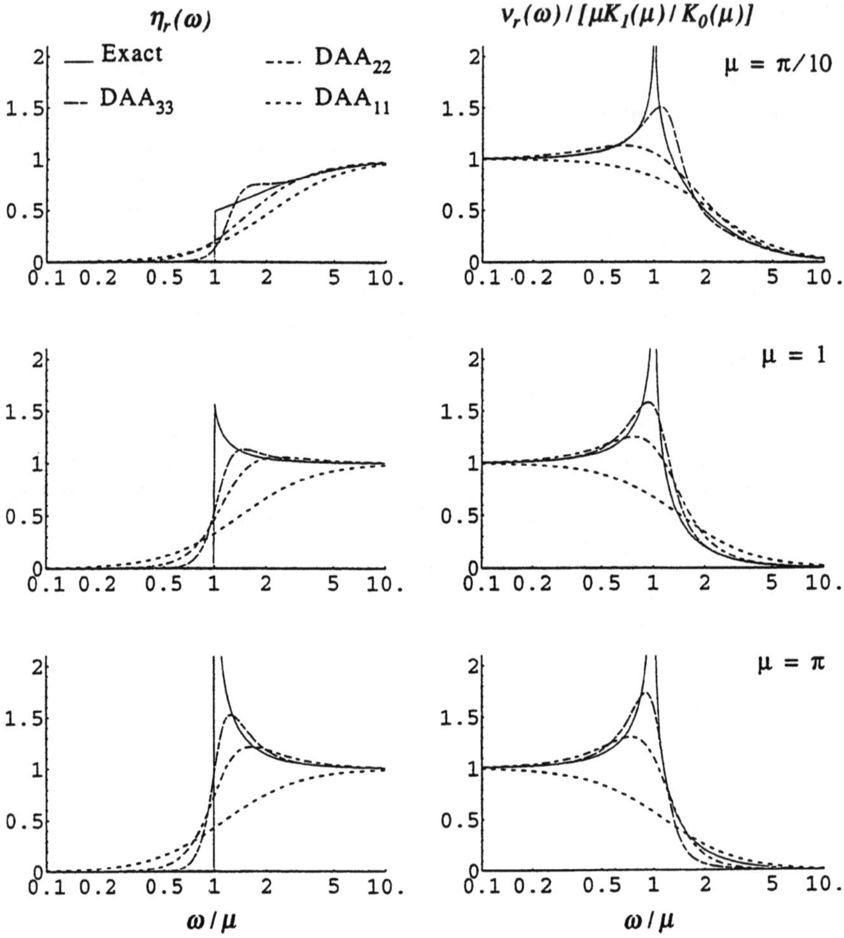

Figure 2. Modal Specific Acoustic Impedance Curves for Axisymmetric Modes of an Infinite Circular Cylindrical Surface

4. Doubly Asymptotic Boundaries

Impedance boundaries based on doubly asymptotic approximations (DAAs) are formulated to be asymptotically exact in both the high- and low-frequency limits. One subset of such boundaries may be formulated on the basis of *operator matching*, which is akin to the method of matched asymptotic approximations [Nicolas-Vullierme, 1991; Geers and Zhang, 1994; Geers and Lewis, 1997]. Theoretically, DAAs of unlimited order may be formulated.

In acoustics, the first-order DAA, the fully-matched, symmetric, second-order DAA, and the fully-matched, symmetric, third-order DAA are given by

$$\text{DAA}_{11}: \quad \dot{p} + P_{10}p = \rho c \dot{u},$$

$$\text{DAA}_{22}: \quad \ddot{p} + P_{21}\dot{p} + P_{20}p = \rho c(\ddot{u} + U_{20}\dot{u}), \tag{4}$$

$$\text{DAA}_{33}: \quad \dddot{p} + P_{32}\ddot{p} + P_{31}\dot{p} + P_{30}p = \rho c(\dddot{u} + U_{31}\ddot{u} + U_{30}\dot{u}),$$

where the P_{ij} and U_{ij} are spatial integral operators. Following boundary-element discretization, the functions p and u become computational column vectors and the operators P_{ij} and U_{ij} become full matrices, so that (4) become ordinary differential equations in time.

The performance of these DAAs for the spherical and infinite cylindrical surface problems of Sections 2 and 3 are demonstrated in Figures 1 and 2, respectively. Figure 1 for the sphere shows that DAA_{11} is exact for dilatation but unsatisfactory for translation and ovalization, DAA_{22} is exact for dilatation and translation and satisfactory for ovalization, and DAA_{33} is exact for dilatation, translation and ovalization. Figure 2 for the infinite cylinder shows that none of the DAAs can follow completely the jump and extreme resonant behavior of the exact curves, but performance improves steadily with increasing DAA order.

5. Perfectly Matched Layer

It is interesting to observe the performance of an excellent absorbing boundary [Berenger, 1994] when applied to a Type 2 problem. The problem is that of Section 2, with the inner radius of a perfectly matched layer (PML) located at several multiples of the spherical body's radius; acoustic medium, of course, lies between. Specific acoustic impedance curves as seen by the body are shown in Figures 3 and 4, which are taken from Qi

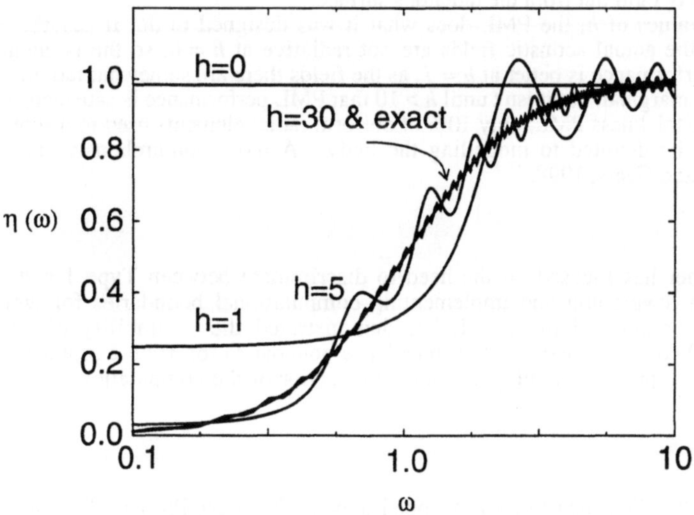

Figure 3. Specific Acoustic Resistance Curves for the Dilatational Motion of a Spherical Surface ($q_r\delta = 6$)

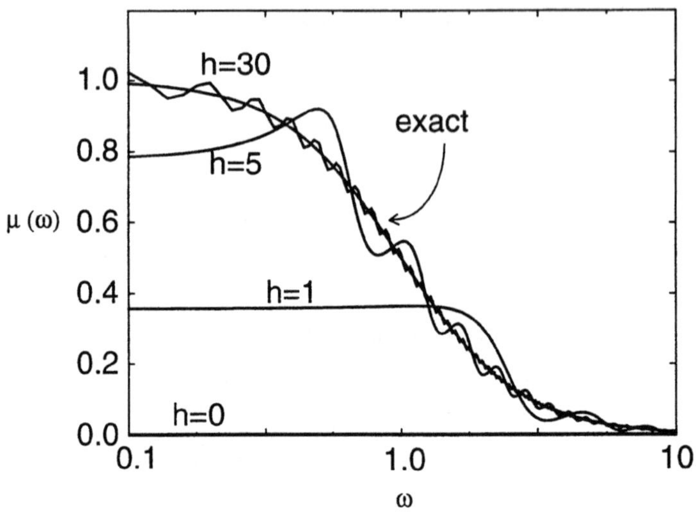

Figure 4. Specific Acoustic Inertia Curves for the Dilatational Motion of a Spherical Surface ($q_r\delta = 6$)

and Geers, 1998. The exact curves for $\eta(\omega)$ and $v(\omega)$ are the non-resonant curves of Figure 1, and the the other curves pertain to placement of the PML's inner surface on ($h = 0$) and at one, five and thirty radii out from the radiator's surface.

For all values of h, the PML does what it was designed to do: it absorbs energy. Unfortunately, the actual acoustic fields are not radiative at $h = 0$, so the boundary fails completely. Performance is better at $h = 1$, as the fields there are somewhat radiative. At $h = 5$, the PML is marginal, but it isn't until $h > 10$ that PML performance is satisfactory. With $h = 8$ and a unit-thickness PML, only 10% of uniform radial elements used to discretize this problem would be devoted to modelling the body. A more comprehensive analysis is presented in Qi and Geers, 1998.

6. Conclusion

This paper has focused on the need to discriminate between Type 1 and Type 2 situations when developing and implementing computational boundaries for wave-field problems in unbounded domains. It has demonstrated the desirability of absorbing boundaries for Type 1 problems and of impedance boundaries for Type 2 problems. For a broadband Type 2 problem, doubly asymptotic behavior of the computational boundary is essential.

7. References

Astley, R.J., 1996, "Transient Wave Envelope Elements for Wave Problems", *J. Sound Vib.*, Vol. 192, pp. 245-261.

Berenger, J.-P., 1994, "A Perfectly Matched Layer for the Absorption of Electromagnetic Waves", *J. Comp. Phys.*, Vol. 114, pp. 185-200.

Felippa, C.A., 1980, "A Top-down Derivation of Doubly Dsymptotic Approximations for Structure-fluid Interaction analysis," Shaw, R.P., et al., eds, *Innovative Numerical Analysis for the Engineering Sciences,* U. Press of Virginia, Charlottesville, pp. 79-88.

Geers, T. L., 1971, "Residual Potential and Approximate Methods for Three-dimensional Fluid-structure Interaction Problems," *J. Acoust. Soc. Am.*, Vol. 49, pp. 1505-1510.

Geers, T. L., 1978, "Doubly Asymptotic Approximations for Transient Motions of Submerged Structures," *J. Acoust. Soc. Am.*, Vol. 64, pp. 1500-1508.

Geers, T.L., and Lewis, B.A., 1997, "Doubly Asymptotic Approximations for Transient Elastodynamics", *Int. J. Solids Struc.*, Vol. 34, pp. 1293-1305.

Geers, T. L., and Zhang, P., 1988, "Doubly Asymptotic Approximations for Electromagnetic Scattering Problems," Tanaka, M. and Cruse, T.A., eds., *Boundary Element Methods in Applied Mechanics,* Pergamon Press, Oxford, pp. 357-369.

Geers, T. L., and Zhang, P., 1994, "Doubly Asymptotic Approximations for Submerged Structures with Internal Fluid Volumes: Formulation," *J. Appl. Mech.* , Vol 61, pp. 893-899.

Givoli, D., 1991, "Non-reflecting Boundary Conditions", *J. Comp. Phys.*, Vol. 94, pp. 1-29.

Junger, M.C., and Feit, D., 1986, *Sound, Structures, and Their Interaction*, 2nd ed., MIT Press, Cambridge, MA.

Kinsler, L.E., et al., 1982, *Fundamentals of Acoustics*, Wiley, New York.

Nicolas-Vullierme, B., 1991, "A Contribution to Doubly Asymptotic Approximations: An Operator Top-down Approach", *J. Vib. Acoust.*, Vol. 113, pp.409-415.

Qi, Q., and Geers, T.L., 1997, "Doubly Asymptotic Approximations for Transient Poro-elastodynamics", *J. Acous. Soc. Am.*, Vol. 102, pp. 1361-1371.

Qi, Q., and Geers, T.L., 1998, "Evaluation of the Perfectly Matched Layer for Computational Acoustics", *J. Comp. Phys.*, Vol. 139, pp.166-183.

Toothaker, B.J., 1997, "A Study of Higher Order Doubly Asymptotic Approximations", M.S. Thesis, University of Colorado, Boulder.

Underwood, P.G., and Geers, T.L., "Doubly Asymptotic Boundary-element Analysis of Dynamic Soil-structure Interaction, *Int. J. Solids Struc.*, Vol. 17, pp. 687-697.

Wolf, J.P., and Song, Ch., 1995, "Doubly Asymptotic Multi-directional Transmitting Boundary for Dynamic Unbounded Medium-structure-interaction Analysis", *Earthquake Eng'g and Struc Dyn.*, Vol. 24, pp. 175-188.

INFINITE ELEMENT METHODS

K. Gerdes
Seminar für Angewandte Mathematik
ETH Zürich
CH-8092 Zürich, Switzerland

1 Introduction

The present work is motivated by the new concept on various infinite elements, [Burnett, 1994], [Astley, Macaulay and Coyette, 1994], [Cremers, Fyfe and Coyette, 1994], [Cremers and Fyfe, 1995], [Gerdes and Demkowicz, 1996] and [Gerdes, 1996]. These new IEM are extensions of the original IEM, which was introduced by Bettess [Bettess, 1992].

The problem of interest deals with scattering of acoustic waves on elastic or rigid objects. The mathematical formulation consists of the Helmholtz equation in the exterior domain accompanied by the Sommerfeld radiation condition and Neumann boundary condition on the boundary of the scatterer (rigid scattering).

Problems of the described type are usually solved using various versions of the Boundary Element Method (BEM) [Demkowicz and Oden, 1994]. The mathematics of the BE approximation (especially in the Galerkin version) is well established and the method delivers reliable results in the whole range of wave numbers. The main drawback of the BEM is its cost - the method becomes prohibitively expensive for large wave numbers. The approach based on the truncation of the infinite domain to a finite one and application of the so called absorbing boundary conditions has always been an alternative technique to solve the problem [Givoli and Keller, 1994], [Kirsch and Monk, 1994]. Another technique is to approximate the Dirichlet-to-Neumann operator by solving an auxiliary Riccati equation [Wolf and Song, 1996]. The drawback is that this method and the exact absorbing boundary conditions correspond to nonlocal operators and result in dense matrices. The local absorbing boundary conditions are not exact but are computationally attractive. The recent versions of the Infinite Elements that are based on multipole expansions fit into the local absorbing boundary condition framework and offer accuracy of arbitrary high order and can be coupled with standard C^0 finite elements. The recent results on convergence of such methods [Babuska and Shirron, 1995], [Demkowicz and Gerdes, 1998] and [Demkowicz and Ihlenburg, 1996] support reliability of such an approach and add to its attractiveness. An essential difference between the various versions of the infinite elements has been recently pointed out in [Babuska and

T.L. Geers (ed.), IUTAM Symposium on Computational Methods for Unbounded Domains, 143–150.

Shirron, 1995] and [Gerdes, 1998]. The difference lies in the fact whether one does or does not use the complex conjugate over the test functions (sesquilinear vs. bilinear form formulation).

In this paper, we investigate the Galerkin FE/IE formulations for a three dimensional rigid scattering problem. The FEM is applied in a finite annular domain obtained by the domain decomposition approach in the exterior of the scatterer. The scatterer is first enclosed by a sphere of finite radius, forming an artificial boundary inside the acoustic medium. In the domain between the scatterer and the sphere, a FEM is used for the numerical solution whereas outside the artificial boundary the solution is approximated with an IEM. This procedure has been described in detail, together with a convergence analysis, for the special case of a spherical scatterer [Gerdes and Demkowicz, 1996]. Here, we account for a general shape of the scatterer, which is located within the unit sphere. In particular, we study the different variational formulations that can be obtained by using a bilinear form or a sesquilinear form in the weak formulation. This corresponds to using or not using the complex conjugate in the weak formulation [Gerdes, 1998]. Additionally, we present the different space settings, that were introduced by Leis [Leis, 1986] and in [Demkowicz and Gerdes, 1998] for the Burnett formulation. These possible space settings result in different convergence properties of the IE schemes.

2 The exterior Helmholtz problem

The goal is to find a function $u = u(\boldsymbol{x})$ satisfying:

- the Helmholtz equation with wave number k in the domain exterior to the scatterer,

$$-\Delta u - k^2 u = 0 \qquad \text{in} \quad \Omega^e \tag{2.1}$$

- a Neumann boundary condition on the scatterer with a prescribed function g on Γ,

$$\nabla_n u = g \qquad \text{for} \quad \boldsymbol{x} \in \Gamma \tag{2.2}$$

- the Sommerfeld radiation condition at infinity,

$$\left| \frac{\partial u}{\partial n} - iku \right| = O\left(\frac{1}{r^2}\right). \tag{2.3}$$

Figure 1 illustrates the exterior scattering problem.

3 The coupled *hp* FE/IE methodology

We need the following weighted Sobolev spaces in order to state the variational formulations:

$$H_w^1(\Omega^e) = \left\{ u : \|u\|_w^1 < \infty \right\} \tag{3.1}$$

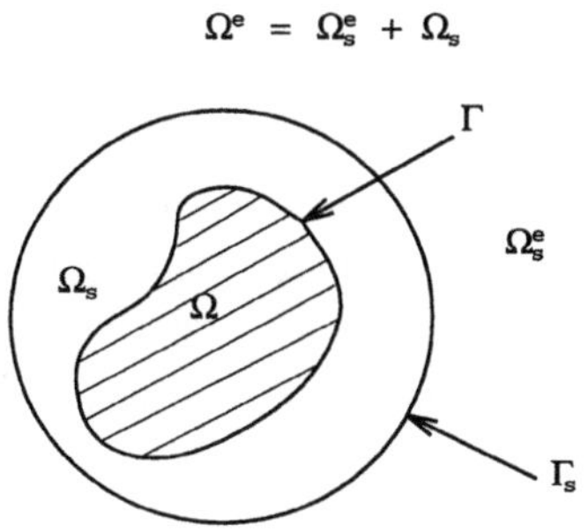

Figure 1: Scatterer within a unit sphere.

with the norm $\|u\|_w^1$ and weights w, w^* defined by

$$(u,v)_w^1 = \int_{\Omega^e} w\, u\,\overline{v} + w\,\nabla u \cdot \nabla \overline{v}\, d\Omega^e + \int_{\Omega^e} \left(\frac{\partial u}{\partial r} - iku\right) \overline{\left(\frac{\partial v}{\partial r} - ikv\right)}\, d\Omega^e, \tag{3.2}$$

$$w(\boldsymbol{x}) = \begin{cases} 1 & \text{for} \quad r = |\boldsymbol{x}| \le 1 \\ \dfrac{1}{r^2} & \text{for} \quad r = |\boldsymbol{x}| > 1 \end{cases}, \qquad w^*(\boldsymbol{x}) = \begin{cases} 1 & \text{for} \quad r = |\boldsymbol{x}| \le 1 \\ r^2 & \text{for} \quad r = |\boldsymbol{x}| > 1 \end{cases}.$$

The variational formulations following the Leis approach [Leis, 1986] read then in the bilinear and sesquilinear case:

$$\begin{cases} \text{Find } u \in H_w^1(\Omega^e) \text{ such that} \\ \displaystyle\int_{\Omega^e} \nabla u \cdot \nabla v\, d\Omega^e - k^2 \int_{\Omega^e} u\, v\, d\Omega^e = \int_{\partial\Omega^e} g\, v\, dS \quad \forall\, v \in H_{w^*}^1(\Omega^e) \end{cases}, \tag{3.3}$$

$$\begin{cases} \text{Find } u \in H_w^1(\Omega^e) \text{ such that} \\ \displaystyle\int_{\Omega^e} \nabla u \cdot \nabla \overline{v}\, d\Omega^e - k^2 \int_{\Omega^e} u\, \overline{v}\, d\Omega^e = \int_{\partial\Omega^e} g\, \overline{v}\, dS \quad \forall\, v \in H_{w^*}^1(\Omega^e) \end{cases}. \tag{3.4}$$

In the Burnett formulations we apply the same space for the test and trial functions, but have to interprete integrals in the Cauchy Principle Value sense. The bilinear and sesquilinear Burnett formulations are:

$$\begin{cases} \text{Find } u \in H_w^1(\Omega^e) \text{ such that } \forall\, v \in H_w^1(\Omega^e) \\ \displaystyle\int_{\Omega^e} \nabla u \cdot \nabla v\, d\Omega^e - k^2 \int_{\Omega^e} u\, v\, d\Omega^e - ik \lim_{\gamma\to\infty} \int_{S_\gamma} uv\, dS_\gamma = \int_{\partial\Omega^e} g\, v\, dS, \end{cases} \tag{3.5}$$

$$\begin{cases} \text{Find } u \in H_w^1(\Omega^e) \text{ such that } \forall\, v \in H_w^1(\Omega^e) \\ \displaystyle\int_{\Omega^e} \nabla u \cdot \nabla \overline{v}\, d\Omega^e - k^2 \int_{\Omega^e} u\, \overline{v}\, d\Omega^e - ik \lim_{\gamma\to\infty} \int_{S_\gamma} u\overline{v}\, dS_\gamma = \int_{\partial\Omega^e} g\, \overline{v}\, dS. \end{cases} \tag{3.6}$$

The separation of variables approach described in [Gerdes and Demkowicz, 1996] motivates the the definition of the IE shape functions. The trial and test functions are defined by

$$\psi_j(r) = \frac{\exp(ikr)}{r^j}, \qquad \tilde{\psi}_j(r) = \frac{\exp(ikr)}{r^{j+m}}, \qquad j \geq 1, \tag{3.7}$$

with $m = 2$ in the Leis formulation and $m = 0$ in the Burnett formulation. Note in the Leis formulation the different powers of r in the denominators. The use of the sesquilinear formulation does eliminate the necessity to integrate the oscillatory component $\exp(ikr)$.

The infinite element shape functions are then given as tensor products of 2D *hp* finite element shape functions and the functions introduced in (3.7), i.e. a typical infinite element trial shape function $N_l(r, \boldsymbol{x})$ is given by

$$N_l(r,\boldsymbol{x}) = N_{l(i,j)}(r,\boldsymbol{x}) = \psi_j(r) \cdot \varphi_i(\boldsymbol{x}), \qquad r > 1,\ \boldsymbol{x} \in \Gamma_s \tag{3.8}$$

In order to minimize the interaction between infinite elements and 3D *hp* finite elements Ω_s, the IE trial shape functions are now modified as follows

$$N_{l(i,j)}(r,\boldsymbol{x}) = \begin{cases} \exp(-ik)\,\psi_j(r)\,\varphi_i(\boldsymbol{x}) & j = 1,\ r > 1,\ \boldsymbol{x} \in \Gamma_s \\ \exp(-ik)\,\left(\psi_j(r) - \psi_1(r)\right)\,\varphi_i(\boldsymbol{x}) & j \geq 2,\ r > 1,\ \boldsymbol{x} \in \Gamma_s \end{cases} \tag{3.9}$$

with an identical modification for the test functions. In this way, all the shape functions corresponding to $j \geq 2$ will contribute to basis functions with support outside of the unit sphere only. Inclusion of the exponential factor $\exp(-ik)$ in the new definition forces the infinite element shape functions to coincide with the standard 3D *hp* finite element shape functions on the surface of the sphere.

It is evident that this definition of IE shape functions applies to the Leis and Burnett formulation in the bilinear and sesquilinear form. Also, the definition is valid for other 2D and 3D master elements. In the following we use triangular prismatic elements in Ω_s and triangular elements on the surface of the unit sphere. It is possible to apply quadrilateral elements on the surface of the unit sphere and cubic elements in Ω_s. A combination of triangular and tetrahedral elements is also possible. The different choices of master elements only affect the definition of functions φ_i in (3.9) but otherwise the IE shape functions $N_{l(i,j)}$ remain the same.

4 Numerical Implementation

The implementation of the 3D finite elements is done in the standard way and the computations for the infinite elements are similar to 2D FE computations. The integration of the IE shape functions in the radial direction involves the following integrals in the conjugated formulations

$$\int_1^\infty \frac{\exp(ikr)\exp(-ikr)}{r^j}\,dr = \frac{1}{j-1}, \quad j \geq 2. \tag{4.1}$$

In the unconjugated versions terms of the form

$$\int_1^\infty \frac{\exp(ikr)\exp(ikr)}{r^j}\,dr = \int_1^\infty \frac{\exp(2ikr)}{r^j}\,dr \tag{4.2}$$

do appear, which can be evaluated using the exponential integral. The following representation is used to obtain a form of the radial integral that can easily be integrated.

$$\int_1^\infty \frac{\exp(2ikr)}{r^n}\,dr = \sum_{l=1}^{n-1} \frac{-(-i2k)^{l-1}\exp(2ik)}{\prod_{\hat{l}=1}^{l} -n+\hat{l}} + \frac{(-i2k)^{n-1}}{\prod_{\hat{l}=1}^{n-1} -n+\hat{l}} \int_1^\infty \frac{\exp(2ikr)}{r}dr, \tag{4.3}$$

where $n \geq 1$, $\sum_{l=1}^{0} := 0$ and $\prod_{\hat{l}=1}^{0} := 1$. The element contributions are then assembled by a generalized assembling procedure which is adapted to handle complex valued matrices and can also deal with general *hp* meshes. The element load vector is calculated similarly. It should be noted that the final global system may not be symmetric.

The weighted H^1-norm is consistent with the trial function space and can be used to compute the error between the exact solution u and the numerical solution u_h in Ω_s^e. Although this norm is consistent with the mathematical theory by Leis for the Helmholtz equation, it should only be used for the conjugated IEM, i.e. in the sesquilinear case. The stability and convergence analysis presented in [Babuska and Shirron, 1995] and [Gerdes, 1998] show that the unconjugated IEM fails to converge in the far field, but converges rapidly in the near field. Therefore, the L^∞-norm is used on the surface of the scatterer. In the sesquilinear formulations we meassure the error in the weighted H^1-norm, where

$$\begin{aligned} \|u-u_h\|_{1,\Omega^e}^2 &= \int_{\Omega_s^e} \frac{1}{r^2}|u-u_h|^2\,d\Omega_s^e + \int_{\Omega_s^e} \frac{1}{r^2}\left|\nabla(u-u_h)\right|^2\,d\Omega_s^e \\ &+ \int_{\Omega_s} |u-u_h|^2\,d\Omega_s + \int_{\Omega_s} \left|\nabla(u-u_h)\right|^2\,d\Omega_s. \end{aligned} \tag{4.4}$$

The rigid scattering of a plane wave on a sphere allows to determine the exact solution, which is the basis for a numerical convergence study. The evaluation of the error is similar to standard FE computations, for details we refer to [Gerdes and Demkowicz, 1996].

In the following discussion we present numerical results for the sesquilinear Leis formulation for different examples of finite-infinite element meshes. It focuses on the question of, how the number of the shape functions in the radial direction affects the approximation of the exact solution. The error is analyzed in context of the order of approximation in the 'angular' direction p and the order of approximation q in the radial direction. The influence of the number of IE shape functions N on the numerical approximation becomes evident from the numerical results. In our numerical results we set the wave number k to 10 and use up to 3 layers of 3D finite elements. Figure 2 shows the p-convergence rates for different q and number of radial shape functions for an incident plane wave. It is evident, that only enough dof in the angular and radial direction, e.g. high p and q, or more finite elements, and the number of dof in radial direction of the infinite element sufficiently large does lead to a satisfactory result, i.e. the error with $p = 5$, $q = 3$, $N = 6$

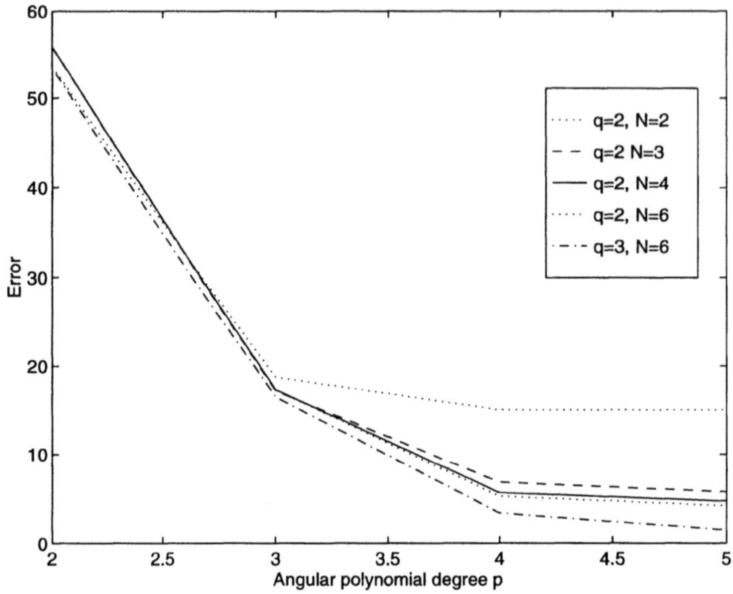

Figure 2: Scattering of a plane wave on a rigid sphere. Experimental p-convergence rates with 72 3D finite elements and 24 infinite elements with $k = 10$.

and three layers of finite elements is below 1.6%, see Figure 2. This clearly indicates that the discretization in the angular and the radial directions of the 3D finite elements is fine enough to obtain a reliable numerical solution. Further, it is evident that the investigated finite-infinite element method works well for wave number $k = 10$ and emphasize that all computations are done on a workstation. The numerical solution is obtained within fifteen minutes for fine meshes. This is a significant advantage compared to the execution time of standard boundary element formulations.

In Figure 3 we show the numerical solution for the rigid scattering on a finite cylinder with spherical incaps. The cylinder is assumed to be within the unit sphere and the Neumann boundary condition on the surface of the cylinder corresponds to an incident plane wave. The IEM is used to compute the numerical solution for the rigid scattering with wave number $k = 10$. The finite element mesh contains three layers of finite elements with 216 finite elements per layer, $p = 3$ and $q = 2$. There are 216 infinite elements with $p = 3$ and 6 radial shape functions. Altogether, 864 elements are used. The computation of the numerical solution with the coupled FE/IE methodology takes about twenty minutes on a workstation. We remark that the high demand of CPU time and memory makes it extremely difficult to obtain the numerical solution with the BEM for wave number $k = 10$.

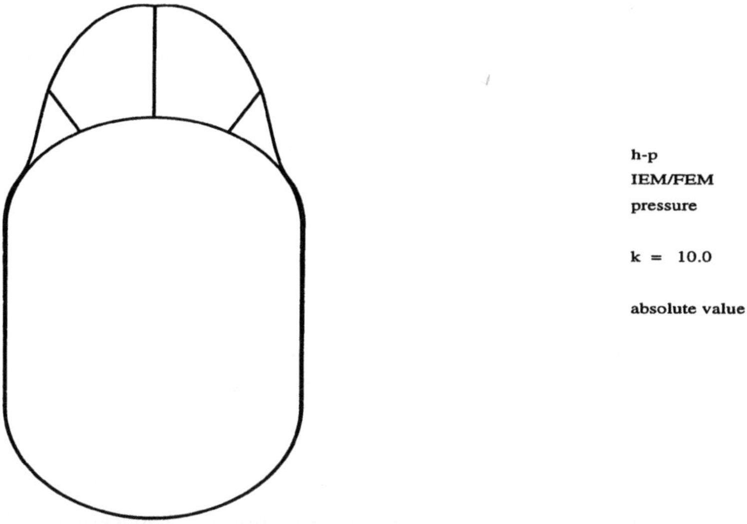

Figure 3: FEM/IEM. Scattering of a plane wave on a rigid cylinder with spherical incaps.

5 Conclusions

Different infinite element formulations for the exterior Helmholtz problem are analyzed. The presented methodologies apply to rigid scattering problems with general shapes of the scatterers. The analyzed formulations combine a variable order *hp*-FE discretization between the scatterer and the unit sphere with an infinite element discretization on the boundary of the unit sphere. The infinite elements are obtained by combining 2D *hp*-FE shape functions with a spectral approximation in the third direction. The radial IE shape functions are motivated from the separation of variables approach. The theoretical results show that the method is reliable and computationally advantageous. The numerical results support this conclusion.

References

[1] Astley, R.J., Macaulay, G.J., and Coyette, J.P., 1994, "Mapped Wave Envelope Elements for Acoustical Radiation and Scattering", *Journal of Sound and Vibration*, Vol. 170, pp. 97-118.

[2] Babuska, I., and Shirron, J.J., 1995, "Solution of Exterior Helmholtz Problems using Finite and Infinite Elements", Dissertation preprint.

[3] Bettess, P., 1992 **Infinite Elements**, Penshaw, Sunderland,UK.

[4] Burnett, D.S., 1994, "A Three-Dimensional Acoustic Infinite Element Based on a Prolate Spheroidal Multipole Expansion", Journal of the Acoustical Society of America, Vol. 96, pp. 2798-2816.

[5] Cremers, L., Fyfe, K.R., and Coyette, J.P., 1994, "A Variable Order Infinite Acoustic Wave Envelope Element", *Journal of Sound and Vibration*, Vol. 171, pp. 483-508.

[6] Cremers, L., and Fyfe, K.R., 1995, "On the use of Variable Order Infinite Wave Envelope Elements for Acoustic Radiation and Scattering", *Journal of the Acoustical Society of America*, Vol. 97, pp. 2028-2040.

[7] Demkowicz, L., and Gerdes, K., 1998, "Convergence of the Infinite Element Methods for the Helmholtz Equation", *Numer. Math.*, Vol. 79, 11-42.

[8] Demkowicz, L., and Ihlenburg, F., 1996, "Analysis of a Coupled Finite-Infinite Element Method for Exterior Helmholtz Problems", *TICAM Report 96-52.*

[9] Demkowicz, L., and Oden, J.T., 1994, "Recent Progress on Application of *hp*-Adaptive BE/FE Methods to Elastic Scattering", *International Journal for Numerical Methods in Engineering*, Vol. 37, pp. 2893-2910.

[10] Gerdes, K., and Demkowicz, L., 1996, "Solutions of 3D-Laplace and Helmholtz Equations in Exterior Domains using hp Infinite Elements", *Comput. Methods Appl. Mech. Engrg.*, Vol. 137, 239-273.

[11] Gerdes, K., 1996, "Solution of the 3D Helmholtz equation in exterior domains of arbitrary shape using *hp*-finite infinite elements", to appear in *Finite Elements in Analysis and Design.*

[12] Gerdes, K., 1998, "The conjugated vs. the unconjugated infinite element method for the Helmholtz equation in exterior domains", *Comput. Methods Appl. Mech. Engrg.* Vol. 152, pp. 125-145.

[13] Givoli, D., and Keller, J.B., 1994, "Special finite elements for use with high-order boundary conditions", *Comp. Methods Appl. Mech. Engrg.* Vol. 119, pp. 199-213.

[14] Kirsch, A., and Monk, P., 1994, "An analysis of the coupling of finite element and Nyström methods in acoustic scattering", *IMA J. Num. Anal.* Vol. 14, pp. 523-544.

[15] Leis, R., 1986, **Initial Boundary Value Problems in Mathematical Physics**, Teubner.

[16] Wolf, J.P., and Song, C., 1996, **Finite-element modelling of unbounded media**, Wiley.

OPTIMAL LOCAL ARTIFICIAL BOUNDARY CONDITIONS

Dan Givoli **Igor Patlashenko**
Department of Aerospace Engineering Asher Space Research Center
Technion — Israel Institute of Technology
Haifa 32000, Israel

1. Introduction

One of the methods commonly used to numerically solve a problem in an infinite domain is the method of *artificial boundary conditions* [1]. For a linear scalar problem, this method may be summarized as follows:

(a) Introduce an artificial boundary $\mathcal{B}$, which divides the original infinite domain into two domains: a finite computational domain Ω and an infinite residual domain D.

(b) By analyzing the problem in D, obtain a relation on $\mathcal{B}$ (exact or approximate) involving the unknown function u and its derivatives.

(c) Use this relation as a boundary condition on $\mathcal{B}$, to obtain a well-posed problem in Ω.

(d) Solve the problem in Ω by the finite element method or some other numerical method.

The relation obtained in step (b) and used as a boundary condition in step (c) is called an *artificial boundary condition* (ABC), or, in the context of wave problems, a *non-reflecting boundary condition* [2]. A standard ABC which might be imposed on $\mathcal{B}$ is simply the condition at infinity. However, in this case Ω must be quite large, or else the ABC would give rise to large inaccuracies. On the other hand, a large computational domain is inefficient, leading to a large number of degrees of freedom. Therefore, the trend in recent work has been to use a more accurate ABC on $\mathcal{B}$, which enables the use of a smaller computational domain.

During the last two decades, various improved ABCs for various problems in infinite domains have been proposed [1, 2]. Most of them are local and approximate. A smaller number of *exact* nonlocal ABCs have been devised for various problems in infinite domains [3]–[9]. For general elliptic linear problems, Keller and Givoli [10, 11] devised an exact ABC on an artificial boundary $\mathcal{B}$ of a simple shape (e.g., a circle or a sphere). In the simplest case, this ABC has the form,

$$\frac{\partial u}{\partial \nu} = -Mu \quad \text{on} \quad \mathcal{B} \,, \tag{1}$$

where $\partial u/\partial \nu$ is the normal derivative of u on $\mathcal{B}$, and M is a nonlocal operator called the Dirichlet-to-Neumann (DtN) map.

T.L. Geers (ed.), IUTAM Symposium on Computational Methods for Unbounded Domains, 151–158.

The DtN boundary condition (1) was incorporated in a finite element scheme, resulting in the *DtN Finite Element method.* Givoli, Keller et al. proposed this method as a general technique to solve elliptic linear problems in infinite domains [1], [11]–[13]. In [14], the method was extended to treat the hyperbolic linear wave equation. Recently, the DtN method has also been extended to elliptic and hyperbolic nonlinear problems.

The DtN method has been further analyzed and improved by Harari and Hughes [15, 16], Grote and Keller [17], and Malhotra and Pinsky [18]. The relation between the DtN method and the mode-matching method has been shown by Astley [19]. Other schemes that use DtN-related ideas, for various problems and configurations, can also be found in [20]–[24].

Despite the fact that exact nonlocal ABCs may be extremely useful in many situations, there are cases where local ABCs may be preferred. A thorough discussion on the relative advantages and disadvantages of local and nonlocal ABCs can be found in [25]. This comparison shows that there is place for both types of ABCs. In this light, in [25]–[27], a procedure has been proposed to *localize* the DtN condition. This localization results in a sequence of local ABCs of increasing order. The local conditions are constructed so that the Nth-order condition exactly annihilates the first N harmonics, and involves tangential derivatives of up to order $2N$. When $N \geq 2$, special finite elements must be used, which possess high-order regularity along $\mathcal{B}$. A hierarchy of such elements in two and three dimensions has been devised in [25, 27].

In this paper, we approach the localization of the DtN condition in a totally different way. Here we consider a local boundary condition with unknown coefficients, typically of *low* order (so that standard finite elements can be used), and ask the question: what is the best choice for the unknown coefficients, so that the local operator is the best approximation of the DtN map, in some norm? We answer this question in a C^∞ framework. We derive an optimal local condition of a given order. This new condition may be of low order but still represent high-order modes in the solution. We show that the localized DtN conditions derived in [25]–[27] are special cases of the new optimal conditions.

2. Optimal Localized DtN Boundary Conditions

Consider a two-dimensional exterior problem. In the DtN method, we introduce a circular artificial boundary $\mathcal{B}$ of radius R, and impose the DtN boundary condition on it. The DtN boundary condition has the form:

$$-\frac{\partial u}{\partial r}(R,\theta) = (Mu)(\theta) \equiv \sum_{n=0}^{\infty}{}' \int_0^{2\pi} m_n(\theta,\theta')\, u(\theta')\, d\theta' \,, \tag{2}$$

where,

$$m_n(\theta,\theta') = \frac{Z_n}{\pi R} \cos n(\theta - \theta') \,. \tag{3}$$

Here the Z_n are constants. See [1], [11]–[13], [25] for further details. We wish to approximate the nonlocal operator M by a Kth-order local operator. Thus, we replace (2) by,

$$-\frac{\partial u}{\partial r}(R,\theta) = (L_K u)(\theta) \equiv \sum_{k=0}^{K} A_k \frac{\partial^{2k} u}{\partial \theta^{2k}} \ . \tag{4}$$

The question is: How do we choose the coefficients A_k, $k = 0, ..., K$, such that L_K is the "best approximation" of M in some sense?

We approach the problem in a C^∞ framework. We consider the Fourier expansion of u,

$$u = \sum_{j=0}^{N}{}'(u_j^c \cos j\theta + u_j^s \sin j\theta) \ . \tag{5}$$

Here, N is a chosen number of modes. The modes $j = 0, ..., N$ are the "important" modes in the problem, that one wishes to represent accurately. Note that generally N is not equal to K. In fact, typically $N > K$.

Now, we apply both operators M and L to the function u whose expansion is given in (5). This yields,

$$(Mu)(\theta) = \sum_{j=0}^{N}{}' Z_j (u_j^c \cos j\theta + u_j^s \sin j\theta) \ . \tag{6}$$

$$(L_K u)(\theta) = \sum_{j=0}^{N}{}' \sum_{k=0}^{K} A_k (-1)^k j^{2k} (u_j^c \cos j\theta + u_j^s \sin j\theta) \tag{7}$$

The difference between (6) and (7) is:

$$(M - L_K)u = \sum_{j=0}^{N}{}' \left[Z_j - \sum_{k=0}^{K} A_k (-1)^k j^{2k} \right] (u_j^c \cos j\theta + u_j^s \sin j\theta) \tag{8}$$

We denote the expression in the brackets by β_j^K, i.e.,

$$\beta_j^K = Z_j - \sum_{k=0}^{K} A_k (-1)^k j^{2k} \ . \tag{9}$$

We also define the vector $\boldsymbol{\beta}^K$ whose entries are the β_j^K, namely,

$$\boldsymbol{\beta}^K = (\beta_0^K \quad \beta_1^K \quad \dots \quad \beta_N^K)^T \ . \tag{10}$$

Thus, (8) becomes,

$$(M - L_K)u = \sum_{j=0}^{N}{}' \beta_j^K (u_j^c \cos j\theta + u_j^s \sin j\theta) \ . \tag{11}$$

We now measure the $L_2(\mathcal{B})$ norm of $(M - L_K)u$. A simple calculation gives:

$$\begin{aligned} ||(M - L_K)u||_0^2 &= ||\sum_{j=0}^{N}{}' \beta_j^K (u_j^c \cos j\theta + u_j^s \sin j\theta)||_0^2 \\ &= \pi \sum_{j=0}^{N}{}' ((u_j^c)^2 + (u_j^s)^2)(\beta_j^K)^2 \end{aligned} \tag{12}$$

We consider two options:

(a) Nothing is know about the exact solution u.
In this case, we get from (12),

$$||(M - L_K)u||_0^2 \leq \pi \left(\sum_{j=0}^{N}{}' ((u_j^c)^2 + (u_j^s)^2) \right) \left(\sum_{j=0}^{N} (\beta_j^K)^2 \right) = ||u||_0^2 \, ||\boldsymbol{\beta}^K||^2 \qquad (13)$$

In the right side of (13), the first norm is the L_2 norm of u, whereas the second norm is the Euclidian norm of the vector $\boldsymbol{\beta}^K$. Thus, in order that $||(M - L_K)u||$ be minimal for an arbitrary u of the form (5), we require that $||\boldsymbol{\beta}^K||$ *be minimal.*

(b) We can estimate the relative importance of the different modes of u.
This information is sometimes available from the given boundary conditions and geometry. In this case, we let W_j be given weights, where,

$$W_j \simeq \pi((u_j^c)^2 + (u_j^s)^2) \qquad (14)$$

(with a factor of 1/2 for W_0). Then (12) gives,

$$||(M - L_K)u||_0^2 \simeq \sum_j W_j^2 (\beta_j^K)^2 \equiv ||\boldsymbol{\beta}^K||_W^2 \qquad (15)$$

Here $||\boldsymbol{\beta}^K||_W$ is the *weighted* Euclidian norm of $\boldsymbol{\beta}^K$.

We can regard option (a) as a special case of option (b), where all the weights have value 1. Thus, the general problem is that of minimizing $||\boldsymbol{\beta}^K||_W$.

A necessary condition for the minimum of $||\boldsymbol{\beta}^K||_W^2$ is $\partial ||\boldsymbol{\beta}^K||_W^2 / \partial A_l = 0$ for $l = 0, ..., K$. This gives,

$$\sum_{j=0}^{N} W_j \sum_{k=0}^{K} (-1)^{(k+l)} j^{2(k+l)} A_k - \sum_{j=0}^{N} W_j Z_j (-1)^l j^{2l} = 0 \ . \qquad (16)$$

We define:

$$P_l = \sum_{j=0}^{N} W_j Z_j (-1)^l j^{2l} \quad , \qquad l = 0, ..., K \qquad (17)$$

$$B_{lk} = \sum_{j=0}^{N} W_j (-1)^{(k+l)} j^{2(k+l)} \quad , \qquad k, l = 0, ..., K \qquad (18)$$

Then (16) becomes the linear system of equations,

$$\boldsymbol{BA} = \boldsymbol{P} \ . \qquad (19)$$

The solution of this system gives the desired coefficients A_k in the local boundary condition (4).

3. A Special Case: $N = K$

Suppose $N = K$. Then (16) becomes,

$$\sum_{j=0}^{N} W_j j^{2l}(-1)^l \left(\sum_{k=0}^{N} A_k(-1)^k j^{2k} - Z_j\right) = 0 . \tag{20}$$

It can be shown that the expression in the parentheses in (20) must vanish for each j separately. We also denote $\alpha_k = A_k(-1)^k$. Then we get,

$$\sum_{k=0}^{N} \alpha_k j^{2k} - Z_j = 0 \qquad , \qquad j = 0, ..., N . \tag{21}$$

But (21) is exactly the system that yields the localized DtN boundary condition of order N, devised in a totally different manner in [25]–[27]! Thus, we conclude that the new optimal local ABCs reduce to the older localized DtN conditions in the special case $N = K$.

4. A Numerical Example

We consider a two-dimensional semi-infinite strip that occupies the domain $x \geq 0$, $0 \leq y \leq b$. In the strip, Laplace's equation governs. The boundary conditions are: on the two rays $y = 0$ and $y = b$ we impose $u = 0$; at infinity the solution must be bounded; at the edge $x = 0$ we impose the Dirichlet condition $u = g(y)$, where $g(y)$ is a given 6th-order polynomial. We set $b = 2$, and introduce an artificial boundary $\mathcal{B}$ at $x = 0.5$. Thus, the computational domain is the rectangle $0 \leq x \leq 0.5$, $0 \leq y \leq 2$. We use a finite element mesh of 15×60 bilinear elements. On $\mathcal{B}$ we use the optimal ABC of order $K = 1$, and we vary the number of represented harmonics N.

The figures show the $L_2(\mathcal{B})$ error, denoted E, in a logarithmic scale, as a function of N, for a number of ABCs on $\mathcal{B}$. These include a simple Neumann condition, the nonlocal DtN condition truncated after the Nth term, and the optimal ABC ($K = 1$) with three types of weights W_i: unit weights, weights that are based on the Fourier coefficients of the function $g(y)$ given on the left boundary, and weights based on $g(y)$ where the exponential decay of the solution in the x direction has also been taken into account. The latter is the best local condition of order 1 that can be obtained.

5. Conclusion

A local artificial boundary condition of a given order has been developed, which is closest in a certain sense to the exact nonlocal DtN boundary condition. This new condition may be of low order but still represent high-order modes.

Undergoing work includes:

- Adapting the optimal local ABCs to a variety of problems, and especially wave problems in various configurations.

- Analyzing the numerical stability of the optimal ABCs.
- Deriving error estimates for the optimal ABCs.
- Performing extensive experimentation with the optimal ABCs.

Progress in these directions will be reported in a future publication.

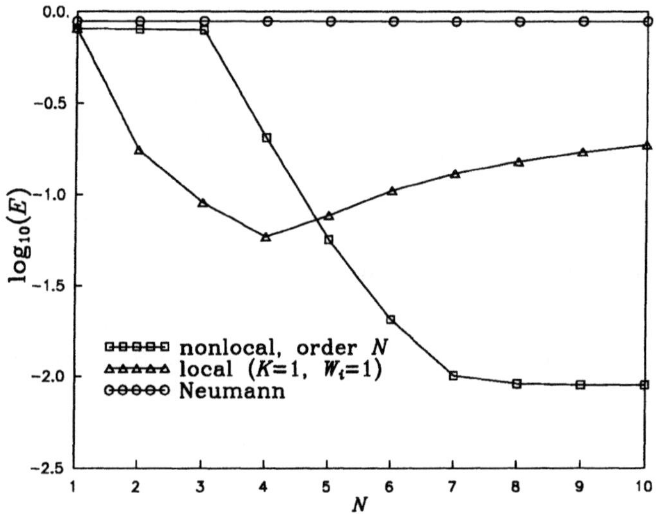

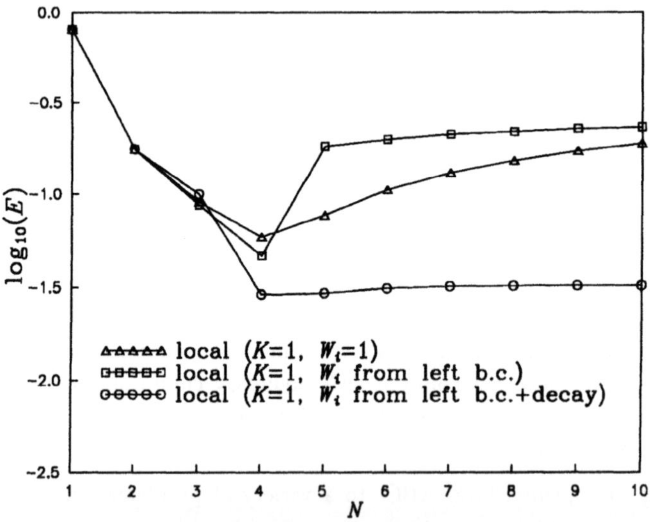

Acknowledgments

This work was partly supported by the U.S.-Israel Binational Science Foundation (BSF), Grant No. 92-00016, and by the Fund for the Promotion of Research at the Technion.

References

[1] D. Givoli, *Numerical Methods for Problems in Infinite Domains*, Elsevier, Amsterdam, 1992.

[2] D. Givoli, "Non Reflecting Boundary Conditions," *J. Comput. Phys.*, 94, pp. 1–29, 1991.

[3] L. Ferm and B. Gustafsson, "A Down-Stream Boundary Procedure for the Euler Equations," *Comp. Fluids*, 10, pp. 261–276, 1982.

[4] T. Hagstrom and H.B. Keller, "Exact Boundary Conditions at an Artificial Boundary for Partial Differential Equations in Cylinders," *SIAM J. Math. Anal.*, 17, pp. 322–341, 1986.

[5] T. Hagstrom, "Conditions at the Downstream Boundary for Simulations of Viscous, Incompressible Flow," *SIAM J. Sci. Stat. Comp.*, 12, pp. 843-858, 1991.

[6] L. Ting and M.J. Miksis, "Exact Boundary Conditions for Scattering Problems," *J. Acoust. Soc. Am.*, 80, pp. 1825–1827, 1986.

[7] D. Givoli and D. Cohen, "Non-reflecting Boundary Conditions Based on Kirchhoff-type Formulae," *J. Comput. Phys.*, 117, pp. 102–113, 1995.

[8] M.J. Grote and J.B. Keller, "Nonreflecting Boundary Conditions for Time Dependent Scattering," *J. of Comput. Phys.*, 127, pp. 52–65, 1996.

[9] S.V. Tsynkov, E. Turkel and S. Abarbanel, "External Flow Computations Using Global Boundary Conditions," *AIAA J.*, 34, pp. 700–706, 1996.

[10] J.B. Keller and D. Givoli, "Exact Non-Reflecting Boundary Conditions," *J. Comput. Phys.*, 82, pp. 172–192, 1989.

[11] D. Givoli and J.B. Keller, "A Finite Element Method for Large Domains," *Comput. Meth. Appl. Mech. Engng.*, 76, pp. 41–66, 1989.

[12] D. Givoli and J.B. Keller, "Non-Reflecting Boundary Conditions for Elastic Waves," *Wave Motion*, 12, pp. 261–279, 1990.

[13] D. Givoli and S. Vigdergauz, "Artificial Boundary Conditions for 2D Problems in Geophysics," *Comput. Meth. Appl. Mech. Engng.*, 110, pp. 87–101, 1993.

[14] D. Givoli, "A Spatially Exact Non-Reflecting Boundary Condition for Time Dependent Problems," *Comput. Meth. Appl. Mech. Engng.*, 95, pp. 97–113, 1992.

[15] I. Harari and T.J.R. Hughes, "Galerkin/Least-squares Finite Element Methods for the Reduced Wave Equation with Non-reflecting Boundary Conditions in Unbounded Domains," *Comput. Meth. Appl. Mech. Engng.*, 98, pp. 411–454, 1992.

[16] I. Harari and T.J.R. Hughes, "Analysis of Continuous Formulations Underlying the Computation of Time-harmonic Acoustics in Exterior Domains," *Comput. Meth. Appl. Mech. Engng.*, 97, pp. 103–124, 1992.

[17] M.J. Grote and J.B. Keller, "On Nonreflecting Boundary Conditions," *J. Comput. Phys.*, 122, pp. 231–243, 1995.

[18] M. Malhotra and P.M. Pinsky, "A Matrix-free Interpretation of the Non-local Dirichlet-to-Neumann Radiation Boundary Condition," *Int. J. Num. Meth. Engng.*, 39, pp. 3705–3713, 1996.

[19] R.J. Astley, "FE Mode-Matching Schemes for the Exterior Helmholtz Problem and Their Relationship to the FE-DtN Approach," *Commun. Num. Meth. Engng.*, 12, pp. 257–267, 1996.

[20] R.C. MacCamy and S.P. Marin, "A Finite Element Method for Exterior Interface Problems," *Int. J. Math. Math. Sci.*, 3, pp. 311–350, 1980.

[21] C.I. Goldstein, "A Finite Element Method for Solving Helmholtz Type Equations in Wave Guides and Other Unbounded Domains," *Math. Comp.*, 39, pp. 309–324, 1982.

[22] M. Masmoudi, "Numerical Solution of Exterior Problems," *Numer. Math.*, Vol. 51, pp. 87–101, 1987.

[23] M. Verriere and M. Lenoir, "Computation of Waves Generated by Submarine Landslides," *Int. J. Num. Meth. Fluids*, 14, pp. 403–421, 1992.

[24] G.N. Gatica and G.C. Hsiao, "The Coupling of Boundary Integral and Finite Element Methods for Nonmonotone Nonlinear Problems," *Numer. Func. Anal. Optim.*, 13, pp. 431–447, 1992.

[25] D. Givoli, I. Patlashenko and J.B. Keller, "High Order Boundary Conditions and Finite Elements for Infinite Domains," *Comp. Meth. Appl. Mech. Engng.*, 143, pp. 13–39, 1997.

[26] I. Patlashenko and D. Givoli, "Nonlocal and Local Artificial Boundary Conditions for Two-dimensional Flow in an Infinite Channel," *Int. J. Numerical Methods for Heat and Fluid Flow*, 6, pp. 47–62, 1996.

[27] I. Patlashenko and D. Givoli, "Non-Reflecting Finite Element Schemes for Three-Dimensional Acoustic Waves," *J. Comput. Acoustics*, 5, pp. 95–115, 1997.

Reduction of the Complexity of the Intrisically Nonlinear Problem of Aerodynamic Sound Generation in an Unbounded Domain

Sheryl M. Grace and Allan D. Pierce

Department of Aerospace and Mechanical Engineering
Boston University
Boston, MA 02215

1 Introduction

In this research, the long standing problem of the sound generated by a complex unbounded flow field is revisited. It is shown that although this problem is inherently nonlinear, a unique splitting of the field quantities into acoustic, vortical, and entropy modes exists. The splitting for the nonlinear case is formally produced with projection operators, and as such, the method described in this paper for determining the acoustic portion of a complex flow field is called the acoustic projection method. In its most general form, the acoustic projection method includes the effect of arbitrary Mach, Reynolds, and Prandtl numbers as well as viscous and thermal diffusivity and refraction effects. It can be shown to yield the common form of Lighthill's analogy by applying the correct assumptions.

The development of the acoustic projection method was partially motivated by a recent article written by members of the computational fluid mechanics division at NASA Langley entitled, "An Acoustic/Viscous Splitting Technique for Computational Aeroacoustics" (Hardin & Pope 1994). This paper called for a new method for making use of existing computational fluid dynamic results, even if they are valid only for very low Mach number, to predict sound. A scheme for calculating the acoustic field of a complex flow field by using results from incompressible fluid dynamical computations would increase the power of conventional CFD as a design tool. The idea behind the acoustic/viscous splitting technique is extremely attractive, although the technique as originally stated is heuristic and needs a more structured theoretical basis.

Thus, the acoustic projection method is being developed as a computational acoustics scheme that takes, as a starting point, results calculated using existing computational fluid dynamic codes. The main focus is on calculating the sound generated when the primary fluid dynamics are modeled well using the incompressible assumption.

T.L. Geers (ed.), IUTAM Symposium on Computational Methods for Unbounded Domains, 159–168.

The other motivation behind the development of the acoustic projection method comes from the classical Lighthill analogy for predicting the sound from turbulence (Lighthill 1952). The analogy has provided a powerful tool for acoustic predictions for many years, but gives information concerning the generated sound only at points external to the primary flow field. In application, however, one may want to compute the sound field within a turbulent region. This cannot be done using the Lighthill analogy or any of its derivatives which have appeared over the years. Another difficulty associated with the analogy is the tendency for those who use the analogy to neglect effects such as viscous and thermal diffusivity as well as refraction effects due to the pressure waves traveling through the complex fluid dynamic field.

By rewriting the Navier-Stokes-Fourier governing equations of motion using the field splitting, one can identify the nonlinear equations used in incompressible CFD analyses as the equations governing the vortical part of the flow field. It can also be shown that the governing equation for the acoustic mode can be written as a wave equation which incorporates both the refraction of the sound by the vorticity mode and nonlinear sources of sound as forcing functions. Therefore, to project out the sound generated by a complex flow field the vorticity mode and nonlinear sources of sound can be calculated by an incompressible CFD and then used as forcing terms or refraction terms in a computation for sound propagation from sources in an inhomogeneous moving medium.

2 Analytical Approach

The discussion below is intended to be suggestive and shows the spirit of the analytical approach to be followed. The full formulation is still being developed. This section should help to explain the underlying assumption of the acoustic projection method that a partitioning of field quantities into vortical, acoustic, and entropy modes can be performed regardless of the strength of the nonlinearities and the manner in which the projection operators are defined.

The analysis starts with the general form of the governing equations of fluid motion. For instance, the velocity field is governed by the Navier-Stokes equation

$$\rho \frac{D\vec{v}}{Dt} = - \nabla p + \mu[\nabla^2 \vec{v} + \frac{1}{3}\frac{\partial}{\partial x_i}(\nabla \cdot \vec{v})] \tag{1}$$

where, $\vec{v}$ is the velocity vector, p is pressure, ρ is the density, and μ is the dynamic viscosity. Including the continuity and energy equations, the full set of governing equations, after some minor algebraic manipulations, can be rewritten as

$$\frac{Dv_x}{Dt} + \dots = 0 \tag{2}$$

$$\frac{Dv_y}{Dt} + \ldots = 0 \quad (3)$$
$$\frac{Dv_z}{Dt} + \ldots = 0 \quad (4)$$
$$\frac{Dp}{Dt} + \ldots = 0 \quad (5)$$
$$\frac{Ds}{Dt} + \ldots = 0 \quad (6)$$

in which the only explicit time differentiation in each equation is that contained in the first term. The terms indicated by ... do not involve time differentiation. This set of governing equations can be subsequently expressed such that all of the nonlinear terms appear only on the right hand side of the equality

$$\frac{\partial v_x}{\partial t} + [\text{Linear terms}] = Q_{v_x} \quad (7)$$
$$\frac{\partial v_y}{\partial t} + [\text{Linear terms}] = Q_{v_y} \quad (8)$$
$$\frac{\partial v_z}{\partial t} + [\text{Linear terms}] = Q_{v_z} \quad (9)$$
$$\frac{\partial p}{\partial t} + [\text{Linear terms}] = Q_p \quad (10)$$
$$\frac{\partial s}{\partial t} + [\text{Linear terms}] = Q_s \quad (11)$$

For instance, for a quiescent background medium where the ambient values of the field quantities are given by $v_o = 0, p_o = const, \rho_o = const$, when the Navier-Stokes-Fourier equations are used, the first three equations become

$$\frac{\partial \vec{v}}{\partial t} + \frac{1}{\rho_o}\nabla p - \frac{\mu}{\rho_o}\nabla^2\vec{v} - \frac{1}{3}\frac{\mu}{\rho_o}\frac{\partial}{\partial x_i}(\nabla\cdot\vec{v}) = -(\vec{v}\cdot\nabla)\vec{v} - \left(\frac{\rho-\rho_o}{\rho_o^2}\right)\left[-\nabla p + \mu[\nabla^2\vec{v} + \frac{1}{3}\frac{\partial}{\partial x_i}(\nabla\cdot\vec{v})]\right] \quad (12)$$

where terms of third and higher order in the field amplitudes have for brevity been omitted.

The acoustic portion of the flow can then be identified if one introduces the field splitting which allows each field variable to be written as the sum of a vorticity mode, an acoustic mode, and an entropy mode. That is

$$\vec{v} = \vec{v}_{vort} + \vec{v}_{ac} + \vec{v}_{ent} \quad (13)$$

This splitting has been used since 1868 when Kirchhoff (1868) used it in his study on the influence of heat conduction in a gas on sound propagation. However, for the most part, the splitting was only used and only thought to be meaningful when the equations were written in linearized form. (The work of Chu and Kovásznay (1958) is an important exception, but the ideas in the paper were never fully exploited by the fluid mechanics community.) However, it can be shown that Eq. (13) holds regardless of how nonlinear the equation is. This is true in part because of the definitions of the modes that is applied in this research. The modes are physically justified and the specific sets of equations that govern each mode are given in (Pierce 1989).

To show how the projection operators, which create the splitting, are obtained, first Eqs. (7) - (11) are written in matrix form by introducing the vector $\{\Psi\}$

$$\{\Psi\} = \left\{ \begin{array}{c} v_x \\ v_y \\ v_z \\ p \\ s \end{array} \right\} \tag{14}$$

Then the governing equations can be written as

$$\frac{\partial}{\partial t}\{\Psi\} + [L(\nabla)]\{\Psi\} = \{Q\} \tag{15}$$

where the square brackets denote a matrix. Eq. (15) can be translated into wave-number space using

$$\{\Psi\} = \int \{\tilde{\Psi}\} e^{i\vec{k}\cdot\vec{x}} d^3k \tag{16}$$

Then, when one makes the connection between ∇ and $i\vec{k}$, Eq. (15) becomes

$$\frac{\partial}{\partial t}\{\tilde{\Psi}\} + [A(\vec{k})]\{\tilde{\Psi}\} = \{\tilde{Q}\} \tag{17}$$

The wave-number vector is only introduced here as an analytical tool in order to avoid treating ∇ as an algebraic entity.

An eigenanalysis of the matrix $[A(\vec{k})]$ gives rise to eigenvalues λ_n with right and left eigenvectors $\{\Psi\}_n$ and $\{\tilde{\Phi}\}_n$ respectively, where the eigenvectors satisfy the relation

$$\{\tilde{\Phi}\}_n^T \{\tilde{\Psi}\}_m = 0 \text{ if } n \neq m \tag{18}$$

At this point the acoustic, vortical, and entropy parts of the fields can be identified. To do this, the properties of the different fields are used. First the vorticity mode must satisfy the requirement that the velocity be perpendicular to the wave number vector or

$$\vec{k} \cdot \vec{v} = 0. \tag{19}$$

this being equivalent to the incompressible flow condition $\nabla \cdot \vec{v} = 0$. Two of the eigenvectors will satisfy this requirement. The other three eigenvectors will satisfy the requirement that the velocity be parallel to the wave-number vector or

$$\frac{\vec{k} \cdot \vec{v}}{|\vec{k}||\vec{v}|} = 1. \tag{20}$$

Of these three vectors, the vector corresponding to the entropy mode will be dominated by its fifth entry. The other two eigenvectors then, must correspond to the acoustic mode. One corresponds to a wave propagating in the direction of $\vec{k}$, the other to a wave propagating opposite to the direction of $\vec{k}$. Hence one can write

$$\begin{array}{ccc} \{\tilde{\Psi}\}_1 & \rightarrow & \{\tilde{\Psi}\}_{vort,1} \\ \{\tilde{\Psi}\}_2 & \rightarrow & \{\tilde{\Psi}\}_{vort,2} \\ \{\tilde{\Psi}\}_3 & \rightarrow & \{\tilde{\Psi}\}_{ac,+} \\ \{\tilde{\Psi}\}_4 & \rightarrow & \{\tilde{\Psi}\}_{ac,-} \\ \{\tilde{\Psi}\}_5 & \rightarrow & \{\tilde{\Psi}\}_{ent} \end{array} \tag{21}$$

Now the projection operator for the n-th mode in the above list can be defined as

$$[P]_n = \frac{1}{\{\tilde{\Phi}\}_n^T \{\tilde{\Psi}\}_n} \{\tilde{\Phi}\}_n^T \tag{22}$$

The it is easily seen that these so defined projection operators have the following properties

$$[P]_n \{\tilde{\Psi}\}_n = \{\tilde{\Psi}\}_n \tag{23}$$

$$[P]_n \{\tilde{\Psi}\}_m = 0 \tag{24}$$

$$[P]_n \{\tilde{\Psi}\} = \{\tilde{\Psi}\}_n \tag{25}$$

$$\sum_n [P]_n = [I] \tag{26}$$

where $[I]$ is the identity matrix. It is important to note that the operators exist and have the above properties regardless of whether nonlinear terms are important in subsequent calculations.

The projection operator formulation can then be used to extract acoustic mode information from the total field by simply noting that $\{\tilde{\Psi}\}_{ac} = \{\tilde{\Psi}\}_{ac,+} + \{\tilde{\Psi}\}_{ac,-}$ and therefore

$$\{\tilde{\Psi}\}_{ac} = ([P]_+ + [P]_-)\{\tilde{\Psi}\} = [P]_{sum}\{\tilde{\Psi}\} \tag{27}$$

If one is interested in the acoustic pressure, the fourth entry of the vector equation is used. For example, for a quiescent background medium, $v_o = 0$, with constant mean properties $p_o = const, \rho_o = const$, if one takes the wave number vector, $\vec{k}$ in the z direction, such that $\nabla \rightarrow ik_z\hat{k}$, the matrix $[A(\vec{k})]$ in Equation (17) becomes

$$\begin{bmatrix} \frac{\mu}{\rho}k_z^2 & 0 & 0 & 0 & 0 \\ 0 & \frac{\mu}{\rho}k_z^2 & 0 & 0 & 0 \\ 0 & 0 & \frac{4}{3}\mu\rho k_z^2 & \frac{ik_z}{\rho} & 0 \\ 0 & 0 & i\rho c_o^2 k_z & (\gamma-1)\frac{\kappa}{\rho c_p}k_z^2 & \frac{c_o^2\beta T_o\kappa}{c_p^2}k_z^2 \\ 0 & 0 & 0 & \frac{\beta\kappa}{\rho^2 c_p}k_z^2 & \frac{\kappa}{\rho c_p}k_z^2 \end{bmatrix} \tag{28}$$

where c_o is the ambient speed of sound, c_p is the specific heat at constant pressure, T is temperature, γ is the ratio of specific heats, κ is the constant of thermal conduction, and β is $\sqrt{1-M^2}$ with M being the Mach number.

One can perform the eigenanalysis to obtain the acoustic projection operator $[P]_{sum}$ and subsequently find the projection $[P]_{sum}\{\tilde{\Psi}\}$. Performing this analysis gives the acoustic pressure mode in x-space as

$$p_{ac} = p + \frac{\beta T_o \rho c^4 \tau_\kappa^2}{c_p}[1 - \frac{4}{3}Pr]\nabla^2 s \tag{29}$$

where $Pr = \mu c_p/\kappa$ is the Prandtl number, and $\tau_\kappa = \kappa/(\rho c^2 c_p)$.

Given the basic definitions of projection operators, it is possible to formulate methods for using field data calculated from an incompressible CFD analysis to determine the associated acoustics. In that which corresponds most closely to Lighthill's acoustic analogy, one operates with $[P]_{ac,+}$ and $[P]_{ac,-}$ on both sides of Equation (17) to obtain

$$\left(\frac{\partial}{\partial t} + [A(\vec{k})]\right)\{\tilde{\Psi}\}_{ac,+} = \left(\frac{\partial}{\partial t} + \lambda_+\right)\{\tilde{\Psi}\}_{ac,+} = [P]_+\{\tilde{Q}\} \tag{30}$$

$$\left(\frac{\partial}{\partial t} + [A(\vec{k})]\right)\{\tilde{\Psi}\}_{ac,-} = \left(\frac{\partial}{\partial t} + \lambda_-\right)\{\tilde{\Psi}\}_{ac,-} = [P]_-\{\tilde{Q}\} \tag{31}$$

Multiplying the first equation by the operator $\left(\frac{\partial}{\partial} + \lambda_-\right)$ and the second by the operator $\left(\frac{\partial}{\partial t} + \lambda_+\right)$ and adding the two equations, one obtains

$$\left(\frac{\partial}{\partial t}+\lambda_+\right)\left(\frac{\partial}{\partial t}+\lambda_-\right)\{\tilde{\Psi}\}_{ac}=[P]_+\left(\frac{\partial}{\partial t}+\lambda_-\right)\{\tilde{Q}\}+[P]_-\left(\frac{\partial}{\partial t}+\lambda_+\right)\{\tilde{Q}\} \tag{32}$$

For the example discussed earlier where the background medium is quiescent and where the z-direction is taken to be that of $\vec{k}$, one can solve Eq. (28) for the eigenvalues and eigenvectors. Substituting these into Eq. (32) gives

$$\left(\frac{\partial^2}{\partial t^2}+c^2k_z^2+[\tau_\mu+(\gamma-1)\tau_\kappa]c^2k_z^2\frac{\partial}{\partial t}\right)\tilde{p}_{ac}=\left(\frac{\partial}{\partial t}+\frac{1}{2}[\tau_\mu+(\gamma-1)\tau_\kappa]c^2k_z^2\right)\tilde{Q}_p$$

$$-ick_z[\rho c\tilde{Q}_{v_z}+\frac{ic}{2}[\tau_\mu-(\gamma-1)\tau_\kappa]k_z\tilde{Q}_p-\frac{i\beta T\rho c^3}{c_p}\tau_\kappa k_z\tilde{Q}_s \tag{33}$$

where $\tau_\mu = 4\mu/(3\rho^2)$.

Now (33) can be translated back to x-space as

$$\left(\frac{\partial^2}{\partial t^2}-c^2\nabla^2-c^2[\tau_\mu+(\gamma-1)\tau_\kappa]\nabla^2\frac{\partial}{\partial t}\right)p_{ac}=\left(\frac{\partial}{\partial t}-\frac{c^2}{2}[\tau_\mu+(\gamma-1)\tau_\kappa]\nabla^2\right)Q_p$$

$$-c\nabla[\rho cQ_{v_z}+\frac{c}{2}[\tau_\mu-(\gamma-1)\tau_\kappa]\nabla Q_p-\frac{\beta T\rho c^3}{c_p}\tau_\kappa\nabla Q_s \tag{34}$$

The full Q vector for this example, but with terms of order of three and higher deleted, for brevity, is

$$Q_v = (\vec{v}\cdot\nabla)\vec{v}-\left(\frac{\rho-\rho_o}{\rho_o^2}\right)\left[-\nabla p+\mu[\nabla^2\vec{v}+\frac{1}{3}\frac{\partial}{\partial x_i}(\nabla\cdot\vec{v})]\right] \tag{35}$$

$$Q_p = -\vec{v}\cdot\nabla p+\frac{c_o^2\beta}{c_p}\phi+(\rho_o c_o(c-c_o)+(\rho-\rho_o)c_o^2)\nabla\cdot\vec{v} \tag{36}$$

$$-\frac{c_o^2\beta}{c_p}\left(\frac{\rho-\rho_o}{\rho_o}+\frac{T-T_o}{T_o}\right)\kappa\nabla^2 T \tag{37}$$

$$Q_s = \vec{v}\cdot s-\frac{1}{T_o\rho_o}\left(\left(\frac{\rho-\rho_o}{\rho_o}\right)\kappa\nabla^2 T+\phi-\left(\frac{T-T_o}{T_o}\right)\kappa\nabla^2 T\right) \tag{38}$$

where $\nabla\cdot\vec{q}=\kappa\nabla^2 T$, $\phi=2\mu[(\frac{\partial v_i}{\partial x_j}-\frac{\partial v_j}{\partial x_i})-\frac{1}{3}\nabla\vec{v}\delta_{ij}]^2$, and $\delta T=\frac{T_o\beta}{\rho_o c_p}\delta p+\frac{T_o}{c_p}\delta s$ (Pierce 1989) have been used. The above equation, although having strong similarities to other governing equations which have appeared in the aeroacoustic literature, is suggested as a starting point for making systematic approximations for numerical calculations. One feature of the utility of this equation is that by arranging the terms it can be shown that both the effects of refraction, due to acoustic vortical interactions, and nonlinear sources, due to vortex-vortex interactions, entropy-vortex interactions etc. are included.

Evaluating the full wave equation source term in Eq. (34) using the results from a conventional computational fluid dynamic code may be very time consuming. But, by using Eq. (34) as a starting point and evaluating the importance of the terms Q_p, Q_v, and Q_s for specific applications, the important interactions can be identified and used to form the source term.

3 First Benchmark

The first benchmark problem to which the method is being applied is the well known spinning vortex pair (Obermeier 1977; Mitchell, Lele, & Moin 1995; Lee & Koo 1995). This problem was chosen because both the fluid dynamic and acoustic results are known analytically. Many simplifications can be made in Eq. (34) for the spinning vortex pair by assuming there is no thermal or viscous diffusion, no entropy fluctuations, and that vortical effects dominate the acoustic effects. The resulting governing equation for the acoustic pressure is

$$(\frac{\partial^2}{\partial t^2} - \nabla^2)p_{ac} = -\rho c^2 \nabla (\vec{v} \cdot \nabla)\vec{v} \tag{39}$$

which is the familiar form found when using Lighthill's acoustic analogy. Because the fluid dynamic result is known analytically, the simplified source term in Eq. (34) is also known analytically.

Currently, Eq. (39) is being solved using a simple second order central differencing scheme in both space and time on a simple, rectangular, uniformly spaced grid. The nonreflecting boundary condition of Bayliss and Turkel (1982) is applied at the outer boundaries of the computational domain. The implementation of the boundary conditions is based on that described by Hariharan et. al. in (1992). The calculations are being carried out on the Silicon Graphics Origin 2000 supercomputer at Boston University.

4 Conclusions

The acoustic projection method is still being developed. It is based on a field splitting theorem which allows fields to be represented as the sum of vortical, acoustic, and entropy modes. The method can theoretically be used to project out of a complex flow the part of the overall field which is acoustic. Computations are under way to validate the method on benchmark problems.

The method has two advantages over the commonly used derivatives of Lighthill's analogy. It offers a way to obtain far and *near* field acoustic information and it is applied using computational fluid dynamic techniques; in the sense that the method is being developed as an easily adaptable extension to CFD codes. Therefore, important acoustic information will be readily obtained from existing CFD codes. Moreover, it will provide an experimental as well as theoretical research tool for fundamental investigations of flow generated sound.

References

Bayliss, A. & Turkel, E. 1982, "Far Field Boundary Conditions for Compressible Flows," *Journal of Computational Physics*, Vol. 48, 182–199.

Chu, B.-T. & Kovásznay, L. S. G. 1958, "Non-linear interactions in a viscous heat-conducting compressible gas," *Journal of Fluid Mechanics*, Vol. 3, 494–514.

Hardin, J. C. & Pope, D. S. 1994, "An Acoustic/Vortical Splitting Technique for Computational Aeroacoustics," *Theoretical and Computational Fluid Dynamics*, Vol. 6, 323–340.

Hariharan, S. I., Ping, Y., & Scott, J. R. 1992, "Time Domain Numerical Calculations of Unsteady Vortical Flows about a Flat Plate Airfoil," *Journal of computational Physics*, Vol. 101, 419–430.

Kirchhoff, G. 1868, "On the Influence of Heat Conduction in a Gas on Sound Propagation," *Annual of Physics and Chemistry*, Vol. 134, 177–193.

Lee, D. J. & Koo, S. O. 1995, "Numerical Study of Sound Generation due to a Spinning Vortex Pair," *AIAA Journal*, Vol. 33, 20–26.

Lighthill, M. 1952, "On Sound Generated Aerodynamically. I. General theory," *Proceedings of the Royal Socety of London*, Vol. A 211, 564–587.

Mitchell, B. E., Lele, S. K., & Moin, P. 1995, "Direct computation of the sound from a compressible co-rotating vortex pair," *Journal of Fluid Mechanics*, Vol. 285, 181–202.

Obermeier, F. 1977, "The application of singular perturbation methods to aerodynamic sound generation,". In *Singular Perturbations and Boundary Layer Theory*, Vol. 594, 400–421. Berlin, Heidelber, New York: Springer-Verlag.

Pierce, A. D. 1989, *Acoustics: An Introduction to Its Physical Principles and Applications*, p. 515. Acoustical Society of America.

TRANSIENT ANALYSIS OF WAVE PROPAGATION IN UNBOUNDED MEDIA: SPACE-TIME METHODS AND CONTINUED-FRACTION IMPLEMENTATIONS

Murthy N. Guddati and John L. Tassoulas

Texas Institute for Computational and Applied Mathematics
The University of Texas at Austin
Austin, TX, USA

1. Introduction

Modeling unbounded domains under transient wave propagation continues to be a challenge to both engineers and scientists. Accurate modeling of this phenomenon is of utmost importance in several fields of engineering ranging from earthquake engineering to electromagnetics. The standard way of analyzing unbounded media is to truncate the domain of interest and apply appropriate conditions on the computational boundary. Several ways of approximating these boundary conditions have been proposed by various researchers in the last two decades (for a survey, see [Givoli, 1992]). Some are local in both space and time. Others are global in space, time, or both. Local methods have case-specific performance, but are economical, whereas, global methods are accurate but expensive.

The methods described in this paper are explained using the problem of two-dimensional scalar wave propagation in homogeneous layer. The aim is to obtain a boundary condi-

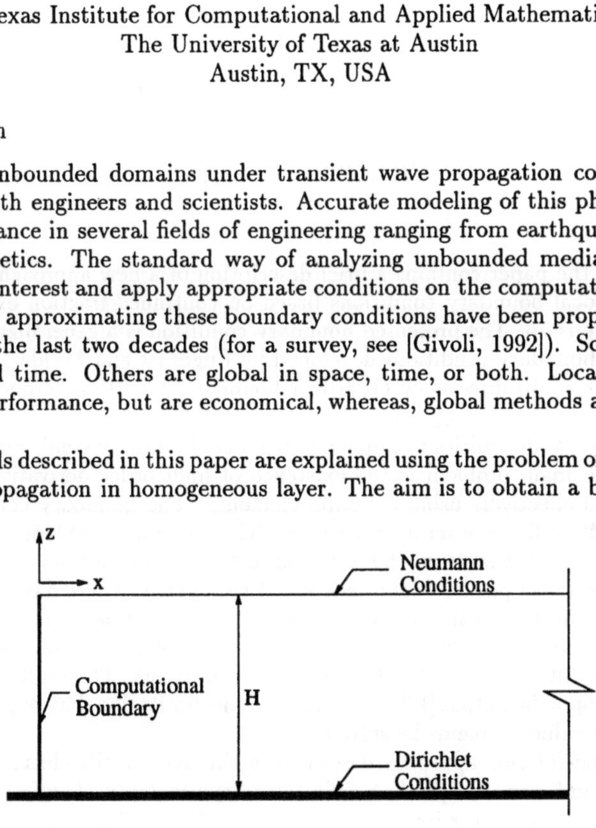

Figure 1: Model problem: Scalar wave propagation in unbounded layer

tion which is a good approximation of the effect of unbounded layer. Without any loss of generality, we assume that the layer is unbounded in the positive x direction and is bounded in the z direction. Dirichlet/ Neumann conditions are applied on the top and bottom boundaries of the layer. To facilitate easy coupling with the finite element method for the interior, it is desirable to obtain a relationship between the traction $(-G\partial u/\partial x)$ and the displacement (u). The problem can thus be summarized as follows:

T.L. Geers (ed.), IUTAM Symposium on Computational Methods for Unbounded Domains, 169–178.

Find the map $DtN : u_\Gamma \rightarrow s$ in which u_Γ is the field variable at the boundary $(x = 0)$ and $s = -Gu_x$ is the traction on the boundary. The field variable satisfies the following differential equation, boundary conditions and initial conditions:

$$\rho u_{tt} - Gu_{xx} - Gu_{zz} = f \quad for \quad x > 0, z \in (-H, 0), t > 0 \tag{1}$$

$$u(x, z, t) = 0 \quad at \quad z = -H \tag{2}$$

$$Gu_z(x, z, t) = 0 \quad at \quad z = 0 \tag{3}$$

$$u(x, z, 0) = u_t(x, z, 0) = 0 \tag{4}$$

This paper contains two parts describing two independent methods to solve the above problem. The first part (Secs. 2-6) deals with a local method and the second part (Secs. 7-10) describes a global method. Each part is independent of each other and can be read separately.

CONTINUED-FRACTION IMPLEMENTATIONS

2. Overview

This part of the paper contains a brief description of a new approach of implementing high-order local boundary conditions based on continued fraction expansion of the dispersion relationship. The presented boundary condition is a generalization of a series of approximate boundary conditions developed by Engquist and Majda [1979]. It is also equivalent to the multi-directional absorbing boundary conditions proposed by Higdon [1986].

There are several difficulties in implementing local boundary conditions of arbitrarily high order. The main problem is the existence of high order derivatives that cannot be approximated effectively using C^0 finite elements. The boundary conditions can be imposed in a finite difference sense, but the resulting bounded problem can have adverse stability properties. Furthermore, the resulting coefficient matrices do not have the desirable properties such as positive-definiteness and symmetry. There has been substantial effort invested in implementing high-order boundary conditions. The implementation proposed by Kallivokas and Bielak [1993] fits well in the context of the finite element method, but is limited to second-order boundary conditions. The multi-stage boundary conditions developed by Safjan [1995] do not have this limitation, but require that several initial-boundary value problems be solved.

The continued-fraction approach described in this section fits elegantly in the finite element setting and results in positive definite and symmetric matrix from the boundary condition. The key idea in the implementation is to add auxiliary *layers* of nodes next to the boundary and reduce the boundary condition to a second-order one. Thus C^0 finite elements suffice. The detailed development, analysis and implementation of this method will be published in the forthcoming special journal issue [Guddati and Tassoulas, 1997b]. The rest of the section contains an outline of the method and its performance.

3. Local boundary conditions based on continued-fractions

We start by examining the Fourier modes for the wave equation. They have the form

$$e^{(ikx + ily)/\sqrt{G} + i\omega t/\sqrt{\rho}}, \tag{5}$$

where k and l are the scaled horizontal and vertical wave numbers, respectively, and ω is the scaled frequency. Then, the wave equation in the Fourier domain is given by:

$$k^2 + l^2 = \omega^2 \tag{6}$$

Our aim is to annihilate waves propagating in the positive x direction. This implies that the ideal boundary condition should be equivalent to the following one-way wave equation (dispersion relationship) in the Fourier domain:

$$ik + i\omega\sqrt{1-\sigma^2} = 0, \tag{7}$$

where $\sigma = il/i\omega$. For traveling waves, σ can be viewed as $sin(\theta)$, where θ is the angle of incidence with respect to the direction normal to the boundary.

To obtain approximate boundary conditions from the dispersion relationship, a rational approximation is obtained for $\sqrt{1-\sigma^2}$ and the approximate dispersion relation contains polynomials in ik, il and $i\omega$. This equation, when transformed back into space and time, results in a local boundary condition containing high-order derivatives.

Engquist and Majda [1979] proposed a series of approximations to $\sqrt{1-\sigma^2}$ based on continued fraction expansions. We generalize their approach and propose a modified continued fraction expansion based on the following iteration:

$$f_1 = cos(\theta_1), \tag{8}$$

$$f_{n+1} = cos(\theta_{n+1}) - \frac{cos^2(\theta_{n+1}) - (1-\sigma^2)}{cos(\theta_{n+1}) + f_n}, \tag{9}$$

where f_n is the nth order continued-fraction approximation of $\sqrt{1-\sigma^2}$ and θ_n are some pre-selected angles. The iterative procedure is a contraction mapping for $0 \le \sigma \le 1$ and thus models the absorption of traveling waves progressively better. It is clear that if $\theta_n = 0$ for every iteration, this procedure is identical to the expansion used by Engquist and Majda.

One way of analyzing the effectiveness of local boundary conditions is to examine the ratio of amplitudes of reflected and incident waves on the boundary. The reflection coefficient for the traveling waves for any approximation of the dispersion relationship is given by:

$$R_n = \frac{f_n - cos(\theta)}{f_n + cos(\theta)}, \tag{10}$$

where θ is the angle of incidence. It can be shown that the reflection coefficient for nth order approximation is

$$R_n = \prod_{k=1}^{k=n} \frac{cos(\theta_k) - cos(\theta)}{cos(\theta_k) + cos(\theta)}. \tag{11}$$

This reflection coefficient is identical to the one for the multi-directional transmitting boundary proposed by Higdon [1986]. The multi-directional boundary conditions have the following form:

$$\prod_{k=1}^{k=n} \left(cos(\theta_k)\frac{\partial}{\partial x} - \frac{\partial}{\partial t} \right) u = 0. \tag{12}$$

Although the proposed local boundary conditions based on Eqs. 8 and 9 may seem more complicated than the multi-directional transmitting boundaries, we show in the following paragraphs that the continued-fraction form is much more amenable to implementation in finite element settings.

4. Mixed form of absorbing boundary conditions

To eliminate the high-order spatial and temporal derivatives, we present a mixed form by defining auxiliary variables. The implementation is restricted to even-order approximations with $\theta_{2k-1} = \theta_{2k}$. Then the iterative process takes the form:

$$f_2 = a_1 \tag{13}$$

$$f_{2n+2} = a_{n+1} - \frac{b_{n+1}^2}{a_{n+1} + f_{2n}}. \tag{14}$$

In the above, a_n and b_n are given by

$$a_n = \frac{cos(\theta_{2n})}{2} + \frac{1-\sigma^2}{2cos(\theta_{2n})} \tag{15}$$

$$b_n = -\frac{cos(\theta_{2n})}{2} + \frac{1-\sigma^2}{2cos(\theta_{2n})}. \tag{16}$$

To satisfy the approximate local boundary condition $(ik + i\omega f_{2n})u = 0$, it is sufficient to satisfy the following matrix equation.

$$\begin{bmatrix} \frac{ik}{i\omega} + a_n & b_n \\ b_n & a_n + f_{2n-2} \end{bmatrix} \begin{bmatrix} u \\ u_1 \end{bmatrix} = \begin{bmatrix} 0 \\ 0 \end{bmatrix}, \tag{17}$$

where u_1 is an auxiliary variable, whose definition is included in the above matrix equation (the second equation). The elimination of u_1 through Schur factorization of Eq. 17 results in the desired approximate boundary condition $(ik + i\omega f_{2n})u = 0$. *This idea of viewing the continued fraction iteration in matrix form is the key to the proposed implementation of high-order boundary conditions.*

The procedure described above can be used recursively to define additional auxiliary variables u_1, u_2, ..., u_n, and to obtain the following matrix form:

$$\begin{bmatrix} \frac{ik}{i\omega} + a_n & b_n & & & \\ b_n & a_n + a_{n-1} & b_{n-1} & & \\ & b_{n-1} & \ddots & \ddots & \\ & & \ddots & \ddots & b_2 \\ & & & b_2 & a_2 + a_1 \end{bmatrix} \begin{bmatrix} u \\ u_1 \\ u_2 \\ \vdots \\ u_{n-1} \end{bmatrix} = \begin{bmatrix} 0 \\ 0 \\ 0 \\ \vdots \\ 0 \end{bmatrix} \tag{18}$$

After some manipulations and transforming back into space and time, we obtain the following boundary condition in terms of the auxiliary variables:

$$\begin{bmatrix} -G\frac{\partial u}{\partial x} \\ 0 \\ \vdots \\ 0 \end{bmatrix} = \sqrt{\rho G}(\mathbf{A} + \mathbf{B})\frac{\partial}{\partial t} \begin{bmatrix} u \\ u_1 \\ \vdots \\ u_{n-1} \end{bmatrix} - G\mathbf{B} \int \frac{\partial^2}{\partial z^2} \begin{bmatrix} u \\ u_1 \\ \vdots \\ u_{n-1} \end{bmatrix} dt. \tag{19}$$

The coefficient matrices **A** and **B** are given by:

$$\mathbf{A} = \frac{1}{2}\begin{bmatrix} c_n & -c_n & & & \\ -c_n & c_n + c_{n-1} & -c_{n-1} & & \\ & -c_{n-1} & \ddots & \ddots & \\ & & \ddots & \ddots & -c_2 \\ & & & -c_2 & c_2 + c_1 \end{bmatrix}, \tag{20}$$

$$\mathbf{B} = \frac{1}{2}\begin{bmatrix} 1/c_n & 1/c_n & & & \\ 1/c_n & (1/c_n) + (1/c_{n-1}) & 1/c_{n-1} & & \\ & 1/c_{n-1} & \ddots & \ddots & \\ & & \ddots & \ddots & 1/c_2 \\ & & & 1/c_2 & (1/c_2) + (1/c_1) \end{bmatrix}, \tag{21}$$

where,

$$c_n = cos(\theta_{2n}). \tag{22}$$

5. Finite element discretization

The discretization of $u_1, \cdots, u_{n-1}$ is chosen to be identical to the discretization of u on the boundary, i.e., in the z direction. We apply Galerkin approximation to the boundary condition. At the layer boundaries, the boundary conditions for the auxiliary variables are chosen to be the homogeneous counterpart of the boundary conditions for u. This choice

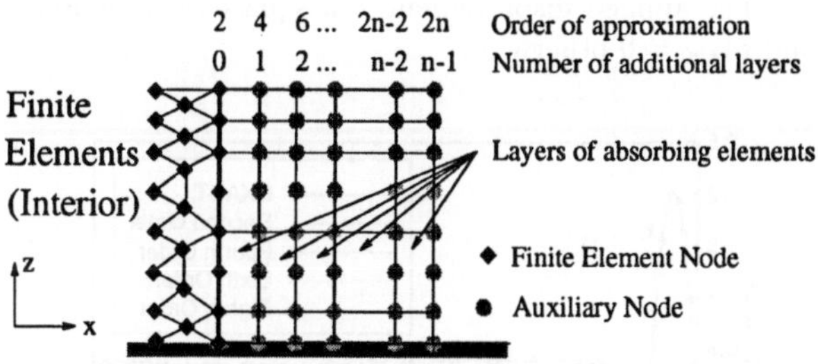

Figure 2: Implementation of higher order absorbing boundary conditions

is appropriate as it approximates the dispersion relationship for the layer in a consistent manner. Under these circumstances, the application of the above boundary condition can be viewed as adding additional nodes and elements to the existing truncated problem. As shown in Fig. 2, we add a layer of elements for every two levels of approximation. The contribution of these *absorbing elements* to the coefficient matrices is symmetric and positive definite. It should also be noted that the connectivity is local and preserves the sparsity of the coefficient matrices. Thus, the analysis of the computational problem in Fig. 2 can be viewed as a standard finite element analysis and all the computational techniques developed for standard analysis such as domain-decomposition techniques apply directly to our computational problem.

The discrete form of evolution equations obtained in this analysis differ from those obtained in standard finite element analysis of wave propagation problems. The interior problem, when augmented with the boundary conditions, results in the following integro-differential equation in time:

$$\mathbf{M}\mathbf{U}_{tt} + \mathbf{C}\mathbf{U}_t + \mathbf{K}\mathbf{U} + \mathbf{R}\int \mathbf{U}dt = \mathbf{F}, \qquad (23)$$

where, matrices **C** and **R** have contributions from the boundary conditions. Care should be exercised when applying conventional time-stepping scheme for this equation. Analysis and design of time-stepping schemes for this system will be presented in the journal paper.

6. Numerical example

To illustrate the effectiveness of high-order boundary conditions and their implementation, a prismatic rod on elastic foundation is analyzed under applied displacements (Fig. 3). The dispersion relationship governing this system is similar to the dispersion relationship of different modes of propagation in acoustic layer/halfspace (see, for e.g.,

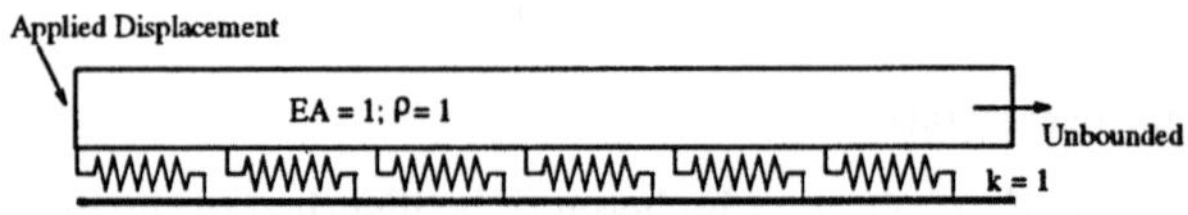

Figure 3: Representative system for unbounded layer/halfspace

[Wolf, 1988]). Hence, this system has been chosen to illustrate the effectiveness of the proposed method. The applied displacement is a squared-sine pulse: $u|_{x=0} = sin^2(\Omega t)$ for $0 \leq \Omega t \leq \pi$ and $u|_{x=0} = 0$ otherwise.

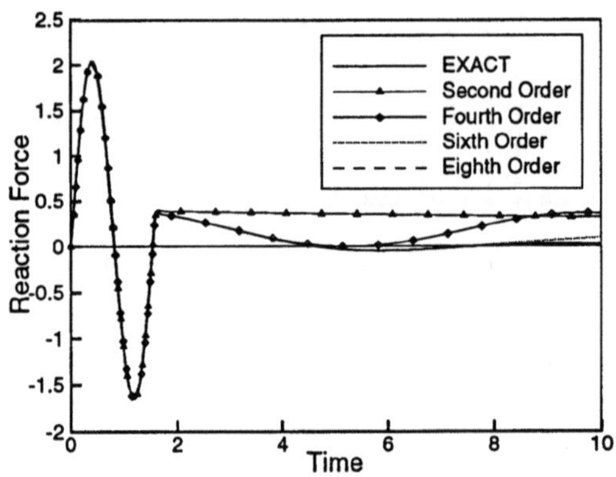

Figure 4: Accuracy of higher order local boundaries

In Fig. 4, we plot the resulting traction at the tip of the rod. We chose $\Omega = 2$ so as to minimize the low-frequency components, i.e., the excitation results in predominantly traveling waves. The proposed implementation is used for the absorbing boundary placed immediately next to the load. The results corresponding to a sequence of increasingly high-order absorbing boundaries are compared with the exact solution obtained by convolution in Fig. 4. It is evident from the figure that the second order and even fourth order boundary conditions are considerably inaccurate. On the other hand, sixth-order and eighth-order boundaries produce remarkably accurate results.

SPACE-TIME METHODS FOR UNBOUNDED MEDIA

7. Overview

It is well known that the local absorbing boundary conditions are effective mainly for traveling waves. When excitation is rich in frequencies lower than fundamental frequency of the unbounded medium, evanescent waves appear and local boundaries can result in erroneous results. The procedures based on impulse response function do not have this limitation, but are computationally very demanding. In this section, we propose a new methodology of analyzing unbounded media. Our approach is based on a two-step procedure. The first step is a semidiscretization process which reduces the spatial dimension of analysis to one. The resulting system of semidiscrete hyperbolic partial differential equations are then solved in the space-time domain in an efficient manner. Two methods have been developed in this direction. The first method, applicable to problems of scalar wave propagation in homogeneous media, is based on characteristics. The second one is a space-time Galerkin technique developed for solving wave propagation problems with multiple wave speeds. In the following paragraphs, we outline the characteristics method, after a brief description of the semidiscretization procedure.

8. Semidiscretization

The layer is discretized in vertical direction such that the discretizations for the interior and exterior are matched at the computational boundary (Fig. 5). Applying the

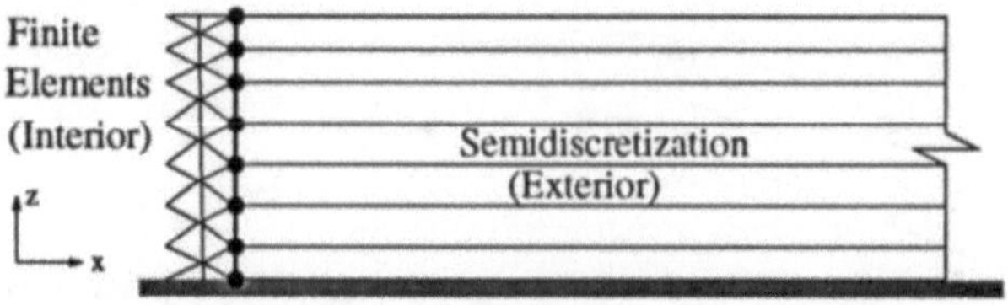

Figure 5: Semidiscretization of an unbounded layer

Galerkin method with this discretization, we arrive at the following semidiscrete problem:

Find the map $\mathbf{DtN} : \mathbf{U}_\Gamma(t) \rightarrow \mathbf{S}(t)$ where $\mathbf{U}_\Gamma$ is the displacement vector at $x = 0$ and $\mathbf{S} = -\mathbf{A}\mathbf{U}_x$ is the corresponding traction vector. We satisfy the following differential equations and initial conditions:

$$\mathbf{A}\mathbf{U}_{xx} - \frac{1}{c^2}\mathbf{A}\mathbf{U}_{tt} - \mathbf{G}\mathbf{U} = \mathbf{0} \quad for \quad x > 0,\ t > 0 \tag{24}$$

$$\mathbf{U} = \mathbf{U}_t = \mathbf{0} \quad at \quad t = 0. \tag{25}$$

In the above, matrix $\mathbf{A}$ is symmetric and positive definite and $\mathbf{G}$ is symmetric positive semidefinite.

9. Characteristics method

The first step in the characteristics method is to decouple to semidiscrete system of equations through spectral decomposition of matrix **G** with with respect to **A**. This results in the following uncoupled form:

$$\hat{\mathbf{U}}_{xx} - \frac{1}{c^2}\hat{\mathbf{U}}_{tt} - \mathbf{\Lambda}\hat{\mathbf{U}} = \mathbf{0}, \tag{26}$$

where $\mathbf{\Lambda}$ is a diagonal matrix. The above equation is transformed into the characteristic variables defined as:

$$\xi = \frac{x + ct}{2} \qquad and \qquad \eta = \frac{x - ct}{2}. \tag{27}$$

The resulting system takes the form:

$$\hat{\mathbf{U}}_{\xi\eta} - \mathbf{\Lambda}\hat{\mathbf{U}} = \mathbf{0}. \tag{28}$$

We integrate the above equation and apply a change of variables to arrive at the following integral formula for $\hat{\mathbf{U}}_x$:

$$\hat{\mathbf{U}}_x = \int [\mathbf{\Lambda}\hat{\mathbf{U}}(\bar{\xi}, \eta)]d\eta - \frac{1}{c}\hat{\mathbf{U}}_t. \tag{29}$$

The characteristics method uses the above equation to compute the traction. The field variable on the corresponding characteristic is computed from the previous characteristics using a cell-centered finite difference technique (Fig. 6). This is done in a cell-by-cell

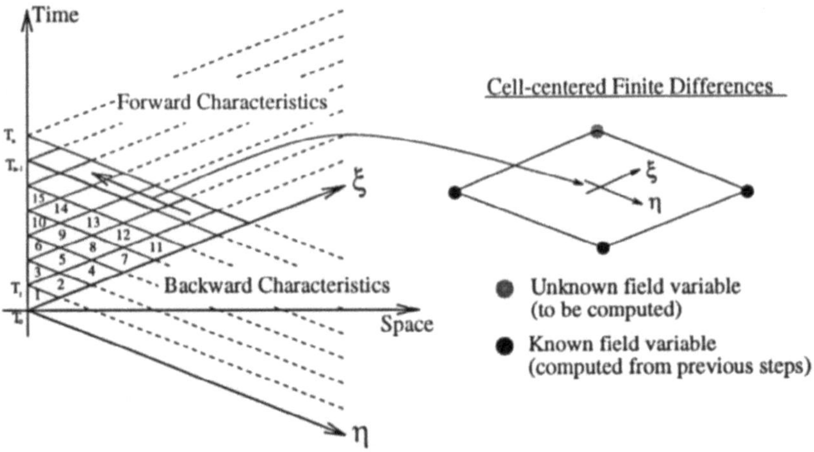

Figure 6: Finite difference grid for the characteristics method

fashion in the order shown in the figure. The equation decoupling and the cell-by-cell computation render the characteristic method highly efficient. In the final form, the boundary condition relating traction and displacement is given by:

$$\mathfrak{I} = \frac{c\Delta T_n}{4}\mathbf{G}\mathbf{U} + \frac{1}{c}\mathbf{A}\frac{\partial \mathbf{U}}{\partial t} + \mathbf{\Pi}, \tag{30}$$

where $\mathbf{H}$ is a vector obtained by integrating the field variable on the η-characteristic corresponding to the current time step.

Note that the coefficient matrices are multiples of matrices $\mathbf{A}$ and $\mathbf{G}$, which are sparse, symmetric and positive semi-definite. Thus, this boundary condition, when applied to the interior problem, preserves all the desirable properties of the coefficient matrices. The computational cost of this method is less than that of comparable global methods by an order of magnitude. The method is proven to be unconditionally stable and optimally accurate.

10. Numerical example

The effectiveness of the characteristics method is demonstrated by analyzing the prismatic rod on elastic foundation described in the first part (Fig. 3). The squared-sine displacement pulse is applied with $\Omega = 1$. The time-stepping for the analysis is chosen such that there is little error in approximating the loading function and the upper bound is chosen considering the cut-off frequency. The results are compared with the exact

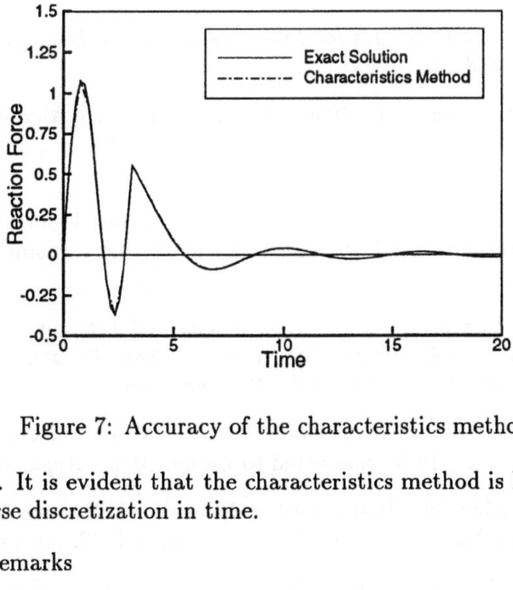

Figure 7: Accuracy of the characteristics method

solution in Fig. 6. It is evident that the characteristics method is highly accurate, even for relatively coarse discretization in time.

11. Concluding remarks

In this paper, we described two new methods for transient analysis of wave propagation in unbounded media. The first method is a continued fraction based local boundary condition facilitating elegant implementation of high-order boundary conditions. The second method is a global method based on space-time discretization. The concluding remarks for these two methods follow.

Continued Fraction Implementations: We presented a new approach to implementing high-order local boundary conditions. A modified version of Engquist-Majda boundaries are shown to be equivalent to the multi-directional boundary conditions and implemented in a mixed form. The key idea behind this approach is to view the continued fraction iteration in a matrix form. The implementation requires only C^0 continuity, and preserves all the desirable properties such as sparsity, symmetry and positive-definiteness of the coefficient matrices. Extensions to curved boundaries and more complicated elastic wave equations will be the focus of further research.

Space-Time Methods: A new two-step approach is presented for the analysis of unbounded media. Semidiscretization is used to reduce the spatial dimension to one and the resulting equations are solved in space-time. The characteristic method is applicable to scalar wave problems in homogeneous media. It contains all the properties that are desirable in an ideal boundary condition. It can also be extended to other geometries of unbounded domains. Local boundary conditions are being developed to facilitate truncation of space-time meshes [Guddati and Tassoulas, 1997c]. For more complicated cases involving multiple wave speeds, a space-time Galerkin method has been developed. The details of that method can be found elsewhere [Guddati and Tassoulas, 1997a].

12. References

Engquist, B. and Majda, A., 1979, "Radiation Boundary Conditions for Acoustic and Elastic Wave Calculations", *Comm. Pure & Appl. Math.*, Vol. 32, pp. 313–357.

Givoli, D., 1992, *Numerical Methods for Problems in Infinite Domains*, Elsevier Science Publishers B.V., Amsterdam.

Guddati, M.N., and Tassoulas, J.L., 1996, "Efficient Numerical Algorithm for Transient Analysis of Exterior Wave Propagation in a Homogeneous Layer", Submitted to *Comp. Meth. Appl. Mech. Engg.*

Guddati, M.N., and Tassoulas, J.L., 1997a, "Space-time Finite Elements for the Analysis of Transient Wave Propagation in Unbounded Layered Media", Submitted to *Int. J. Sol. and Struct.*

Guddati, M.N., and Tassoulas, J.L., 1997b, "Continued Fraction Implementations of Local Absorbing Boundary Conditions for Transient Wave Propagation Problems", to be submitted to the *special journal issue for this conference.*

Guddati, M.N., and Tassoulas, J.L., 1997c, "Characteristics Method for Transient Analysis of Unbounded Media", To be submitted to *Comp. Meth. Appl. Mech. Engg.*

Higdon, R.L., 1986, "Absorbing Boundary Conditions for Difference Approximations to the Multi-Dimensional Wave Equation", *Math. Comp.*, Vol. 47, pp. 437–459.

Kallivokas, L.F., and Bielak, J., 1993, "Time-domain analysis of transient structural acoustics problems based on the finite element method and a novel absorbing boundary element", *J. Acous. Soc. Am.*, Vol. 94, pp. 3480–3492.

Safjan, A., 1995, *Highly Accurate Non-Reflecting Boundary Conditions for Finite Element Simulation of Transient Acoustics Problems*, TICAM Report 95-11, Texas Inst. for Comp. and Appl. Math., Univ. of Texas, Austin, TX.

Wolf, J.P., 1988, *Soil-Structure-Interaction in Time Domain*, Prentice-Hall, Englewood Cliffs, N.J.

EXACT AND HIGH-ORDER BOUNDARY CONDITIONS IN THE TIME DOMAIN

Thomas Hagstrom

Department of Mathematics and Statistics
The University of New Mexico
Albuquerque, NM 87131, USA

1 Introduction

The problem of imposing accurate radiation boundary conditions at the artificial boundary is generally the primary accuracy bottleneck in the numerical solution of scattering problems in the time domain. An impressive number of schemes have been proposed to overcome this difficulty. Nonetheless, the ideal of solving the problem on a fixed domain and achieving convergence by some systematic improvement of the boundary treatment and/or simple mesh refinement seems rarely to have been used. In this talk we advocate such an approach for the basic constant coefficient equations of wave propagation - the wave equation, Maxwell's equations, the convective wave equation and the compressible Euler equations linearized about a uniform flow. We first develop representations of the exact boundary conditions for special boundaries using separation of variables. The difficult, that is temporally nonlocal, part of the exact condition is expressed as a temporal convolution operator. Approximate boundary treatments, both new and old, are viewed as approximations to the exact convolution kernel, allowing a fairly easy identification of convergence rates and the development of complexity estimates.

For the sake of simplicity, we present our analysis for the scalar wave equation, though most results also hold for all the equations listed above. Besides theoretical analyses, we present some recent results using our favorite method and discuss some important open problems.

T.L. Geers (ed.), IUTAM Symposium on Computational Methods for Unbounded Domains, 179–186.

2 Exact and Approximate Conditions

We begin with the case of a flat boundary. Consider the wave equation in a half-space:

$$u_{tt} = \nabla^2 u + f, \quad t > 0, \quad x = (x_1, y) \in (0, \infty) \times R^{n-1}, \tag{2.0.1}$$

$$u(x, 0) = g(x), \quad B_0 u(0, y, t) = g_0(y, t). \tag{2.0.2}$$

We suppose, for some $L, \delta > 0$, that $f = g = 0$ for $x_1 \geq L - \delta$. Let $\hat{u}(x_1, k, s)$ be the Fourier-Laplace transform of u with respect to y and t. Then $\hat{u}$ satisfies the exact boundary condition at $x_1 = L$:

$$\frac{\partial \hat{u}}{\partial x_1} + \left(s^2 + |k|^2\right)^{1/2} \hat{u} = 0. \tag{2.0.3}$$

This exact condition is expressed in terms of u in the following way: Let $\mathcal{F}$ denote Fourier transformation with respect to y and $\mathcal{F}^{-1}$ be its inverse. Let:

$$\mathcal{K}(t) = \frac{J_1(t)}{t} = \mathcal{L}^{-1}\left(\left(s^2 + 1\right)^{1/2} - s\right). \tag{2.0.4}$$

Inverting the transform we finally have the exact condition at $x_1 = L$:

$$\frac{\partial u}{\partial x_1} + \frac{\partial u}{\partial t} + \mathcal{F}^{-1}\left(|k|^2 \mathcal{K}(|k|t) * (\mathcal{F}u(x_1, \cdot, t))\right) = 0. \tag{2.0.5}$$

Similar formulations are possible on cylindrical and spherical boundaries, assuming the support of the forcing and initial data to be restricted to a distance $R - \delta$ from the coordinate origin. Using polar coordinates in two dimensions, we place the artificial boundary at $r = R$ and expand the solution, u, in a Fourier series in θ:

$$u(r, \theta, t) = \sum_l \bar{u}_l e^{il\theta}. \tag{2.0.6}$$

The exact boundary condition is then given by:

$$\frac{\partial \bar{u}_l}{\partial r} + \frac{\partial \bar{u}_l}{\partial t} + \frac{1}{2R}\bar{u}_l + \frac{1}{R^2} A_l(t/R) * \bar{u}_l = 0, \tag{2.0.7}$$

$$\hat{A}_l(z) = -z\left(\frac{K_l'(z)}{K_l(z)} + 1 + \frac{1}{2z}\right). \tag{2.0.8}$$

(Here, $K_l(z)$ is the modified Bessel function.)

In three dimensions we use spherical coordinates and again take $r = R$ as our artificial boundary. The solution, u, is now expanded in spherical harmonics:

$$u(r, \theta, \phi, t) = \sum_l \bar{u}_l Y_l(\theta, \phi), \tag{2.0.9}$$

where

$$\nabla^2_{\text{sphere}} Y_l = -l(l+1)Y_l. \tag{2.0.10}$$

In analogy with the two-dimensional case, the exact condition is related to the inverse Laplace transform of the logarithmic derivative of spherical Bessel functions. In particular we have:

$$\frac{\partial \bar{u}_l}{\partial r} + \frac{\partial \bar{u}_l}{\partial t} + \frac{1}{R}\bar{u}_l + \frac{1}{R^2} S_l(t/R) * \bar{u}_l = 0, \tag{2.0.11}$$

$$\hat{S}_l(z) = -z\left(\frac{(z^{-1/2}K_{l+1/2}(z))'}{z^{-1/2}K_{l+1/2}(z)} + 1 + \frac{1}{z}\right) = \frac{P_l(z)}{Q_l(z)}, \tag{2.0.12}$$

where, for $l \neq 0$,

$$P_l(z) = \sum_{k=0}^{l-1} \frac{(2l-k)!}{k!(l-k-1)!}(2z)^k, \quad Q_l(z) = \sum_{k=0}^{l} \frac{(2l-k)!}{k!(l-k)!}(2z)^k, \tag{2.0.13}$$

and $S_0 = 0$. Observe that $S_l(z)$ is rational and, therefore, (2.0.11) may be localized in time.

In each case the part of the exact boundary operator which is difficult to apply takes the form:

$$T^{-1}G_l * (Tu), \tag{2.0.14}$$

where T is a spatial transform along the boundary. In the cases listed, there are reasonably fast algorithms for applying T and T^{-1}. Approximate boundary conditions can generally be expressed as:

$$T^{-1}Q_l * (Tu), \tag{2.0.15}$$

and approximation errors depend on:

$$\|G_l - Q_l\|_{L_1(0,T)}, \tag{2.0.16}$$

as well as on stability constants. In standard approaches, Q_l is a rational function of s so that the time convolution may be localized. Spatial localization depends on the behavior of Q_l as a function of the spatial transform variables. It is also possible to analyze absorbing layer techniques, such as PML, in this framework.

An interesting problem is the extension of (2.0.14) to other domains. It is unclear to me as yet how to usefully express the exact condition for rectangular or spheroidal boundaries, which could be more efficient for high aspect ratio scatterers.

3 Comparison of High-Accuracy Methods

To compare the computational complexity of different techniques, we consider the following situation: we seek a solution with an error ϵ in a three-dimensional domain of unit size over a time T. Assuming a pth order discretization, we expect the number of mesh points to scale like $\epsilon^{-3/p}$ and the number of time steps like $T\epsilon^{-1/p}$. Hence the total work and storage for the interior scheme is:

$$\text{Work}_i \propto T\epsilon^{-4/p}, \quad \text{Storage}_i \propto \epsilon^{-3/p}. \tag{3.0.17}$$

Note that for many schemes the theoretical error estimates are incomplete, and we will list the additional assumptions made as we go along.

No Boundary Condition

The simplest approach is to take advantage of the finite speed of propagation and move the boundary far enough so that the solution in the domain of interest is unaffected by it. We then require additional work and storage no more than:

$$\text{Work}_b \propto T^4\epsilon^{-4/p}, \quad \text{Storage}_b \propto T^3\epsilon^{-3/p}. \tag{3.0.18}$$

(These bounds might be improvable by considering a graded mesh.) Clearly, the additional cost is significant for moderate to large T.

Direct Rational Approximations

In collaboration with Alpert and Greengard [2], rational approximations to the transforms of the convolution kernels have been constructed. For the three-dimensional case, the approximation is found to require q poles with

$$q \propto (\ln 1/\epsilon)^2. \tag{3.0.19}$$

This yields:

$$\text{Work}_b \propto T\epsilon^{-4/p} + qT\epsilon^{-3/p}, \quad \text{Storage}_b \propto q\epsilon^{-2/p}. \tag{3.0.20}$$

We note that the dominant work, which is comparable to that of the interior scheme, comes from the spherical harmonic transformations. If a fast spherical harmonic transform were devised, we'd find the boundary work to be completely dominated by the work associated with the interior scheme. For cylindrical boundaries or periodic problems this is the case.

Engquist-Majda-Padé Approximants

In their pioneering two decades old work [5], Engquist and Majda proposed a sequence of local approximate boundary conditions on a flat boundary based on Padé approximants of the symbol of the exact condition. They are easily (ignoring corner problems!) implemented to any order using auxiliary functions. Recently, Xu and I [13] have developed a sequence of conditions including the Padé scheme as a special case and analyzed them for half-space and periodic problems. (See also [10, 11].) Assuming that similar estimates hold for a rectangular domain we find we require q auxiliary functions where:

$$q \propto \max\left(\mathrm{T}, \frac{\ln 1/\epsilon}{\ln \ln 1/\epsilon}\right), \tag{3.0.21}$$

$$\mathrm{Work}_b \propto qT\epsilon^{-3/p}, \quad \mathrm{Storage}_b \propto q\epsilon^{-2/p}. \tag{3.0.22}$$

We note that, as $\epsilon \to 0$ for T fixed the work and storage associated with the boundary condition is certainly dominated by the work associated with the interior scheme, though the large T limit is less favorable.

Grote-Keller Conditions and Generalizations

Recently, Grote and Keller [8, 9] proposed a sequence of temporally local boundary conditions which may be made exact for any number of spherical harmonics. The origin of these may be seen in (2.0.12). Recently, Hariharan and I [12] have improved these somewhat using progressive wave expansions, removing the necessity of a spherical harmonic expansion. Assuming the condition is made exact for all spherical harmonics which can be represented on the mesh we find for both formulations:

$$\mathrm{Work}_b \propto T\epsilon^{-4/p}, \quad \mathrm{Storage}_b \propto \epsilon^{-3/p}. \tag{3.0.23}$$

(The constant is smaller in the new formulation.) Here the work and storage is always comparable to that associated with the interior scheme.

Conditions Based on Kirchoff's Formula

Ting and Miksis [15] proposed a different formulation of exact boundary conditions with the apparently attractive property of involving only a finite time history. These were later implemented by Givoli and Kohen [7]. However, a fast method for the spatial integrations is lacking and we find:

$$\mathrm{Work}_b \propto T\epsilon^{-5/p}, \quad \mathrm{Storage}_b \propto \epsilon^{-3/p}. \tag{3.0.24}$$

Asymptotic Boundary Conditions

Also nearly twenty years ago, Bayliss and Turkel [3] proposed conditions based on progressive wave expansions. As mentioned above, we have found it possible to reformulate the expansion so as to produce exact conditions. Stopping at a finite number of terms, one obtains an error proportional to R^{-m} for some m. Ignoring the dependence of the error constants on T (which is probably not justifiable) we find:

$$\text{Work}_b \propto T\epsilon^{-4/p-3/m}, \quad \text{Storage}_b \propto \epsilon^{-3/p-3/m}, \qquad (3.0.25)$$

which is already unfavorable in comparison with other techniques.

Perfectly Matched Layer

Recently Berenger [4] proposed a new absorbing layer for the Maxwell system. He called it "perfectly matched" because no waves are reflected off the layer interface. We have formulated the PML for the scalar wave equation and expressed it as an equivalent boundary condition at the layer interface. We have not, however, yet completed an error analysis. Moreover, it is known that the PML equations are not strongly well-posed (see [1]), which will effect the error bounds. In the best case, we expect the layer width to scale like $\ln 1/\epsilon$ which implies:

$$\text{Work}_b \propto T\epsilon^{-4/p}(\ln 1/\epsilon)^3, \quad \text{Storage}_b \propto \epsilon^{-3/p}(\ln 1/\epsilon)^3. \qquad (3.0.26)$$

This is somewhat worse than the results for direct approximations to exact conditions, but only by logarithmic factors.

4 Conclusions and Problems

Our results clearly show that efficient, high-accuracy solutions of time-domain scattering problems are possible, and that there are a number of reasonable methods for treating the boundary conditions. We slightly prefer the methods based on direct rational approximations to the convolution kernel, but in the absence of a faster spherical harmonic transformation, conditions based on progressive wave expansions are likely to be as efficient. Moreover, for relatively short time computations with high accuracy requirements, conditions based on Padé approximants may be most efficient, although their analysis for exterior problems is incomplete.

A great challenge to the theory is its extension to variable coefficients and nonlinearities. The only hard results I know of are those due to Halpern and Rauch [14] for low-order Engquist-Majda conditions. These do not say anything about convergence as the order is increased. Many of the other conditions we have discussed have no obvious generalization to variable coefficient problems.

References

[1] S. Abarbanel and D. Gottlieb, A mathematical analysis of the PML method, preprint.

[2] B. Alpert, L. Greengard and T. Hagstrom, in preparation.

[3] A. Bayliss and E. Turkel, Radiation boundary conditions for wave-like equations, *Comm. Pure and Appl. Math.*, 33 (1980), 707-725.

[4] J.-P. Berenger, A perfectly matched layer for the absorption of electromagnetic waves, *J. Comp. Phys.*, 114 (1994), 185-200.

[5] B. Engquist and A. Majda, Absorbing boundary conditions for the numerical simulation of waves, *Math. Comp.*, 31 (1977), 629-651.

[6] B. Engquist and A. Majda, Radiation boundary conditions for acoustic and elastic wave calculations, *Comm. Pure and Appl. Math.*, 32 (1979), 313-357.

[7] D. Givoli and D. Kohen, Nonreflecting boundary conditions based on Kirchoff-type formulae, *J. Comp. Phys.*, 117 (1995), 102-113.

[8] M. Grote and J. Keller, Exact nonreflecting boundary conditions for the time-dependent wave equation, *SIAM J. Appl. Math.* 55 (1995), 280-298.

[9] M. Grote and J. Keller, Nonreflecting boundary conditions for time-dependent scattering, *J. Comp. Phys.*, 127 (1996), 52-80.

[10] T. Hagstrom, On the convergence of local approximations to pseudodifferential operators with applications, *Mathematical and Numerical Aspects of Wave Propagation Phenomena*, G. Cohen, E. Bécache, P. Joly, J. Roberts, eds., SIAM (1995), 474-482.

[11] T. Hagstrom, On high-order radiation boundary conditions, *IMA Volume on Computational Wave Propagation*, B. Engquist and G. Kriegsmann, eds., Springer (1996), 1-21.

[12] T. Hagstrom and S. Hariharan, in preparation.

[13] T. Hagstrom and L. Xu, in preparation.

[14] L. Halpern and J. Rauch, Error analysis for absorbing boundary conditions, *Num. Math.*, 51 (1987), 459-467.

[15] L. Ting and M. Miksis, Exact boundary conditions for scattering problems, *J. Acoust. Soc. Am.*, 80 (1986), 1825-1827.

5 Acknowledgements

Collaborators in the work described herein include B. Alpert, L. Greengard, S.I. Hariharan and L. Xu. We'd also like to acknowledge support from NSF Grant DMS-9600146 and ICOMP, NASA Lewis Research Center, Cleveland, OH.

A VARIATIONAL FORMULATION FOR PARTITIONED EXTERIOR PROBLEMS

Isaac Harari

Department of Mechanics, Materials and Structures
Tel-Aviv University
69978 Ramat Aviv, ISRAEL

1 Exterior Boundary-value Problem of Acoustics

We consider a boundary-value problem related to acoustic radiation and scattering governed by the Helmholtz equation in a d-dimensional unbounded region $\mathcal{R} \subset \mathbb{R}^d$ with internal boundary Γ. The unbounded domain $\mathcal{R}$ is partitioned by a smooth artificial boundary Γ_R into a bounded inner domain Ω^i and its unbounded outer complement Ω^o (Fig. 1).

Following [Hughes, 1995] the solution is decomposed into the inner field u^i and outer field u^o:

$$u = u^i + u^o \tag{1}$$

where $u^i|_{\Omega^o} = 0$ and $u^o|_{\Omega^i} = 0$. Continuity across Γ_R is enforced weakly. We assume that f vanishes in the outer domain.

The variational form of the boundary-value problem is stated in terms of sets of trial solutions, $\mathcal{S}^i$ and $\mathcal{S}^o$ for u^i and u^o, respectively. Dirichlet boundary conditions on Γ must be satisfied by functions in $\mathcal{S}^i$. The treatment of $\mathcal{S}^o$ is less conventional, namely

$$\mathcal{S}^o = \left\{ u^o \,|\, \mathcal{L}u^o = 0 \text{ in } \Omega^o,\ \lim_{r\to\infty} r^{\frac{d-1}{2}} \left(\frac{\partial u^o}{\partial r} - iku^o \right) = 0 \right\} \tag{2}$$

Here $\mathcal{L}u := \Delta u + k^2 u$ is the Helmholtz operator, Δ is the Laplace operator and $k \in \mathbb{C}$ is the wave number, $\operatorname{Im} k \geq 0$; $i = \sqrt{-1}$ is the imaginary unit; and r is the distance from the origin. Thus we seek $u^i \in \mathcal{S}^i$ and $u^o \in \mathcal{S}^o$ that render $\Pi(u)$ stationary, where

$$\Pi(u) = \frac{1}{2}a(u^i, u^i) - L(u^i) + \left(\frac{\partial u^o}{\partial n^o}, u^i - \frac{1}{2}u^o \right)_{\Gamma_R} \tag{3}$$

$$a(w, u) = \int_{\Omega^i} \left(\nabla w \cdot \nabla u - wk^2 u \right) d\Omega \tag{4}$$

$$(w, u)_{\Gamma_R} = \int_{\Gamma_R} w\, u\, d\Gamma \tag{5}$$

T.L. Geers (ed.), IUTAM Symposium on Computational Methods for Unbounded Domains, 187–196.

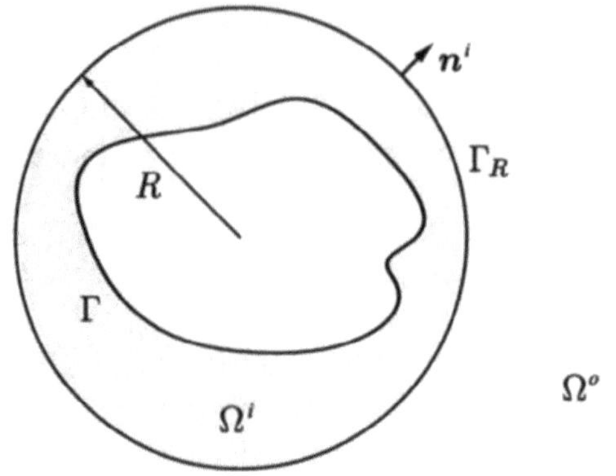

Figure 1: The partitioned domain.

Here $a(\cdot,\cdot)$ and $(\cdot,\cdot)_{\Gamma_R}$ are symmetric bilinear forms which are *not* inner products, and $L(w)$ contains contributions from source terms and natural boundary data. This functional is based on a complex-valued mixed formulation in which continuity across the artificial boundary is enforced by Lagrange multipliers [Harari, Barbone and Montgomery, 1997]. The Euler-Lagrange equations are employed to eliminate the Lagrange multipliers and obtain the form shown here. Continuity is enforced by energy flux-type terms, reminiscent of the radiation condition that is being replaced. In this formulation there is integration only over the bounded domain and the artificial boundary.

The stationary point of the functional (3) is obtained by setting its first variation equal to zero. In partitioned form we have

$$\begin{aligned} a(w^i,u^i) &+ \left(w^i,\frac{\partial u^o}{\partial n^o}\right)_{\Gamma_R} &= L(w^i) \\ \left(\frac{\partial w^o}{\partial n^o},u^i\right)_{\Gamma_R} &- \left(\frac{\partial w^o}{\partial n^o},u^o\right)_{\Gamma_R} &= 0 \end{aligned} \tag{6}$$

Here, $w^i \in \mathcal{V}^i$ and $w^o \in \mathcal{V}^o$ are the arbitrary variations of u^i and u^o, respectively. Homogeneous counterparts of Dirichlet boundary conditions on Γ must be satisfied by functions in $\mathcal{V}^i$. In the outer field $\mathcal{V}^o = \mathcal{S}^o$.

2 Absorbing Boundary Conditions, DtN

Formulations of exterior problems that are amenable to domain-based computation are often derived by introducing an artificial boundary into the problem to bound a computational domain Ω^i. Correct far-field behavior is enforced by specifying non-reflecting (or absorbing) boundary conditions on this boundary. A noteworthy class of non-reflecting boundary conditions is obtained by the DtN method. The method in its general form was developed by [Givoli and Keller, 1989].

We now re-derive the DtN boundary conditions in the framework of our formulation

(6). We denote the homogeneous Dirichlet Green's function in Ω^o by $g(\boldsymbol{x}, \boldsymbol{x}')$, i.e., $\mathcal{L}g = -\delta(\boldsymbol{x} - \boldsymbol{x}')$, $g|_{\Gamma_R} = 0$, and g is outgoing. Consequently,

$$u^o(\boldsymbol{x}) = K(\boldsymbol{x}, \boldsymbol{x}')u^o(\boldsymbol{x}'), \qquad \boldsymbol{x} \in \Omega^o, \boldsymbol{x}' \in \Gamma_R. \tag{7}$$

Here, $Ku^i = \int_{\Gamma_R} \frac{\partial g}{\partial n^o} u^i \, d\Gamma$. Using g as the weighting function in the second of equations (6) together with (7) yields

$$u^o(\boldsymbol{x}) = K(\boldsymbol{x}, \boldsymbol{x}')u^i(\boldsymbol{x}'), \qquad \boldsymbol{x} \in \Omega^o, \boldsymbol{x}' \in \Gamma_R. \tag{8}$$

Equation (8) is a relation for the outer field in terms of the inner field on the interface.

The normal derivative of u^o is required for the first of equations (6)

$$\frac{\partial u^o}{\partial n^o}(\boldsymbol{x}) = M(\boldsymbol{x}, \boldsymbol{x}')u^i(\boldsymbol{x}'). \qquad \boldsymbol{x} \in \Omega^o, \boldsymbol{x}' \in \Gamma_R \tag{9}$$

Here $M = \frac{\partial K}{\partial n^o}$ is the standard DtN map. Substituting this expression into the first equation of (6) yields

$$a(w^i, u^i) + \left(w^i, Mu^i\right)_{\Gamma_R} = L(w^i) \tag{10}$$

Equation (10) is identical to the variational equation for u^i in the computational domain obtained by the Givoli-Keller derivation. It is equivalent to specifying the boundary condition

$$\frac{\partial u^i}{\partial n^i} = -Mu^i, \qquad \text{on } \Gamma_R \tag{11}$$

in the strong form of the inner-field problem, where $\boldsymbol{n}^i$ denotes the outward normal with respect to Ω^i (Fig. 1).

3 Piecewise Outer Field Representations

Formulation (6) is generalized to account for derivative discontinuities across infinite element boundaries, typical of standard infinite element approximations (e.g., [Astley, Macaulay, and Coyette, 1994], [Bettess, 1997], [Burnett, 1994], [Gerdes, and Demkowicz, 1996], and [Zienkiewicz, Bando, Bettess, Emson, and Chiam, 1985]). Therefore, we consider a partition of the outer domain Ω^o into n_{sec} non-overlapping sectors Ω^s with boundaries Γ^s, $s = 1, \ldots, n_{sec}$. For example, Fig. 2 shows the partition in a two-dimensional configuration with a circular interface. We denote the union of sector interiors by

$$\tilde{\Omega}^o = \bigcup_{s=1}^{n_{sec}} \Omega^s. \tag{12}$$

The interior sector boundaries are

$$\Gamma^o_{int} = \left(\bigcup_{s=1}^{n_{sec}} \Gamma^s\right) \setminus \partial\Omega^o, \tag{13}$$

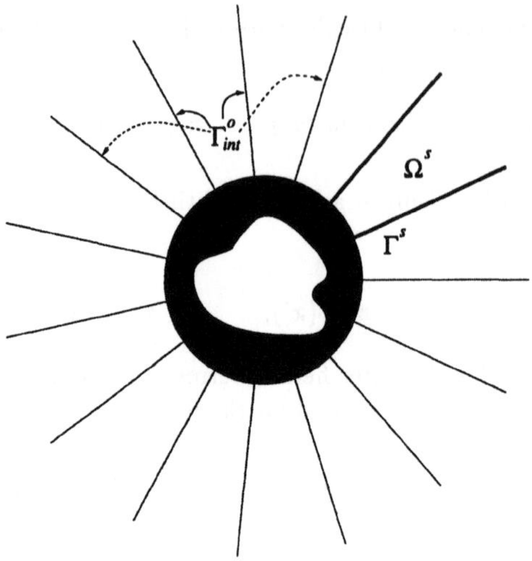

Figure 2: Partition in a two-dimensional configuration with a circular interface.

where $\partial\Omega^o$ is the boundary of Ω^o.

One side of Γ^o_{int} is designated as "+" and the other side is "−." Allowing for discontinuities across Γ^o_{int}, we define the jump across Γ^o_{int} as:

$$\left[w^o \frac{\partial u^o}{\partial n}\right] = \left(w^o \frac{\partial u^o}{\partial n^+}\right)\bigg|_{\boldsymbol{x}^+} + \left(w^o \frac{\partial u^o}{\partial n^-}\right)\bigg|_{\boldsymbol{x}^-} \tag{14}$$

Here $\boldsymbol{x}^+$ and $\boldsymbol{x}^-$ are obtained by approaching Γ^o_{int} from the + and − sides, respectively. Similarly, $\boldsymbol{n}^+$ and $\boldsymbol{n}^-$ are unit normals to Γ^o_{int} pointing out of the + and − sides, respectively. The jump is invariant to the arbitrary + and − designations.

For continuous, piecewise-smooth representations of the outer field the set of trial solutions is modified

$$\mathcal{S}^o = \left\{u^o \,|\, u^o \in C^0(\Omega^o),\ \mathcal{L}u^o = 0 \text{ in } \tilde{\Omega}^o,\ \lim_{r\to\infty} r^{\frac{d-1}{2}}\left(\frac{\partial u^o}{\partial r} - iku^o\right) = 0\right\} \tag{15}$$

and $\mathcal{V}^o = \mathcal{S}^o$. The weak form must now account for possible jumps in derivatives across Γ^o_{int}. Thus we seek $u^i \in \mathcal{S}^i$ and $u^o \in \mathcal{S}^o$ such that

$$\begin{aligned} a(w^i, u^i) \quad &+ \quad \left(w^i, \frac{\partial u^o}{\partial n^o}\right)_{\Gamma_R} &&= L(w^i) \\ \left(\frac{\partial w^o}{\partial n^o}, u^i\right)_{\Gamma_R} &- \left(\frac{\partial w^o}{\partial n^o}, u^o\right)_{\Gamma_R} + \left(w^o, \left[\frac{\partial u^o}{\partial n}\right]\right)_{\Gamma^o_{int}} &&= 0 \end{aligned} \tag{16}$$

The formulation (16) is a generalization of (6) that accounts for possible discontinuities across infinite element boundaries. We may now employ infinite element basis

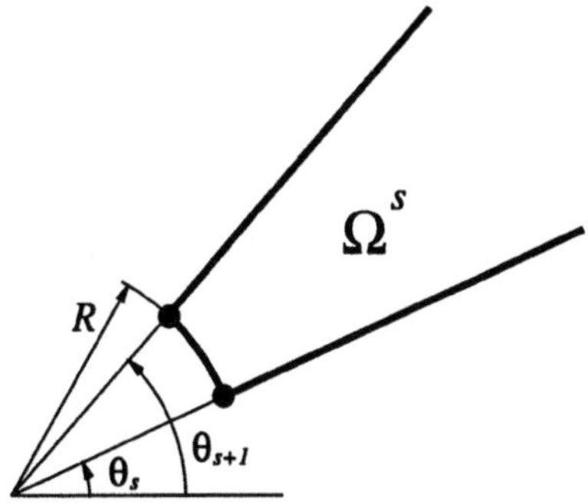

Figure 3: A typical sector in a two-dimensional configuration with a circular interface.

functions that exactly satisfy the Helmholtz equation within each sector and are continuous across infinite element boundaries. In most cases, the circumferential behavior of these functions on the artificial boundary differs from piecewise polynomial variation of finite element shape functions. Weak enforcement of continuity across the artificial boundary is then performed naturally within the framework of our formulation (16). Consequently, the infinite element mesh need not be the restriction of the finite element mesh to the artificial boundary.

To fix ideas consider the two-dimensional configuration with a circular interface shown in Fig. 2. The outer domain Ω^o is partitioned into non-overlapping sectors

$$\Omega^s =]R, \infty[\times]\theta_s, \theta_{s+1}[, \qquad s = 1, \ldots, n_{sec}. \tag{17}$$

Here $\theta_s < \theta_{s+1}$ and $\theta_{n_{sec}+1} = \theta_1 + 2\pi$. A typical sector is shown in Fig. 3. The infinite element approximation is based on interpolation of values on the arc $r = R$, $\theta_s < \theta < \theta_{s+1}$.

4 Linear Infinite Elements

The basic two-noded infinite element is based on the representation

$$u^o \approx N_1 u^o(R, \theta_s) + N_2 u^o(R, \theta_{s+1}), \qquad \theta_s < \theta < \theta_{s+1} \tag{18}$$

where

$$N_1 = \frac{H_0^{(1)}(kr)}{H_0^{(1)}(kR)} \frac{\theta_{s+1} - \theta}{\theta_{s+1} - \theta_s} \qquad \text{and} \qquad N_2 = \frac{H_0^{(1)}(kr)}{H_0^{(1)}(kR)} \frac{\theta - \theta_s}{\theta_{s+1} - \theta_s} \tag{19}$$

are shape functions that satisfy the homogeneous Helmholtz equation within the sector.

The infinite element arrays obtained from the terms $-\left(\frac{\partial w^o}{\partial n^o}, u^o\right)_{\Gamma_R}$ and $\left(w^o, \left[\frac{\partial u^o}{\partial n}\right]\right)_{\Gamma_{int}^o}$ in the second equation in (16) require the evaluation of α_{00}, a special case of a term that arises in higher-order approximation

$$\alpha_{\nu\mu} = \int_R^\infty \frac{H_\nu^{(1)}(kr) H_\mu^{(1)}(kr)}{H_\nu^{(1)}(kR) H_\mu^{(1)}(kR)} \frac{dr}{r} \tag{20}$$

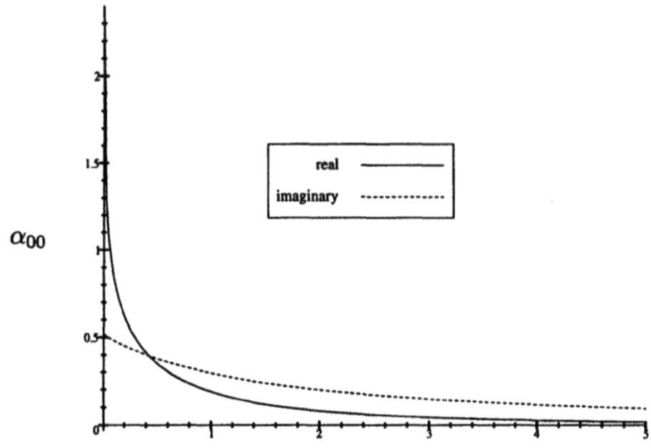

Figure 4: The coefficient required for two-noded infinite elements.

These integrals are evaluated analytically based on properties of solutions of the homogeneous Bessel equation, yielding

$$\alpha_{\nu\mu} = \frac{kR}{\mu^2 - \nu^2} \left(\frac{H_\mu^{(1)\prime}(kR)}{H_\mu^{(1)}(kR)} - \frac{H_\nu^{(1)\prime}(kR)}{H_\nu^{(1)}(kR)} \right) \tag{21}$$

The prime denotes differentiation with respect to the argument. In practice, the unequal-order case is not integrated, it is evaluated by symmetry considerations. For equal-order functions we approximate the limit

$$\alpha_{\mu\mu} = \lim_{\nu \to \mu} \alpha_{\nu\mu} \tag{22}$$

at small values of ν. Numerical experience indicates that this limit converges without difficulty for arithmetic of sufficient precision. The dependence of α_{00} on kR is shown in Fig. 4, indicating the diminishing importance of the jump terms for a given problem, as the location of the artificial boundary is moved outwards.

Numerical results are presented for an exterior problem bounded internally by an infinite circular cylinder of radius a. Soft boundary conditions are specified on the wet surface to represent a pressure-release cylinder. A circular interface Γ_R is located at $R = 2.5\,a$. Finite elements are used in the inner domain Ω^i and infinite elements in the outer domain Ω^o. The mesh of infinite elements coincides with the finite element mesh on the interface (although this is not required by the formulation). The inner domain Ω^i is discretized by 3×36 bilinear quadrilateral finite elements (Fig. 5).

We consider the non-uniform radiation from an infinite circular cylinder with a constant inhomogeneous value on an arc $(-\alpha < \theta < \alpha)$ and vanishing elsewhere, so that there are two points of discontinuity in the boundary data. We select $\alpha = 20°$. Fig. 6(a) shows the real part of the analytical and numerical solutions. The low-amplitude oscillations of the *analytical* solution in the vicinity of the wet surface are merely an artifact of the truncated series representation (60 terms) of the discontinuity in the boundary

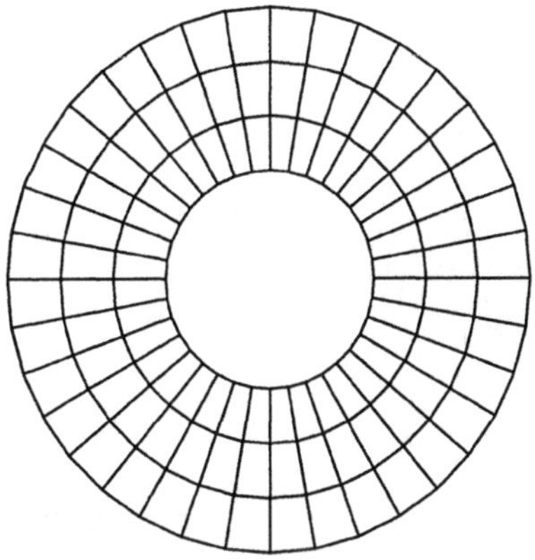

Figure 5: The computational domain exterior to a cylinder of radius a, with a circular interface at $R = 2.5\,a$, discretized by 3×32 linear quadrilateral finite elements.

data, and are not relevant to the validation of the numerical results. The numerical solution captures the essential physics of the problem, without visible reflection from the interface. The error relative to the interpolation of the analytic solution, measured in the energy norm, is

$$\frac{|u^h - u^I|_1}{|u^I|_1} = 0.52\% \tag{23}$$

Fig. 6(b) depicts the solutions along the interface. The performance of the infinite elements is quite good.

5 High-order Infinite Elements

Several extensions of this continuous, piecewise smooth representation to higher-order approximation are possible. The basic interpolation in (18) can be enriched by a bubble $H^{(1)}_{\nu_1}(kr)\sin(\nu_1(\theta - \theta_s))$ where $\nu_n = n\pi/(\theta_{s+1} - \theta_s)$. The added coefficient is a generalized degree of freedom rather than a nodal value. The matrix equations are statically condensed back to the size of the original two-noded element so that no additional computational effort is required in the solution. Again considering the non-uniform radiation problem, Fig. 7(a) shows the real part of the analytical and enriched numerical solutions. The error relative

$$\frac{|u^h - u^I|_1}{|u^I|_1} = 0.49\% \tag{24}$$

shows improvement over the linear infinite element. Fig. 7(b) depicts the solutions along the interface. The improvement over the linear infinite elements is apparent.

Alternatively, the outer field may be based on a truncation of the series representation

$$u^o = {\sum_{n=0}^{\infty}}' H^{(1)}_{\nu_n}(kr)\left(a_n \cos(\nu_n(\theta - \theta_s)) + b_n \sin(\nu_n(\theta - \theta_s))\right) \tag{25}$$

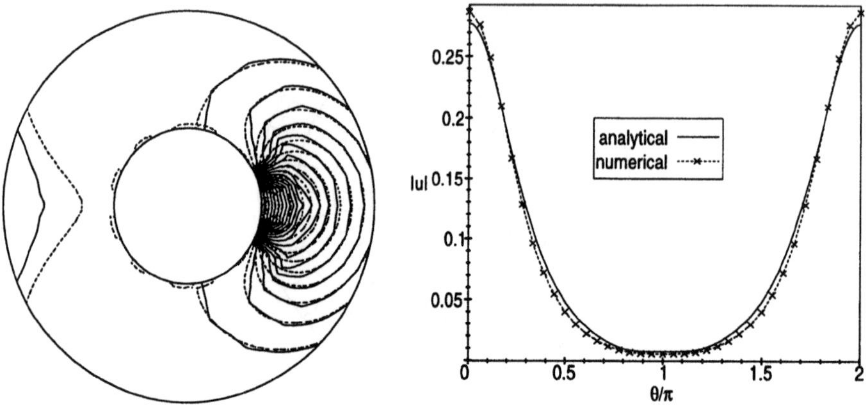

Figure 6: Radiation from a sector of a cylinder, $ka = 1$, ˜13 points per wavelength, linear infinite elemnts; (a) dashed contours denote analytical solution, solid contours denote numerical solution; (b) along the interface $R = 2.5\,a$.

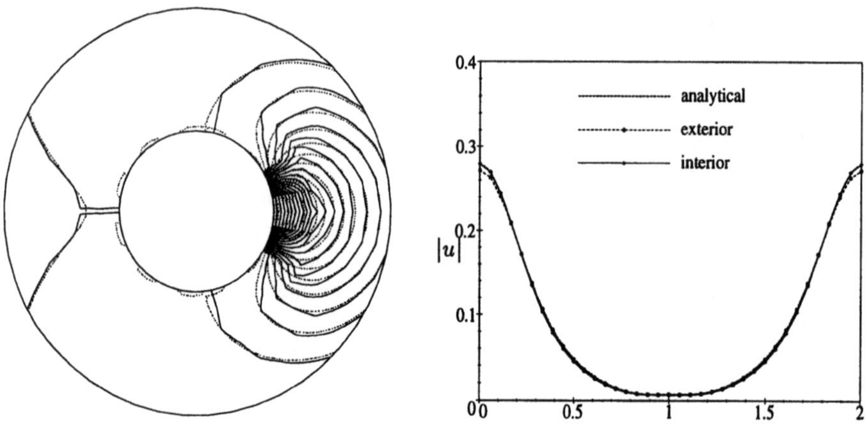

Figure 7: Radiation from a sector of a cylinder, $ka = 1$, ˜13 points per wavelength, sine-enriched linear infinite elements (a) dashed contours denote analytical solution, solid contours denote numerical solution; (b) along the interface $R = 2.5\,a$.

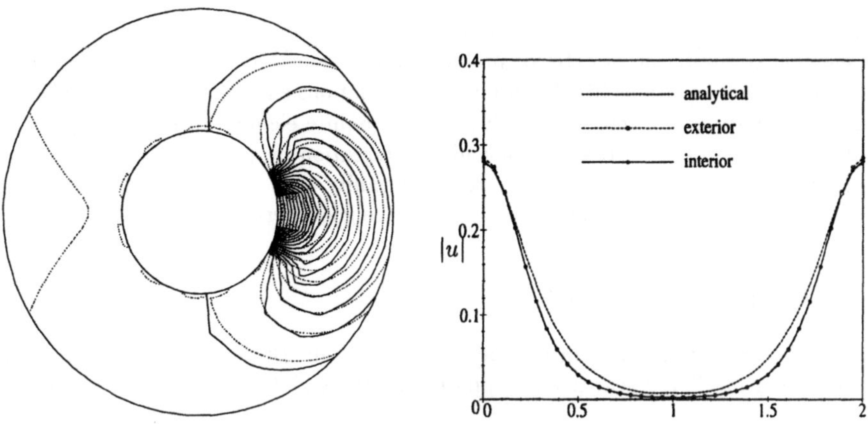

Figure 8: Radiation from a sector of a cylinder, $ka = 1$, ~13 points per wavelength, trigonometric infinite elements (a) dashed contours denote analytical solution, solid contours denote numerical solution; (b) along the interface $R = 2.5\,a$.

The prime on the sum indicates that the first term is halved. In particular, a three-degree-of-freedom element combines the interpolatory shape functions

$$N_1 = \frac{1}{2}\left(\frac{H_0^{(1)}(kr)}{H_0^{(1)}(kR)} + \frac{H_{\nu_1}^{(1)}(kr)}{H_{\nu_1}^{(1)}(kR)}\cos\left(\nu_1(\theta-\theta_s)\right)\right) \tag{26}$$

$$N_2 = \frac{1}{2}\left(\frac{H_0^{(1)}(kr)}{H_0^{(1)}(kR)} - \frac{H_{\nu_1}^{(1)}(kr)}{H_{\nu_1}^{(1)}(kR)}\cos\left(\nu_1(\theta-\theta_s)\right)\right) \tag{27}$$

with the bubble $H_{\nu_1}^{(1)}(kr)\sin\left(\nu_1(\theta-\theta_s)\right)$, again associated with a generalized degree of freedom. Due to the orthogonality of these functions, terms multiplying equal-order radial integrals $\alpha_{\mu\mu}$ vanish, so that no radial integration whatsoever is required. Again, static condensation of internal degrees of freedom is employed. Considering the non-uniform radiation problem, as before, Fig. 8(a) shows the real part of the analytical and numerical solutions. The error relative

$$\frac{|u^h - u^I|_1}{|u^I|_1} = 0.74\% \tag{28}$$

is a degradation from the linear infinite element. Fig. 8(b) depicts the solutions along the interface.

6 Conclusions

A general framework for developing finite element formulations for solving exterior problems is presented in this work, with particular application to time-harmonic acoustics.

By selecting specific representations of the outer field, several methods underlying finite element computation of exterior problems are obtained as specific instances of the mixed variational equations (6). Examples of formulations employing DtN non-reflecting boundary conditions are reproduced. Employing infinite element interpolations for the outer field yields methods that are devoid of integration over the unbounded domain, simplifying the task of discretization. Strong continuity with finite element interpolation is not a prerequisite. Numerical results validate the good performance of this scheme.

7 Acknowledgment

Parts of this work were done in collaboration with Paul Barbone, Rami Shalom, and Michael Slavutin. This research was supported by the U.S. Office of Naval Research.

8 References

Astley, R.J., Macaulay, G.J., and Coyette, J.-P., 1994, "Mapped Wave Envelope Elements for Acoustical Radiation and Scattering," *J. Sound Vib.*, Vol. 170, pp. 97–118.

Bettess, P., 1977, "Infinite Elements," *Int. J. Numer. Methods Eng.*, Vol. 11, pp. 53–64.

Burnett, D.S., 1994, "A Three-dimensional Acoustic Infinite Element Based on a Prolate Spheroidal Multipole Expansion," *J. Acoust. Soc. Am.*, Vol. 96, pp. 2798–2816.

Gerdes, K., and Demkowicz, L., 1996, "Solution of 3D-Laplace and Helmholtz Equations in Exterior Domains Using *hp*-infinite Elements," *Comput. Methods Appl. Mech. Eng.*, Vol. 137, pp. 239–273.

Givoli, D., and Keller, J.B., 1989, "A Finite Element Method for Large Domains," *Comput. Methods Appl. Mech. Eng.*, Vol. 76, pp. 41–66.

Harari, I., Barbone, P.E., and Montgomery, J.M., 1997, "Finite Element Formulations for Exterior Domains: Application to Hybrid Methods, Non-reflecting Boundary Conditions and Infinite Elements," *Int. J. Numer. Methods Eng.*, Vol. 40, In press.

Hughes, T.J.R.H., 1995, "Multiscale Phenomena: Green's Functions, the Dirichlet-to-Neumann Formulation, Subgrid Scale Models, Bubbles, and the Origins of Stabilized Methods," *Comput. Methods Appl. Mech. Eng.*, Vol. 127, pp. 387–401.

Zienkiewicz, O.C., Bando, K., Bettess, P., Emson, C., and Chiam, T.C., 1985, "Mapped Infinite Elements for Exterior Wave Problems," *Int. J. Numer. Methods Eng.*, Vol. 21, pp. 1229–1251.

A SYSTEMATIC APPROACH FOR CONSTRUCTING ASYMPTOTIC BOUNDARY CONDITIONS FOR WAVE-LIKE EQUATIONS

S. I. Hariharan
Department of Mathematical Sciences
The University of Akron
Akron, OH, USA
and National Science Foundation

Thomas Hagstrom
Department of Mathematics and Statistics
The University of New Mexico
Albuquerque, NM, USA

1 Introduction

In this work, we investigate some computational and theoretical issues for a class of problems in compressible flows. The issues under consideration are mainly un-addressed in the literature, except for some relevant ideas. Our work is driven by two classes of problems which we propose as benchmark problems. They are

1. Time-dependent transonic flows in open domains.
2. Time-dependent three-dimensional spherically non-symmetric flows in open domains. process.

This paper addresses possible approaches to these problems. These approaches include combination of asymptotic and numerical treatments to examine solutions as well as simulations. These are computationally intensive problems. Theoretical issues related to computations, namely useful estimates that relate to anticipated solution behavior, will also be investigated. We motivate the treatment by two model problems: the two-dimensional wave equation and a sound diffraction problem governed by the non-homogeneous linearized Euler equations. In both cases asymptotic decay of errors are obtained, and suggest possible implementations to other classes of problems.

The problems considered are in areas in which the authors have made at least partial progress, however, the issues considered are not fully understood. These are continuations of the work which appeared in [1], [5], [7] and [8]. Even though these problems fall into the category of well studied compressible flows, in multi-dimensions, both the theory as well as *time dependent simulations* lack details, despite the advanced state of numerical methods for hyperbolic conservation laws.

T.L. Geers (ed.), IUTAM Symposium on Computational Methods for Unbounded Domains, 197–206.

2 Time Dependent Transonic Flows

This part of the work deals with a prototype problem one can address in the context of challenging multi-dimensional hyperbolic problem. The specific problem consists of solving the Euler equations past an impulsively started cylinder. This is a rarely studied problem in the literature due to the lack of proper boundary conditions. Ten years ago there was a workshop which took place in the Netherlands to study this problem [9] . It still remains unresolved, due to the lack of systematic procedures in prescribing accurate boundary conditions. A thorough procedure for prescribing this condition will lead to the solution of other classes of problems governed by hyperbolic equations, such as Maxwell's equations. Early results of this work for shorter times are reported in [4]. We are close to bringing this subject to a conclusion. Another contender for the solution of the problem is the popular perfectly matched layer (PML) method. However, it behaves rather non uniformly and requires very large layers for practical applications (see [6]). In addition, private communications with S. S. Abarbanel (Tel Aviv University) indicate that the PML methods are only marginally stable.

The problem is described below:

PROBLEM: Solve for $\mathbf{U} = (m_r, m_\theta, \rho, e_t)^T$, where $\mathbf{U}$ satisfies

$$\frac{\partial \mathbf{U}}{\partial t} + \frac{\partial \mathbf{F}(\mathbf{U})}{\partial r} + \frac{1}{r}\frac{\partial \mathbf{G}(\mathbf{U})}{\partial \theta} + \frac{1}{r}\mathbf{H}(\mathbf{U}) = \mathbf{0}$$

$$\mathbf{F}(\mathbf{U}) = \left[p + \frac{m_r^2}{\rho}, \frac{m_r m_\theta}{\rho}, m_r, \frac{m_r(e_t+p)}{\rho}\right]^T$$

$$\mathbf{G}(\mathbf{U}) = \left[\frac{m_r m_\theta}{\rho}, p + \frac{m_\theta^2}{\rho}, m_\theta, \frac{m_\theta(e_t+p)}{\rho}\right]^T$$

$$\mathbf{H}(\mathbf{U}) = \left[\frac{m_r^2 - m_\theta^2}{\rho}, \frac{2m_r m_\theta}{\rho}, m_r, \frac{m_r(e_t+p)}{\rho}\right]^T$$

$$p = (\gamma - 1)\left(e_t - \frac{1}{2}(m_r^2 + m_\theta^2)\rho\right)$$

SUBJECT TO:

$$m_r = 0 \quad \text{at } r = 1$$

$$\mathbf{U} \to (\rho_\infty U_\infty \cos\theta, -\rho_\infty U_\infty \sin\theta, \rho_\infty, e_\infty) \quad \text{as } r \to \infty.$$

This nonlinear problem contains the following features:

- Time-varying transonic region
- Time-dependent shock waves

- Anisotropy in the wave propagation

These features are worth investigation by themselves. A work that is closest to this problem has been investigated by Abarbanel et. al. [18] in the context of low Mach number, viscous flows past a cylinder. They used characteristic boundary conditions, which have fixed *a priori* errors both in space and time, and the procedure is easy to implement if one is only interested in short term solutions. In contrast, the current work consists of inviscid flows (numerically more difficult), larger subsonic Mach numbers (no satisfactory results in the literature), and the long time behavior (both theory and numerics are nonexistent in the literature). It should be noted that the numerical treatment of time dependent shock waves are just under way in the literature (Arora and Roe [11], LeVeque [10]).

Treatment Using Progressive Wave Expansions

In our long term collaboration, we have investigated boundary conditions for compressible flows governed by Euler equations for both linear and nonlinear problems, and isotropic as well as anisotropic cases. The emerging interests in electromagnetic computations give sufficient opportunities to complete a theory and numerical procedures for these problems using progressive wave expansions, which is one of the ingredients of our past work. A key feature in the derivations is that the resulting family of boundary conditions involve a single derivative in the direction normal to the open boundary. The hierarchy of these conditions contain even order tangential derivatives. Here, the focus will be to construct progressive wave expansions for general hyperbolic systems and to use them to derive a class of boundary conditions that involve only time derivatives or time and tangential derivatives. Preliminary results are found in [1].

Here we provide a brief description of the development pertinent to two-dimensional wave equations. The three-dimensional form of progressive wave expansions are established in [3], but in the context of constructing far field solutions. We established in [1], that for two-dimensional wave equation, the series representing outgoing solutions is of the form:

$$u = \sum_{n=0}^{\infty} a_n(\theta) f^n(r,t),$$

$$f^n = \sum_{k=0}^{\infty} r^{-k-1/2} f_k^n(t-r),$$

where,

$$(f_{k+1}^n)' = -\frac{(k+\frac{1}{2})^2 - n^2}{2(k+1)} f_k^n, \quad a_n'' = -n^2 a_n.$$

Using this expansion and the differential equations satisfied by a_n and f_k^n, we obtain a sequence of conditions that involve time derivatives and even order tangential derivatives using a sequence of auxiliary functions w_j as follows:

$$\frac{\partial u}{\partial t} + \frac{\partial u}{\partial r} + \frac{1}{2r} u = w_1,$$

$$\frac{\partial w_1}{\partial t} + \frac{1}{r} w_1 = \frac{1}{8r^2} u + \frac{1}{2r^2} \frac{\partial^2 u}{\partial \theta^2} + w_2,$$

$$\frac{\partial w_j}{\partial t} + \frac{j}{r} w_j = \frac{(j-\frac{1}{2})^2}{4r^2} w_{j-1} + \frac{1}{4r^2}\frac{\partial^2 w_{j-1}}{\partial \theta^2} + w_{j+1}, \quad j = 2, \ldots$$

$$w_j = O(r^{-2j})u = O(r^{-2j+(1/2)}).$$

Using p auxiliary functions involves setting $w_{p+1} = 0$, which leads to an expected accuracy of $O(r^{-2p-2})$. There is a corresponding estimate in time as well. It should be noted that setting $w_1 = 0$, one obtain a basic condition that is popular in the field of computational wave propagation. These conditions are numerically approximated to obtain excellent results with errors far less than 1% with increasing order of conditions (the results are forthcoming in [2]. This procedure can be extended to hyperbolic systems, namely, to the linearized Euler equations. The authors suggest to use this methodology to a) obtain increasingly higher order conditions in both time and space, b) solve the "impulsively started cylinder" problem with these conditions.

We have derived analogous conditions for the Maxwell's equations as well as for the Euler equations. A sample result for the linearized Euler equations is discussed below.

3 Linearized Euler Equations and a Model problem

In this section we begin with the linearized Euler equations with mean velocity convection.

$$\frac{\partial p}{\partial t} + u_0\frac{\partial p}{\partial x} + v_0\frac{\partial p}{\partial y} + \frac{\partial u}{\partial x} + \frac{\partial v}{\partial y} = 0$$

$$\frac{\partial u}{\partial t} + u_0\frac{\partial u}{\partial x} + v_0\frac{\partial u}{\partial y} + \frac{\partial p}{\partial x} = 0$$

$$\frac{\partial v}{\partial t} + u_0\frac{\partial v}{\partial x} + v_0\frac{\partial v}{\partial y} + \frac{\partial p}{\partial y} = 0$$

Here $(u_0(x,y), v_0(x,y))$ is the steady mean flow velocity, $(u(x,y,t), v(x,y,t))$ is the perturbation velocity and $p(x,y,t)$ is the perturbation pressure. $g1$ and $g2$ model Gaussian spatial momentum source (quadrupole sound source) with time dependent amplitude $A(t)$.

The goal is to compute unsteady flow fields excited over the given mean flow field. A candidate problem considered here consists of an incompressible inviscid mean flow with a vortex dipole located at the origin. This is known as the *Lamb Vortex* in the literature. The mean flow is given by $u_0 = -M\frac{\partial\psi}{\partial y}$ and $v_0 = M\frac{\partial\psi}{\partial x}$ where

$$\psi = \left\{ \begin{array}{cc} J_1(kr)\sin\theta & r < 1 \\ \frac{k}{2}J_1'(k)(r - \frac{1}{r})\sin\theta & r > 1 \end{array} \right\}$$

and k is the first zero of $J_1'(z) = 0$. In the far field $u_0 = M + O(\frac{1}{r^2})$ and $v_0 = O(\frac{1}{r^2})$, where M is the speed at infinity. For the specific problem considered above the acoustic fields were excited by a source whose components have the form:

$$\mathbf{u}_s = A(t)\sin(\omega t)\begin{pmatrix} 0 \\ g_1(x,y) \\ g_2(x,y) \end{pmatrix}$$

where $g1$ and $g2$ model Gaussian spatial momentum source and $A(t)$ is an amplitude function which both oscillates sinusoidally and decays algebraically in time.

Test Case $\frac{MaxError}{MaxSolution}$

$$\delta x = \delta y = 0.5 \quad M = 0.5 \quad (x_0, y_0) = (-1.5, 0)$$

$$T = 60 \text{ periods}$$

$$A(t) = 150 \sin(2\pi t)$$

Boundary Condition

L	1	2	3
10	6.7%	10.1%	4.7%
15	6.3%	5.6%	2.0%
20	6.3%	3.9%	1.2%

Graphical results for the decay of errors are found in Figure 1. The first sub-figure indicates the solution at time $t = 80$ and the other three sub-figures show the behavior of the errors for different boundary conditions. In this figure (as well as in the table above), the first condition is the standard characteristic boundary condition analogous to the condition proposed in [18], first asymptotic boundary condition, and the second asymptotic boundary condition respectively. Note there is no decay of the error with increasing domain size for the characteristic condition.

4 Pulsating, Squishing and Oscillating Sphere in a Gas

The second problem that we wish to investigate is a three-dimensional nonlinear wave propagation problem, which describes the acoustic field in an inviscid, isentropic fluid generated by a solid body. This problem is of interest in studying the sound wave structure in under-water acoustics, aerodynamics, and astrophysics. This problem has been analyzed by asymptotic methods in the literature (Crow [12], Whitham [13] and more recently by Geer and Pope [14]) in several limiting asymptotic regimes. In particular, Geer and Pope considered three cases for low Mach numbers: pulsating, oscillating and squishing spherical bodies. The problem is as follows: Solve for ρ, z, m and q, where these quantities are density, radial momentum, angular momentum and the azimuthal momentum respectively.

$$\frac{\partial \rho}{\partial t} + \frac{\partial z}{\partial r} + \frac{1}{r}\frac{\partial m}{\partial \theta} + \frac{1}{r \sin\theta}\frac{\partial q}{\partial \phi} = g_1,$$

$$\frac{\partial z}{\partial t} + \frac{\partial}{\partial r}\left(\frac{z^2}{\rho} + f(\rho)\right) + \frac{1}{r}\frac{\partial}{\partial \theta}\left(\frac{mz}{\rho}\right) + \frac{1}{r\sin\theta}\frac{\partial}{\partial \phi}\left(\frac{qz}{\rho}\right) = g_2,$$

$$\frac{\partial m}{\partial t} + \frac{\partial}{\partial r}\left(\frac{mz}{\rho}\right) + \frac{1}{r}\frac{\partial}{\partial \theta}\left(\frac{m^2}{\rho} + f(\rho)\right) + \frac{1}{r\sin\theta}\frac{\partial}{\partial \phi}\left(\frac{mq}{\rho}\right) = g_3,$$

$$\frac{\partial q}{\partial t} + \frac{\partial}{\partial r}\left(\frac{qz}{\rho}\right) + \frac{1}{r}\frac{\partial}{\partial \theta}\left(\frac{qm}{\rho}\right) + \frac{1}{r\sin\theta}\frac{\partial}{\partial \phi}\left(\frac{q^2}{\rho} + f(\rho)\right) = g_4,$$

where, $g_1 = -\frac{2z}{r} - \frac{m\cot\theta}{r},$, $g_2 = -\frac{2z^2 - m^2 - q^2}{\rho r} - \frac{mz\cot\theta}{\rho r}$, $g_3 = -\frac{3mz}{\rho r} - \frac{\cot\theta}{\rho r}(m^2 - q^2)$, $g_4 = -\frac{3qz}{\rho r} - \frac{2qm\cot\theta}{\rho r}$.
and where the physical boundary conditions are:

$u(a, \theta, \phi) = M \sin\omega t$ - pulsating sphere

$u(a, \theta, \phi) = M \sin\omega t \cos\theta$ - oscillating sphere

$u(a, \theta, \phi) = M \sin\omega t \sin^2\theta \sin 2\phi$ - squishing sphere.

Here, M is the Mach number which arises in the nondimensionalization of the problem. It turns out this is a very computationally challenging problem. Hariharan and Johnson considered the first case of a pulsating sphere [8]. The results matched the asymptotic theory and a complete simulation was carried out. Upon the success of this case, the the authors in collaboration with David Thompson considered the other two cases. One has cylindrical symmetry and the other has no symmetry. In these problems, the difficulty is not the boundary condition on the artificial boundary (accurate nonlinear boundary conditions for this problem are derived by the authors in [19]), but rather on the coordinate transformation and resulting singularities. In particular, when the Euler equations are cast into spherical coordinates, there are singularities on the poles (at $\theta = 0$, and at $\theta = \pi$). In our several attempts, even though we were able to treat these singularities by a variety of methods, such as cutting off a small slice of the domain at those singular points to local Taylor approximations of the solutions, we saw instabilities evolving in long time. Typical simulation is given in Figure 2, sub-figures 1, 2 and 3 which describe all three cases. The propagation of singularities is seen in the later two cases mainly at (at $\theta = 0$, and at $\theta = \pi$). For this sample simulation, we used 33 points in the r direction and 18 and and 36 points in the θ and ϕ directions respectively.

5 Conclusions

As far as boundary conditions for the open domain problems are concerned, specifically for istropic linear problems such as second order wave equations and Maxwell's equations, we have a solid development of *easily implementable and higher order* classes of boundary boundary conditions. For nonisotropic problems such as linear acoustics the same procedure applies, but the resulting conditions do not share the mathematical easiness in developing higher order conditions as in the case of isotropic equations, but in principle are derivable to arbitrary orders. For nonlinear isotropic problems, nonlinear boundary conditions are available. However, singularities resulting from coordinate transformations are still a computational challenge as reflected by the squishing and oscillating sphere in a gas problem. What is of interest is the second class of problems in the effect of coordinate mappings and resulting singularities. In particular, spherical geometry poses challenges and is a recent topic of investigation (see [15], [16] and [17]).

Acknowledgement: S. I. Hariharan was supported by a NASA grant NCC - 3 - 283 and by NSF grant No. DMS-9528105, and T. Hagstrom was supported in part by ICOMP, NASA Lewis Research Center, Cleveland, OH and NSF Grant No. DMS-9600146.

Bibliography

1. T. Hagstrom, S. I. Hariharan, "Progressive Wave Expansions and Open Boundary Problems", With T. Hagstrom, *IMA Volumes in Mathematics and its Applications - Computational Wave Propagation*, Vol. 86, 1996, Eds. B. Engquist and G. A. Kriegsmann, Springer Verlag, pp. 23-43.

2. T. Hagstrom, S. I. Hariharan and D. Watkins, "Easily Implementable Higher Order Conditions for Wave-Like Problems" (in preparation)

3. F.G. Friedlander, "On the Radiation Field of Pulse Solutions of the Wave Equation", *Proc. Roy. Soc.* A, 279, (1964), pp. 386-394.

4. S. I. Hariharan and T. Hagstrom, "Far Field Expansion for Anisotropic Wave Equations," *Proceedings of the Second IMACS Symposium on Computational Acoustics*, Vol. 2, North Holland, Eds. D. Lee, A. Cakmak and R. Vichnevetsky, (1990), pp. 283-294.

5. S. I. Hariharan, "Long Time Behavior of Unsteady Flow Computations," *Unsteady Aerodynamics, Aeroacoustics, and Aeroelasticity of Turbomachines and Propellers*, Springer Verlag, Ed. H. M. Atassi, (1993), pp. 73-90.

6. J. De Moerloose and M. Stuchly," Behavior of Berenger's ABC for evanescent Waves", *IEEE Microwave and Guided Wave Lett.*, Vol. 5, No. 10, 1995, pp.344.

7. T. Hagstrom, S. I. Hariharan and R. C. MacCamy, "On the Accurate Long-Time Solution of the Wave Equation in Exterior Domains: Asymptotic Expansions and

Corrected Boundary Conditions", *Mathematics of Computation*, Vol. 63, No. 208, pp. 507-539 (1994). Also a supplement to this article, pp. s7-s10.

8. S. I. Hariharan and D. K. Johnson, "A Framework for Evaluating Boundary Conditions", *Journal of Computational Acoustics*, Vol. 3, No. 3, 1995, pp. 241-259.

9. "Numerical simulation of Compressible Euler Flows", *Proceedings of the GAMM Workshop*, Rocquencourt, June 10-13, 1986, Eds. A. Dervieux, B. van Leer, J. Periaux and A. Rizzi.

10. R. J. LeVeque, "Wave Propagation Algorithm for Multidimensional Hyperbolic Systems", *Journal of Computational Physics*, Vol. 131, 1997, pp. 327-353.

11. M. Arora and P. L. Roe, "On Postshock Oscillations Due to Shock Capturing Schemes in Unsteady Flows", Vol. 130, 1997, pp. 25-40.

12. S. C. Crow, "Aerodynamic Sound Emission as a Singular Perturbation Problem", *Studies in Applied Mathematics*, Vol. 49, pp. 21-44.

13. G. B. Whitham, "Linear and Nonlinear Waves", !974, Wiley, New York.

14. J. F. Geer and D. S. Pope, "A multiple Scales Approach to Sound Generation by Vibrating Bodies", ICASE Report No. 92-73 (1992).

15. B. Fornberg, "A pseudo-spectral Approach for Polar and Spherical Geometries, *SIAM Journal of Scientific and Statistical Computing*, Vol. 16, No. 5, 1995, pp. 1071-1081.

16. A. Tomboulides, "Direct and Large-Eddy Simulation of Wake Flows: Flow Past a Sphere", Ph.D. thesis, Princeton University (1993).

17. A. Tomboulides, S. A. Orszag and G. E. Karnikadis, "Direct and Large-Eddy Simulation of Axisymmetric Wakes", AIAA 93-0546 (1993).

18. S. S. Abarbanel, W. S. Don, D. Gottlieb, D. H. Rudy and J. C. Townsend, "Secondary Frequencies in the Wake of a Circular Cylinder with Vortex Shedding", *Journal of Fluid Mechanics*, Vol. 225, 1991, pp. 557-574.

19. T. Hagstrom and S. I. Hariharan, "Accurate Boundary Conditions for Exterior Problems in Gas Dynamics", *Mathematics of Computation*, Vol. 51, No. 184, (1988), pp. 581-597.

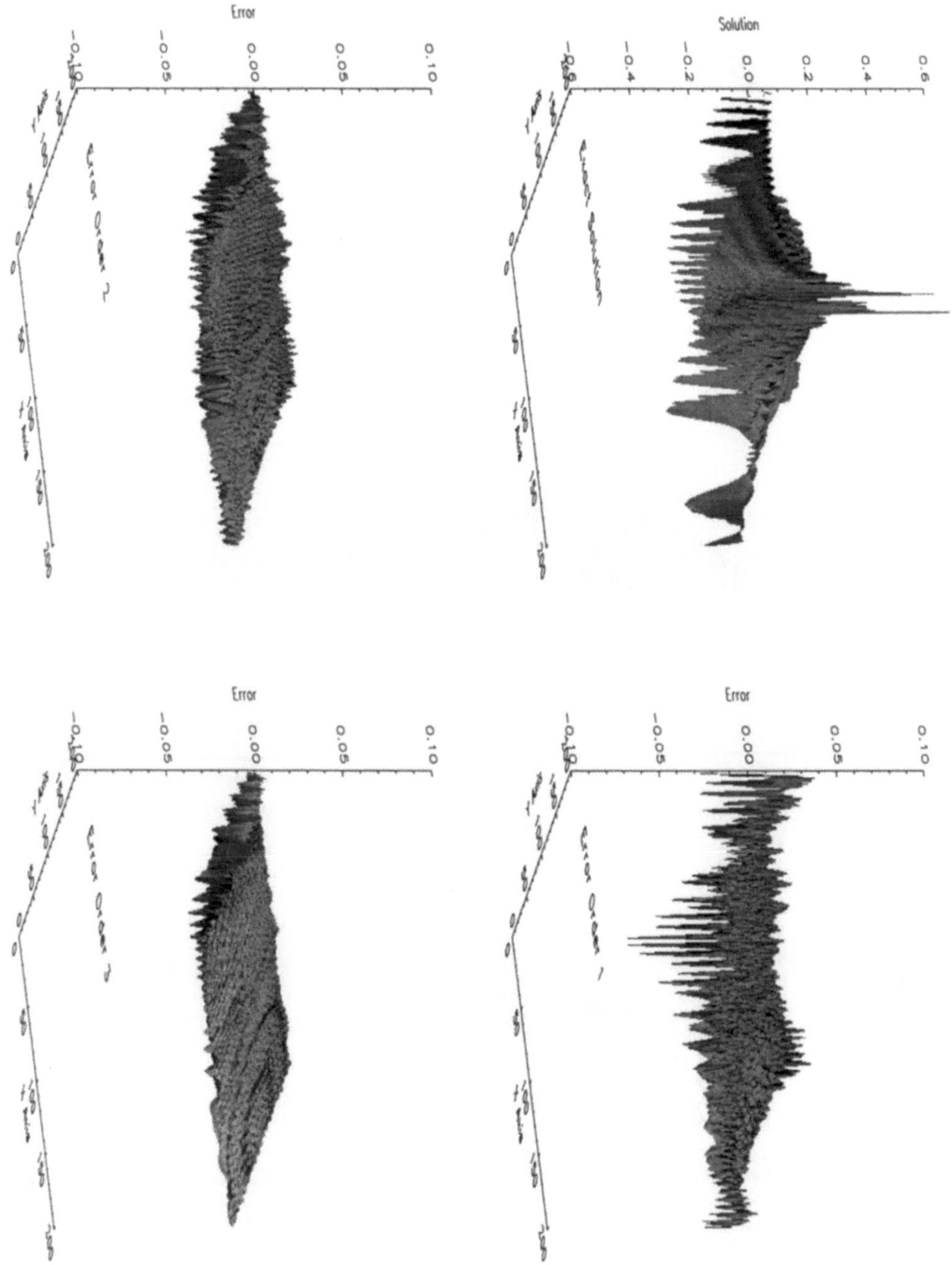

Solution
Exact Solution
Error
Error Order 1
Error
Error Order 2
Error
Error Order 3

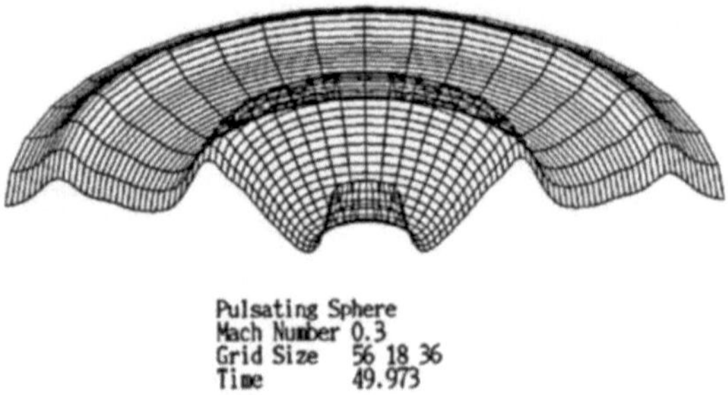
Pulsating Sphere
Mach Number 0.3
Grid Size 56 18 36
Time 49.973

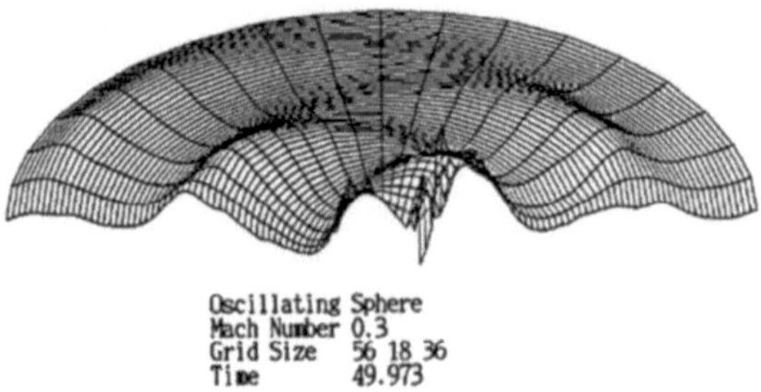
Oscillating Sphere
Mach Number 0.3
Grid Size 56 18 36
Time 49.973

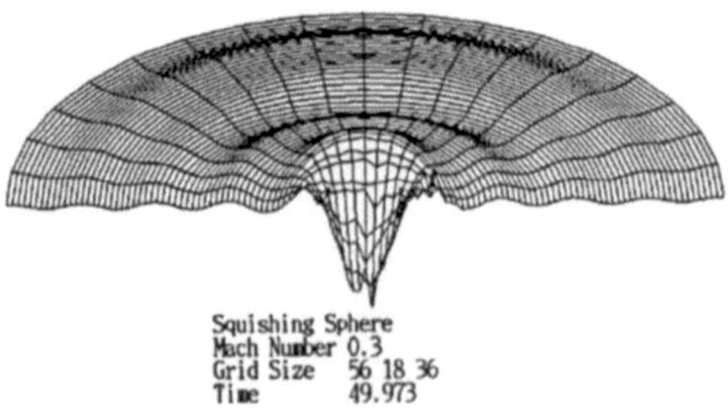
Squishing Sphere
Mach Number 0.3
Grid Size 56 18 36
Time 49.973

EXPERIENCE WITH PML BOUNDARY CONDITIONS IN FLUID-FLOW COMPUTATIONS

M. Ehtesham Hayder
Institute for Computer Applications in Science and Engineering
NASA Langley Research Center, Hampton, VA 23681

Harold L. Atkins
NASA Langley Research Center, Hampton, VA 23681

Abstract

We review our experiences with absorbing layer boundary conditions in fluid flow. The absorbing layer is constructed by splitting the temporal derivative and applying independent damping in each split equation. This methodology follows the basic principle outlined by Berenger for the solution of Maxwell's equations and is known as the Perfectly Matched Layer (PML) boundary condition. We consider several physical problems and numerical algorithms. The results indicate that the PML boundary condition efficiently minimize reflections and are suitable for aeroacoustic computations.

1. Introduction

The proper treatment of computational boundaries is crucial in the numerical solution of equations that govern any physical problem. Reviews and evaluations of many boundary treatments may be found in literature [see for example, Givoli (1991), Hayder and Turkel (1995)]. Numerical boundary conditions based on the characteristics of linearized governing equations and the asymptotic solutions at the far field have been widely used. However, such boundary conditions are not satisfactory if the outflow is nonlinear or involves multi-directional waves. An alternative approach of a buffer zone abutting the computational boundary, which absorbs the incident waves, has been proposed in such situations. In the buffer zone, the governing equations are modified to preclude or dampen upstream propagation of waves into the domain of computation. However, the buffer zone solutions are not physical. Various types of buffer zone techniques have been used in flow simulations. For example, Colonius et al.(1993) used buffer zones in which the solutions were filtered. In a different approach, governing equations are modified to change the dispersion relation

T.L. Geers (ed.), IUTAM Symposium on Computational Methods for Unbounded Domains, 207–216.

for the equations near the boundary such that no wave reflects from the outer boundary of the buffer zone [see for example, Streett and Macaraeg (1989), Ta'asan and Nark (1995)]. Recently, Berenger(1994) proposed a very effective PML technique for the Maxwell equations. In this approach, solution variables inside the layer are split into subcomponents and outgoing waves are absorbed almost perfectly. Following Berenger, Hu (1996a) developed an analogous technique for the linearized Euler equations, and provided analytical results for the case of uniform flow. We have applied this splitting principle in various fluid flow and aeroacoustic computations [Hayder et al. 1996, 1997; Atkins 1997]. Here we present a summary of our experiences with PML boundary condition in computations of various physical problems.

2. Numerical Models

2.1 Shock-vortex Interaction

Hayder et al. (1996, 1997) used the numerical model of Erlebacher et al. (1997) to examine the effectiveness of PML boundary condition for the problem of shock-vortex interactions governed by the compressible Euler equations. The numerical methodology consisted of a sixth-order compact-difference discretization for spatial derivatives, and a fourth-order Runge-Kutta scheme for temporal derivatives. The shock formed a computational boundary and was fitted to preclude Gibb's phenomena. This physical problem provides mechanisms of turbulence alternation caused by interaction with a shock wave. This complex problem is important as it relates to noise generation in many practical flows. Hayder et al. (1996, 1997) presented results for the fully nonlinear compressible Euler equations. The time dependent Rankine-Hugoniot conditions were used to obtain time evolution equation for the shock acceleration; this shock acceleration equation was solved simultaneously with the Euler equations.

To construct the absorbing layer for this problem, they split the solution variables in computational space. For the pressure equation which is written as

$$\frac{\partial p}{\partial t} = -a_1\frac{\partial p}{\partial X} - a_2\frac{\partial p}{\partial Y} - \gamma p(\frac{\partial w_1}{\partial X} + \frac{\partial w_2}{\partial Y}),$$

in the interior domain, two PML equations

$$\frac{\partial p_1}{\partial t} = -a_1\frac{\partial p}{\partial X} - \gamma p\frac{\partial w_1}{\partial X} - \sigma_X p_1$$

and

$$\frac{\partial p_2}{\partial t} = -a_2\frac{\partial p}{\partial Y} - \gamma p\frac{\partial w_2}{\partial Y} - \sigma_Y p_2$$

are used in the absorbing layer. In the above equations w_1 and w_2 are velocity components in the x and y directions, a_1 and a_2 are contravariant velocity components (which include the effect of grid motion) in computational space, and $p = p_1 + p_2$. Locally one-dimensional equations for the other variables were constructed in a similar manner. The damping factor σ_X was zero in a layer parallel to the X direction; similarly σ_Y was zero in a layer parallel to the Y direction. However, in the corner region both damping factors were positive.

Damping is introduced gradually to minimize any reflections caused by the change of equations inside the absorbing layer. A large domain solution was used as the reference solution, and error was computed as the deviation of the solution on the standard domain from the reference solution. They studied a simple wave given by

$$u - U_1 = A\frac{k_y}{k}cos(k_x x + k_y y - u_0 t)$$

$$v = -A\frac{k_x}{k}cos(k_x x + k_y y - u_0 t)$$

$$\rho = p = T = 1$$

as the upstream condition ahead of the shock; U_1 was the upstream mean velocity normal to the undisturbed shock, $k_y = k\ sin\theta$, $k_x = k\ cos\theta$ where k was the upstream wave number, A was the amplitude of the wave and θ was the angle between the y axis and the wave vector. The damping factor (σ_X) was equal to $C\alpha d^2$, where d was the distance from the interface between the interior and the buffer layer, C was constant and α was a function of mean quantities. Now consider the same conditions for the linearized Euler equations. The fluctuation of pressure from the solution at mid height of the computational domain at time equal to 20 is shown in figure 1. The solution with 25 points in the buffer layer is visually indistinguishable from the reference (large domain) solution. The strength of the absorbing layer can be changed by varying the size of the layer and by changing the value of the constant C. Figure 2 shows the effectiveness of a 15 point buffer layer for different values of C. The damping is not adequate with C equal to 100 and both phase and amplitude errors are noticeable. With C equal to 300, the same size buffer layer gives much better results. Although inadequate damping yields poor results, rapid increase in damping or a large amount of damping may lead to large reflections and even numerical instability. When the computations are carried on for a long time, a portion of the reflection from the outflow boundary propagates to the shock and to other computational boundaries, which then may be amplified and/or reflected back to interior. Errors in pressure for solution of the linearized Euler equations with different layer size and damping levels are shown in figures 3 and 4 respectively. Hayder et al. (1996, 1997) found that the PML boundary condition was satisfactory for solution of the nonlinear Euler equations. This approach was more effective at lower angle of incidence and wave number. For an angle of incidence of 10° the error in pressure at time equal to 20 was about 0.05 per cent. The error increased as the angle of incidence increased and reached approximately 1 per cent for a 45° angle of incidence.

2.2 Free Shear Layer

Hu (1996b) studied the evolution of a Kelvin-Helmholtz instability wave in the free shear layer as it propagates downstream and impinges on the absorbing layers. He used the PML boundary condition for the solution of the linearized Euler equations. Hayder et al. (1997) revisited the problem in which they solved both the linearized and nonlinear Euler equations. They used a 4th-order 7-point explicit central difference scheme for the spatial discretization in Cartesian coordinates. Backward differences were used at the boundaries. Time integration was carried out by a 4th order Low-Dissipation and Low-Dispersion Runge-Kutta scheme (Hu et al., 1996). Splitting of the variables in the absorbing layer was done in coordinate directions. The PML boundary condition was very effective in reducing numerical reflection for the linearized Euler equation. With 10 points in the buffer layer the error was less that 0.03 per cent. For the nonlinear Euler solution, error as high as 4 percent were noted in their simulations.

2.3 Axisymmetric Jet

Hayder et al. (1996, 1997) considered the PML method for computations of an axisymmetric jet. The computation of such flow fields has attracted considerable attention because of the need to suppress the noise radiated by jets. Both the linearized and nonlinear Euler equations were solved by using the fourth order MacCormack scheme (Gottlieb and Turkel, 1976). This scheme uses predictor and corrector steps to compute time accurate solutions. One sided differences (forward or backward) were used to compute spatial derivatives at each predictor or corrector step. This problem was solved in polar coordinates and the governing equation $Q_t + F_z + G_r = S$ was split into two one-dimensional equations. The PML boundary condition was applied to these split equations by splitting conservation variables. At all exterior boundaries extrapolation is used to obtain the fluxes outside the computational domain. The inflow was excited by $\epsilon Re(qe^{i\omega t})$, where q was the set of eigenfunctions given by the linear stability theory. The error became quasi periodic after initial transients. effectiveness of the PML layer for this case was comparable to the free shear layer case.

2.4 Acoustic scattering

Atkins (1997) applied the discontinuous Galerkin method (Atkins 1997; Atkins and Shu 1996) with PML boundary condition to compute the sound field scattered off a circular cylinder. The linear Euler equations were solved by using a fifth-order discontinuous Galerkin method on unstructured grids with third-order Runge-Kutta time marching. The discontinuous Galerkin method is well suited to the buffer domain type of approach in that abrupt changes in the governing equations can be accurately treated. One benefit of this feature is that the damping does not need to be introduced gradually as it is in a finite-difference method. Instead the damping coefficient can be abruptly set to any desired value and held constant throughout the PML zone. Another benefit of the compact form of the discontinuous Galerkin method is that the direction used for splitting the equations can be chosen locally within each element, and can be allowed to vary from element to element. Thus, when treating cylindrical geometries there is no need to transform the equations to a cylindrical or spherical form before splitting.

Extensive numerical tests indicate that a relatively thin layer (three cells) can reduce reflections by an order of magnitude over that obtained using characteristic boundary conditions. Figure 5 shows the difference between solutions obtained using a PML layer of varying thickness and a solution obtained on a larger domain with the same spatial resolution. The plots give the maximum difference between solutions measured at $r = 5$ for $0 < \theta < \pi$, $t < 12$. Similar reductions in the non-physical reflections can be obtained with either constant or variable damping coefficients.

To place these results in perspective, figure 6 gives a similar set of results for a Gaussian pulse in a semi-infinite duct – a problem that is ideally suited to the PML method. Again, comparisons are made between solutions obtained using a PML layer of varying thickness and a solution in which the duct was extended to approximately 161 with the same resolution. The measurements were taken at $x = 48$ for $-50 < y < 50$, $t < 200$. Here, the reductions in the reflections are only slightly better than those shown in figure 5; this result suggests that the PML method performed well for the cylinder-scatter problem even though the direction for splitting was chosen locally and varied from one element to the next.

Although both variable and constant damping factor appear to give similar results, a closer examination at the data suggests otherwise. If the PML layer were to function perfectly, then a wave should pass from the physical domain to the PML zone with no reflection and should continue to propagate across the PML zone. Although the wave is dissipated as it propagates across the PML zone, a small reflection off the outer boundary of the PML zone is expected. The reflection should be reduced as the thickness of the PML layer is increased. In addition, the time at which the reflection appears should increase with the layer thickness because the travel time across the zone increases. This last feature is not observed in cases for which the damping is introduced gradually. Figures 7(a) and 7(b) show the time history of the solution differences with variable and constant damping coefficients, respectively. For cases in which a variable damping coefficient was used, a reduction is noted in the measured differences as the thickness of the layer increases, however, no other distinct differences were noted in the history of the data. These results are typical of published results. In contrast, when the damping coefficient is constant throughout the PML zone, a distinct delay is noted in the time at which the solution differences begin to increase.

3. Conclusion

Absorbing layers were constructed by using the operator splitting (Perfectly Matched Layer condition) technique for computations of several physical problems (with different numerical algorithms). Our numerical studies and also those of Hu (1996a, 1996b) indicate that this methodology is highly effective and is suitable for aeroacoustic computations.

References

Atkins, H. L., 1997, "Continued Development of the Discontinuous Galerkin Method for Computational Aeroacoustic Applications", AIAA paper 97-1581.

Atkins, H. L. and Shu, Chi-Wang, 1996, "Quadrature Free Implementation of the Discontinuous Galerkin Methods for Hyperbolic Equations", AIAA paper 96-1683.

Berenger, J-P, 1994,"A Perfectly Matched Layer for the Absorption of Electro-magnetic Waves", JCP, **114**, pp 185-200.

Colonius, T., Lele, S. K. and Moin, P., 1993, "Boundary Conditions for Direct Computation of Aerodynamic Sound Generation", AIAA J., **31**, pp 1574-1582.

Erlebacher, G., Hussaini, M. Y. and Shu, C., 1997, "Interaction of a Shock with a Longitudinal Vortex", ICASE report 96-31, *to appear in JFM.*

Givoli, D., 1991, "Non-reflecting Boundary Conditions", JCP, **94**, pp 1-29.

Gottlieb, D. and Turkel, E., 1976, "Dissipative Two-Four Methods for Time Dependent Problems", Math. Comp, **Vol 30**, pp 703-723.

Hayder, M. E. and Turkel, E., 1995,"Nonreflecting Boundary Conditions for Jet Flow Computations", AIAA Journal, **33**, No. 12, pp 2264-2270.

Hayder, M. E., Hu, F. Q. and Hussaini, M. Y., 1996, "Towards Perfectly Absorbing Boundary Layer as a Non-Reflecting boundary", Presented in the 15th ICNMFD, Monterey, CA.

Hayder, M. E., Hu, F. Q. and Hussaini, M. Y., 1997, "Towards Perfectly Absorbing Boundary Conditions for Euler Equations", ICASE report 97-25, AIAA paper 97-2075.

Hu, F. Q., 1996a, "On Absorbing Boundary Conditions for Linearized Euler Equations by a Perfectly Matched Layer", JCP, **129**, pp 201-219.

Hu, F. Q., 1996b, "On Perfectly Matched Layer As An Absorbing Boundary Condition", AIAA paper 96-1664.

Hu, F. Q., Hussaini, M. Y. and Manthey, J. L., 1996,"Low-Dissipation and Low-Dispersion Runge-Kutta Schemes for Computational Acoustics" JCP, **124**, pp 177-191.

Streett, C. L. and Macaraeg, M. G., 1989, "Spectral multi-domain for large scale fluid dynamics simulations", Int. J. of Applied Num. Math., **6**, pp 123-139.

Ta'asan, S. and Nark, D. M., 1995, "An Absorbing Buffer Zone Technique for Acoustic Wave Propagation", AIAA paper 95-0164.

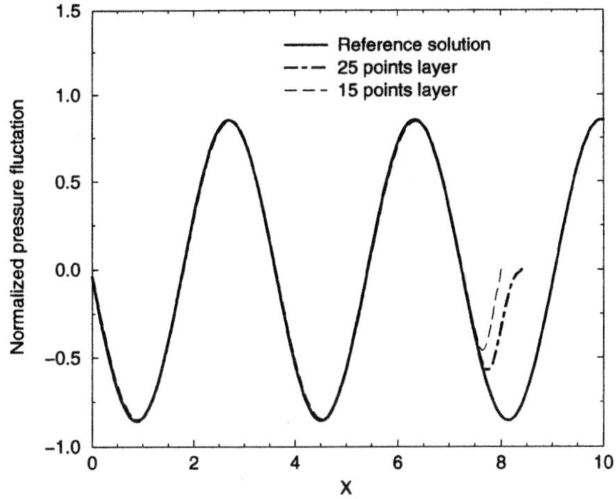

Figure 1: Effect of buffer layer.

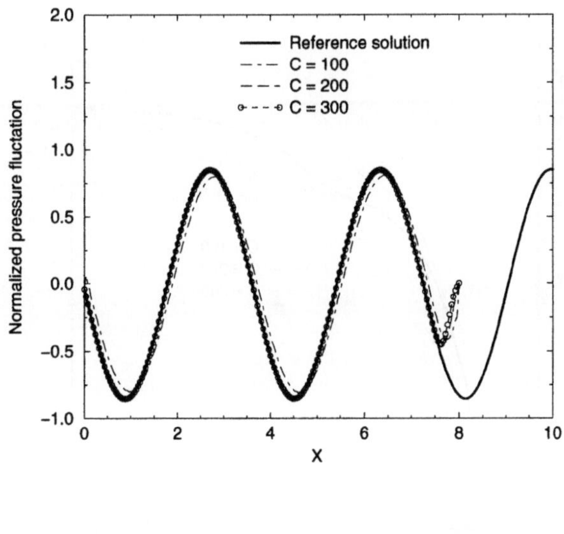

Figure 2: Influence of damping coefficient.

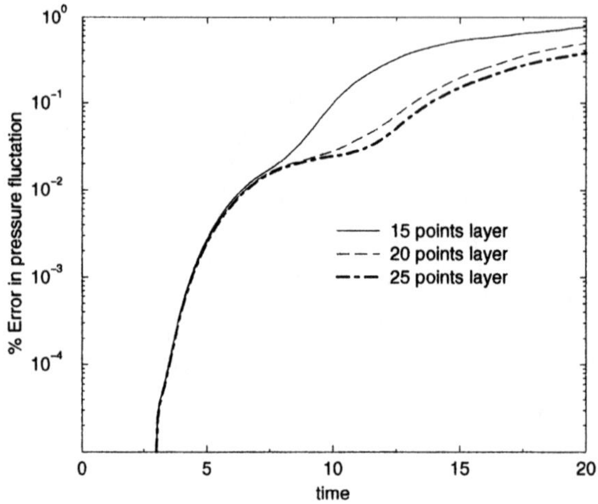

Figure 3: Error for different layer thickness.

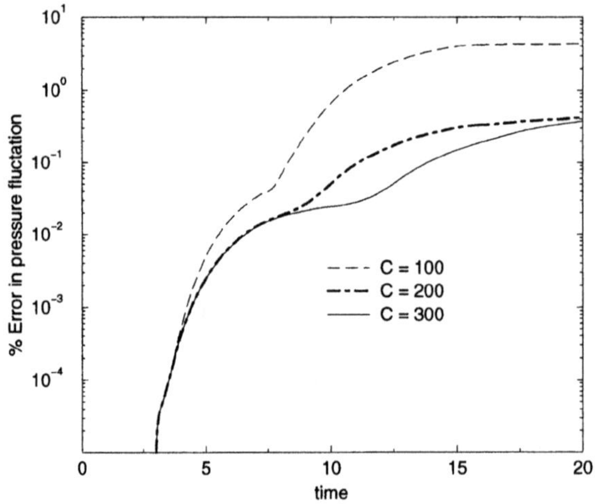

Figure 4: Error for different damping levels.

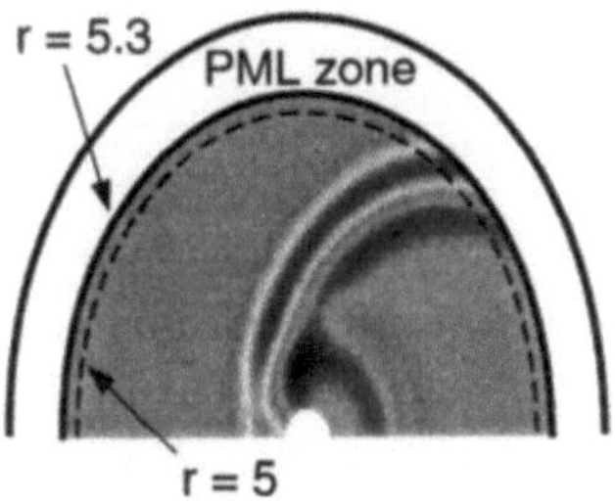

a) Pressure at $t = 4$.

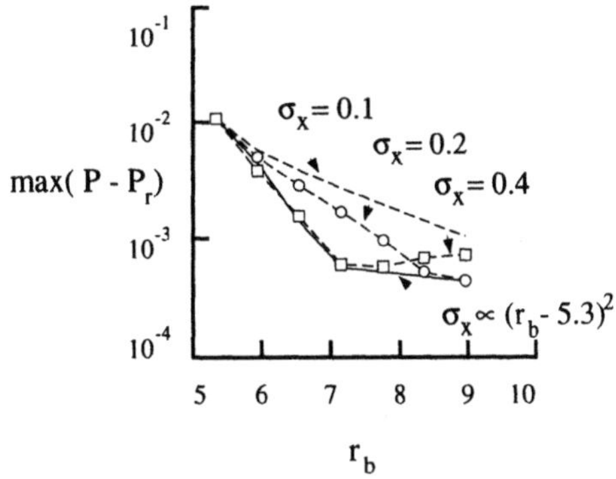

b) Influence of PML layer thickness damping coefficient, σ.

Figure 5: PML method applied to acoustic scattering off cylinder.

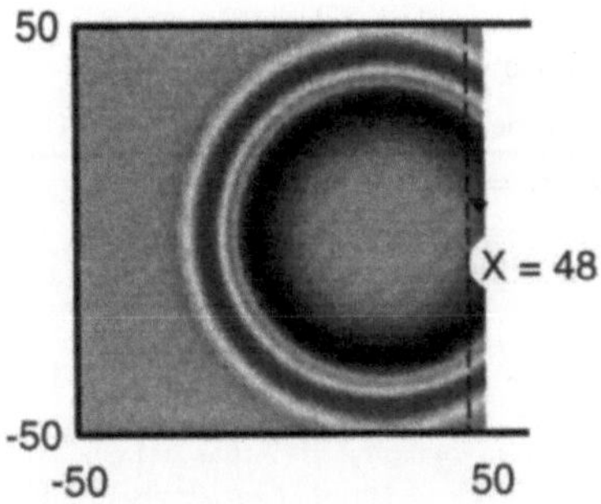

a) Pressure at $t = 40$.

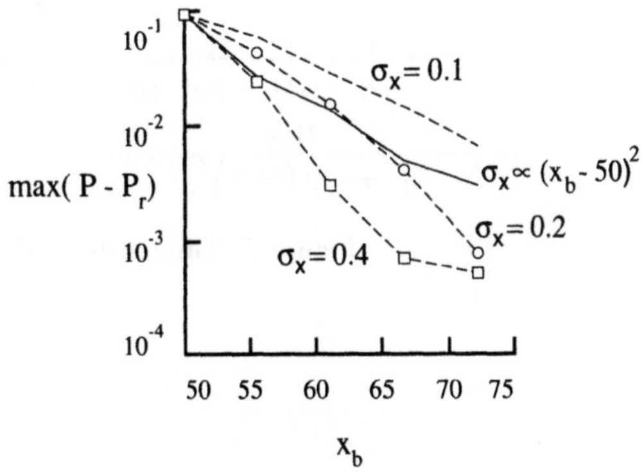

b) Influence of PML layer thickness and damping coefficient, σ.

Figure 6: PML method applied to Gaussian pulse in semi-infinite duct.

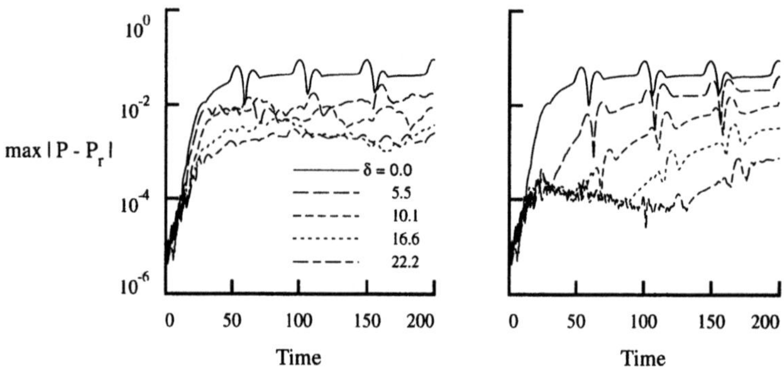

a) $\sigma = 2\left(\frac{x-x_b}{\delta}\right)^2$. b) $\sigma_x = 0.2$; constant.

Figure 7: Time history of solution differences.

ABSORBING BOUNDARY CONDITIONS FOR DISPERSIVE WAVES

Robert L. Higdon

Department of Mathematics
Oregon State University
Corvallis, Oregon 97331-4605

1. Introduction

The goal of an absorbing boundary condition is to simulate the outward radiation of energy at an artificial computational boundary. To that end, the boundary condition should make a valid statement about outgoing waves, at least in an approximate sense, but this statement should not require apriori knowledge of the pointwise values of the wave field. Typically, such a boundary condition involves a wave velocity. However, in the case of dispersive waves there are two kinds of velocity, phase velocity and group velocity, and each of the velocities varies with wavenumber and frequency. This variation of velocity is a central issue in the construction of absorbing boundary conditions for dispersive waves. The present abstract summarizes a technique developed by [Higdon, 1994].

2. Formulation of boundary conditions

An example of dispersive waves is given by solutions of the equation $u_{tt} = c^2 u_{xx} - f^2 u$. A function $u(x,t) = e^{ikx-i\omega t}$ is a solution if and only if the dispersion relation $\omega^2 = c^2k^2 + f^2$ is satisfied. Monochromatic solutions of this form propagate with phase velocity $c_p = \omega/k$. However, suppose that a more general solution is constructed by superposition of such waves. Over time, such a solution evolves into a slowly varying wavetrain that is locally sinusoidal; for a large time $t > 0$, the distance traveled by wavenumber k is approximately $c_g(k)t$, where $c_g(k) = d\omega/dk$ is the group velocity. See, e.g., [Lighthill, 1978]. For the example $u_{tt} = c^2u_{xx} - f^2u$, the dispersion relation is symmetric about $k = 0$, and the group velocity and phase velocity always have the same sign.

Suppose that the computational domain is defined by $x > 0$, in either one or several space dimensions. For a scalar field u, we use absorbing boundary conditions of the form

$$\left[\prod_{j=1}^{m} \left(\frac{\partial}{\partial t} - c_j \frac{\partial}{\partial x}\right)\right] u = 0 \tag{1}$$

at the boundary $x = 0$. Typically, one might use $m = 2$ or $m = 3$. The parameters $c_j > 0$ need to be chosen by the user; for one-dimensional problems, these parameters can be interpreted as phase velocities of waves that are absorbed exactly. The selection of parameters is discussed in Section 4. Boundary conditions similar to (1) can also be used for acoustic and elastic waves [Higdon, 1992].

T.L. Geers (ed.), IUTAM Symposium on Computational Methods for Unbounded Domains, 217–220.

An operator $\partial/\partial t - c_j \partial/\partial x$ can be approximated by a difference operator involving shifts with respect to t and x. A composition of such operators then gives a discretization of the boundary condition (1). Details are given by [Higdon, 1994]. The stencil of the discrete boundary condition is one-dimensional, which is convenient for implementation near the corners of a rectangular domain.

3. Analysis of absorption properties

We now examine the performance of the boundary condition (1). For simplicity, the analysis is stated in terms of the example $u_{tt} = c^2 u_{xx} - f^2 u$. However, the analysis applies more generally, subject to conditions on the symmetry and monotonicity of the dispersion relation which are stated at the end of this section.

In the following analysis, we assume that a known wave is leaving the computational domain and then determine the reflected wave (if any) that is generated at the boundary. In particular, we consider the linear combination

$$a_{in} e^{ik_{in}x - i\omega t} \; + \; a_{out} e^{ik_{out}x - i\omega t}, \tag{2}$$

where a_{out} is a given amplitude of a wave moving out of the domain $x > 0$, and a_{in} is a coefficient for an incoming wave that is to be determined. We would like to have $a_{in} = 0$ or at least very small. The configuration of the analysis is illustrated in Figure 1; in that graph, the solid curve is the graph of the dispersion relation $\omega^2 = c^2k^2 + f^2$ for $\omega > 0$.

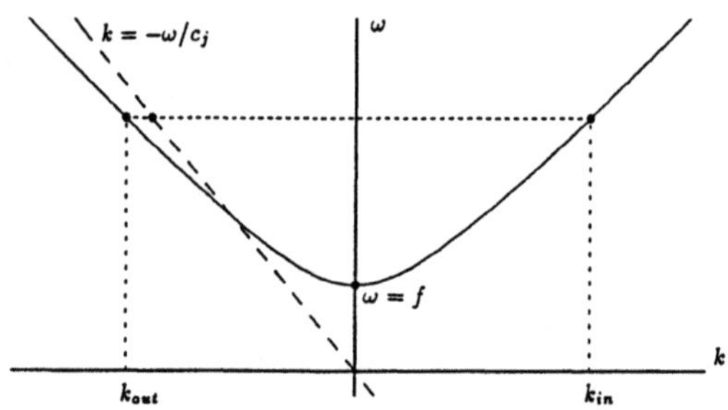

Figure 1.

When the linear combination (2) is inserted into the boundary condition (1), the result is

$$a_{in} \prod_{j=1}^{m} (-i\omega - c_j i k_{in}) \; + \; a_{out} \prod_{j=1}^{m} (-i\omega - c_j i k_{out}) \;\; = \;\; 0.$$

The reflection coefficient is then

$$R \; = \; \left| \frac{a_{in}}{a_{out}} \right| \; = \; \prod_{j=1}^{m} \left| \frac{-i\omega - c_j i k_{out}}{-i\omega - c_j i k_{in}} \right| \; = \; \prod_{j=1}^{m} \left| \frac{-\omega/c_j - k_{out}}{-\omega/c_j - k_{in}} \right|.$$

The j'th factor in the reflection coefficient can be interpreted geometrically. For fixed ω, the numerator is the distance from the line $k = -\omega/c_j$ to the outgoing part of the graph of the dispersion relation, and the denominator is the distance from $k = -\omega/c_j$ to the incoming part of the dispersion relation. As seen in Figure 1, this ratio is always less than 1, and it is zero for the wavenumber and frequency at which the line intersects the dispersion relation. The reflection coefficient R is a product of such ratios, so $R << 1$ with $R = 0$ at the points of exact absorption.

The preceding analysis can be applied more generally. For problems in one space dimension, it suffices to assume that ω is a symmetric function of k and that ω is an increasing (decreasing) function of k when ω and k have the same (opposite) signs. For multi-dimensional problems, assume that the preceding holds for each value of the tangential wavenumber. These hypotheses are satisfied, for example, by the multi-dimensional equation $u_{tt} = c^2\nabla^2 u - f^2 u$.

4. Selection of parameters

The factor $\partial/\partial t - c_j\partial/\partial x$ in the boundary condition (1) leads to the j'th factor in the reflection coefficient R. Each factor in R is less than 1 and is equal to zero for a certain wavenumber and frequency. As suggested by the preceding analysis, the selection of the parameters c_j is equivalent to choosing the wavenumbers and frequencies of exact absorption.

In order to make these choices, it is advisable to obtain a rough estimate of the range of dominant wavenumbers or frequencies in the problem. This estimate could be obtained by physical or heuristic means; it is not necessary to use a Fourier transform to obtain precise information about the spectrum of the solution. For a boundary condition of order m, one then chooses m wavenumbers or frequencies for which waves are to be absorbed exactly. The precise choices of these parameters are not critical, as long as they are chosen within the ranges of dominant wavenumbers or frequencies. Numerical experiments show that the boundary condition (1) gives good results even when the problem is so strongly dispersive that the group and phase velocities differ by orders of magnitude.

5. An equivalent approach

Absorbing boundary conditions can also be derived by a method based on rational approximations to the outgoing part of the dispersion relation. This method was used by [Engquist and Majda, 1979] and by [Trefethen and Halpern, 1986] to derive boundary conditions for the nondispersive wave equation $u_{tt} = c^2\nabla^2 u$. We now show that this approach is essentially equivalent to the method developed above, when applied to the dispersive equation $u_{tt} = c^2\nabla^2 u - f^2 u$. Analogous results also hold for the nondispersive case; see, e.g., [Higdon, 1992].

First consider an example. For the one-dimensional equation $u_{tt} = c^2 u_{xx} - f^2 u$, the dispersion relation is $\omega^2 = c^2k^2 + f^2$. For notational convenience, we use the nondimensional form $\tilde{\omega}^2 = \tilde{k}^2 + 1$, where $\tilde{\omega} = \omega/f$ and $\tilde{k} = ck/f$. For a fixed frequency ω, the possible wavenumbers are defined by $\tilde{k} = \pm\tilde{\omega}\sqrt{1 - 1/\tilde{\omega}^2}$; a wave leaving the domain

$x > 0$ is then characterized by $\tilde{k} = -\tilde{\omega}\sqrt{1 - 1/\tilde{\omega}^2} = -\tilde{\omega}\left(1 - 1/2\tilde{\omega}^2 + O\left(1/\tilde{\omega}\right)^4\right)$. An approximation to this relation is $\tilde{k} \approx -\tilde{\omega} + 1/2\tilde{\omega}$, or $-\tilde{\omega}^2 - \tilde{k}\tilde{\omega} + \frac{1}{2} \approx 0$. The correspondences $-i\tilde{\omega} \leftrightarrow (1/f)\partial/\partial t$ and $i\tilde{k} \leftrightarrow (c/f)\partial/\partial x$ then yield a boundary condition $u_{tt} - cu_{xt} + \frac{1}{2}f^2 u = 0$ that is approximately satisfied by outgoing waves. However, this boundary condition can be written in the form (1). First use the interior equation $u_{tt} = c^2 u_{xx} - f^2 u$ to eliminate $f^2 u$ from the boundary condition. A re-arrangement then yields $\left(\frac{\partial}{\partial t} - c\frac{\partial}{\partial x}\right)^2 u = 0$, which is an example of the form (1). A generalization of this result is the following.

Theorem. If a boundary condition for $u_{tt} = c^2\nabla^2 u - f^2 u$ on the domain $x > 0$ is derived from a symmetric rational approximation to the outgoing part of the dispersion relation, then the boundary condition is either

(a) equivalent to the boundary condition (1), for suitable $c_j > 0$; or

(b) unstable, in the sense of violating the stability criterion of [Kreiss, 1970]; or

(c) not optimal, in the sense that the coefficients in the boundary condition can be changed so as to reduce the reflection coefficient for all frequencies for which the group velocity has a nonzero normal component.

6. Summary

The boundary condition (1) is based on a composition of first-order operators, each of which yields exact absorption for a specified wavenumber or frequency and yields good absorption otherwise. The first-order operators work together, in the sense that the reflection coefficient for (1) is the product of the reflection coefficients for the individual first-order operators. High absorption can thus be obtained over broad ranges of wavenumbers or frequencies. The points of exact absorption are chosen by the user; this generally requires a rough estimate of the range of dominant wavenumbers or frequencies in the problem, but it is not necessary to employ a Fourier transform of the solution.

7. References

Engquist, B., and Majda, A., 1979, "Radiation Boundary Conditions for Acoustic and Elastic Wave Calculations", *Comm. Pure. Appl. Math.*, Vol. 32, pp. 313–357.

Higdon, R. L., 1992, "Absorbing Boundary Conditions for Acoustic and Elastic Waves in Stratified Media", *J. Comput. Phys.*, Vol. 101, pp. 386–418.

Higdon, R. L., 1994, "Radiation Boundary Conditions for Dispersive Waves", *SIAM J. Numer. Anal.*, Vol. 31, pp. 64–100.

Lighthill, J., 1978, *Waves in Fluids*, Cambridge University Press, Cambridge, U.K.

Kreiss, H.-O., 1970, "Initial Boundary Value Problems for Hyperbolic Systems", *Comm. Pure Appl. Math.*, Vol. 23, pp. 277–298.

Trefethen, L. N., and Halpern, L., 1986, "Well-Posedness of One-Way Wave Equations and Absorbing Boundary Conditions", *Math. Comp.*, Vol. 47, pp. 421–435.

ABSORBING BOUNDARY CONDITIONS OF ARBITRARY SHAPE FOR THE THREE-DIMENSIONAL WAVE EQUATION[1]

Loukas F Kallivokas†, Jacobo Bielak†, Richard C. MacCamy‡
†*Computational Mechanics Laboratory,*
Department of Civil and Environmental Engineering
‡*Department of Mathematical Sciences*
Carnegie Mellon University, Pittsburgh, PA 15213, USA

1. Introduction

In this article we describe the theory leading to the development of a family of three-dimensional absorbing boundary conditions for the scalar wave equation; the conditions are applicable to either the transient or harmonic steady-state regimes. The development of the conditions follows an earlier line of investigation on the two-dimensional case [Barry et al, 1987 and Kallivokas et al, 1991]. The introduction of a global coordinate system in the region exterior to the truncation boundary allows for the conditions to be prescribed on convex boundaries of arbitrary shape, eg, on ellipsoidal surfaces. The latter is of importance if scattering problems involving long and slender obstacles were to be efficiently tackled.

An important issue that often arises in practice is the ease with which absorbing boundary conditions can be coupled with the numerical method used for the associated computational problem. Here, we favor finite elements, and demonstrate how a second-order absorbing boundary condition leads to a novel surface-only isoparametric element by virtue of simple decomposition scheme of the absorbing boundary operator. The new element, termed absorbing boundary element (ABE), is completely defined by a pair of time-invariant stiffness and damping matrices. That allows for the ready incorporation of the element, and thus the associated second-order condition, into existing codes.

In the next section we discuss the theoretical development of the absorbing boundary conditions in the context of an exterior initial and boundary value problem for the wave equation.

2. Development

Let Γ be a closed surface with exterior $\Omega \subset \Re^3$ (Fig. 1a). We consider the radiation problem in which Γ is subjected to a prescribed velocity. Then, the exterior problem becomes: find $p(\boldsymbol{x}, t)$ such that:

$$\ddot{p}(\boldsymbol{x},t) = c^2 \Delta p(\boldsymbol{x},t), \quad \boldsymbol{x} \in \Omega, \quad t \geq 0, \tag{1a}$$

$$p_\nu(\boldsymbol{x},t) = f_N(\boldsymbol{x},t), \quad \boldsymbol{x} \in \Gamma, \quad t > 0 \tag{1b}$$

$$\lim_{r\to\infty} r(p_r + \frac{1}{c}\dot{p}) = 0, \quad \text{and} \tag{1c}$$

$$p(\boldsymbol{x},0) = 0, \quad \dot{p}(\boldsymbol{x},0) = 0, \quad \boldsymbol{x} \in \bar{\Omega}. \tag{1d}$$

In these equations p denotes pressure; $\boldsymbol{x}$ is the position vector, t is time; $\boldsymbol{\nu}$ is the outward

[1]Portions of this article have appeared in *Comp. Methods Appl. Mech. Engrg.*, 147 (1997) 235–262; reprinted with permission from Elsevier Science.

T.L. Geers (ed.), IUTAM Symposium on Computational Methods for Unbounded Domains, 221–228.

unit normal on Γ; c is the velocity of wave propagation; Δ is the Laplace operator, and an overdot denotes derivative with respect to time. p_ν denotes the normal derivative of the pressure p, f_N is a prescribed function. Condition (1c) is the Sommerfeld radiation condition, and the system is taken to be initially at rest.

With the introduction of the truncation boundary Γ_a (Fig. 1b) that limits the computational domain to Ω_a, a condition need be prescribed on Γ_a to allow for a mathematically well-posed problem. Such an exact condition exists and is of the form:

$$p_\nu(\boldsymbol{x},t) = \mathcal{F}[p^t(\cdot,\cdot)](\boldsymbol{x}), \quad \boldsymbol{x} \in \Gamma_a, \tag{2}$$

where the p^t denotes the time history of p and is defined as:

$$p^t(t) = p(t-\tau), \; \forall\, \tau: \; 0 \leq \tau \leq t. \tag{3}$$

The dots following p^t indicate dummy variables, and $\mathcal{F}$ is an integral operator that depends on p^t.

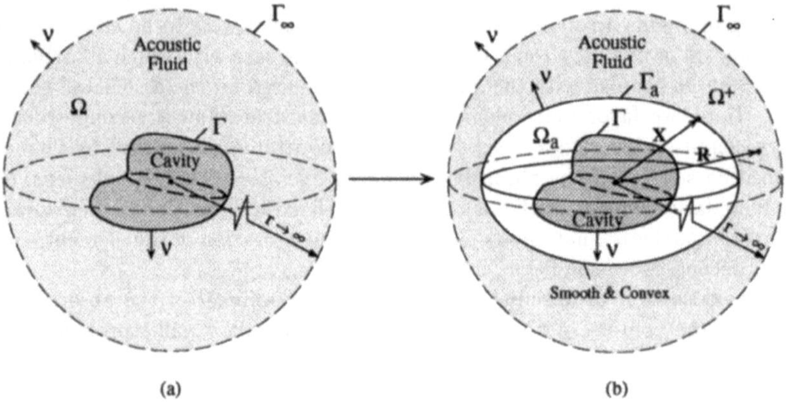

Figure 1: (a) Model of cavity surrounded by infinite acoustic fluid; (b) Reduced model with finite fluid region Ω_a and absorbing boundary.

To determine the exact form of the operator $\mathcal{F}$ we consider the following problem associated with the original problem (1). We Laplace-transform (1a) with respect to time to obtain:

$$s^2\hat{p}(\boldsymbol{x},s) = c^2\Delta\hat{p}(\boldsymbol{x},s), \quad \boldsymbol{x} \in \Omega, \tag{4}$$

where s is the Laplace transform variable and a caret over a variable such as p denotes its Laplace transform. Suppose now that the surface Γ_a is smooth and convex, and let Ω^+ be the exterior of Γ_a (Fig. 1b). We focus on $\hat{p}$ in Ω^+ and formulate the following auxiliary Dirichlet problem in Ω^+ for an auxiliary field $\hat{P}$:

$$s^2\hat{P}(\boldsymbol{x},s;t) = c^2\Delta\hat{P}(\boldsymbol{x},s;t), \quad \boldsymbol{x} \in \Omega^+, \tag{5a}$$

with the boundary condition,

$$\hat{P}(\boldsymbol{x}, s; t) = p(\boldsymbol{x}, t), \quad \boldsymbol{x} \in \Gamma_a. \tag{5b}$$

Then, by virtue of Duhamel's principle [7], one can show that:

$$\hat{p}(\boldsymbol{x}, s) = \int_0^\infty e^{-st} \hat{P}(\boldsymbol{x}, s; t)\, dt, \quad \boldsymbol{x} \in \bar{\Omega}^+, \tag{6a}$$

and hence also,

$$\hat{p}_\nu(\boldsymbol{x}, s) = \int_0^\infty e^{-st} \hat{P}_\nu(\boldsymbol{x}, s; t)\, dt, \quad \boldsymbol{x} \in \Gamma_a. \tag{6b}$$

Equation (6a) together with equations (5) ensure that the solution $\hat{p}$ in Ω^+ will coincide with the solution $\hat{p}$ in Ω from (4). From (5b) it can also be seen that the normal derivative $\hat{P}_\nu$ on Γ_a will be a linear functional of $p(\cdot, t)$ on Γ_a. Let us denote this dependency by:

$$\hat{P}_\nu(\boldsymbol{x}, s; t) = \hat{\mathcal{F}}[p(\cdot, t)](\boldsymbol{x}, s), \quad \boldsymbol{x} \in \Gamma_a, \tag{7a}$$

where $\hat{\mathcal{F}}$ denotes a functional. By substituting (7a) into (6b) there results:

$$\hat{p}_\nu(\boldsymbol{x}, s) = \hat{\mathcal{F}}[\hat{p}(\cdot, s)](\boldsymbol{x}, s), \quad \boldsymbol{x} \in \Gamma_a. \tag{7b}$$

Then, translation of (7b) back to the time-domain yields:

$$p_\nu(\boldsymbol{x}, t) = \mathcal{F}[p^t(\cdot, \cdot)](\boldsymbol{x}), \quad \boldsymbol{x} \in \Gamma_a. \tag{7c}$$

The nonlocal character of the exact $\mathcal{F}$ makes it unsuitable for implementation in the context of most numerical methods, eg, the finite element or finite difference methods. We seek approximations to the exact $\mathcal{F}$ aiming primarily at reducing the temporal nonlocality. It will be shown that this procedure also reduces automatically the spatial nonlocality. To this end, we turn again to the auxiliary Dirichlet problem defined by (5) and introduce, borrowing from geometrical optics, an asymptotic expansion for $\hat{P}$ of the form:

$$\hat{P}(\boldsymbol{x}, s; t) \sim e^{-\frac{s}{c}\chi(\boldsymbol{x})} \sum_{k=0}^{\infty} \left[\frac{c}{(s+\gamma)a}\right]^k A^{(k)}(\boldsymbol{x}, t), \quad \boldsymbol{x} \in \Omega^+, \tag{8a}$$

where a is a characteristic length of the absorbing boundary (e.g. in the case of a spherical Γ_a, a is the radius), $\chi(\boldsymbol{x})$ and $A^{(k)}(\boldsymbol{x}, t)$ are as yet unknown functions, and γ is an arbitrary nonnegative parameter which is introduced for stability. We require that the functions $\chi(\boldsymbol{x})$ and $A^{(k)}(\boldsymbol{x}, t)$ satisfy the following conditions on Γ_a:

$$\chi(\boldsymbol{x}) = 0, \quad A^{(0)}(\boldsymbol{x}, t) = p(\boldsymbol{x}, t), \quad \text{and} \quad A^{(k)}(\boldsymbol{x}, t) = 0, \text{ for } k \geq 1, \quad \boldsymbol{x} \in \Gamma_a. \tag{8b}$$

Equations (8) ensure that $\hat{P}$ is outgoing and that (5b) is satisfied automatically for any functions $\chi(\boldsymbol{x})$ and $A^{(k)}(\boldsymbol{x}, t)$. Equation (8a), by virtue of (8b), yields for $\hat{P}_\nu$ on Γ_a:

$$\hat{P}_\nu(\boldsymbol{x}, s; t) = -\frac{s}{c}\chi_\nu A^{(0)}(\boldsymbol{x}, t) + \sum_{k=0}^{\infty} A_\nu^{(k)}(\boldsymbol{x}, t) \left[\frac{c}{(s+\gamma)a}\right]^k, \quad \boldsymbol{x} \in \Gamma_a. \tag{9a}$$

Substitution of (9a) in (6b) will therefore yield:

$$\hat{p}_\nu(\boldsymbol{x},s) = -\frac{s}{c}\chi_\nu\,\hat{A}^{(0)}(\boldsymbol{x},s) + \sum_{k=0}^{\infty}\hat{A}_\nu^{(k)}(\boldsymbol{x},s)\left[\frac{c}{(s+\gamma)a}\right]^k, \quad \boldsymbol{x}\in\Gamma_a. \tag{9b}$$

From (9b) and (6b) it can be seen that, once the unknown functions $\chi(\boldsymbol{x})$ and $A^{(k)}(\boldsymbol{x},t)$ are determined, an expression for the normal derivative p_ν on Γ_a will be possible by translating (9b) back to the time-domain. To determine $\chi(\boldsymbol{x})$ and $A^{(k)}(\boldsymbol{x},t)$, we proceed by introducing (8a) into (5a) and matching the coefficients of the non-negative powers of s and of the negative powers of the monomials $(s+\gamma)$. There results:

$$\begin{aligned}
& s^2\frac{1}{c^2}\left[|\nabla\chi|^2-1\right]A^{(0)} \\
& +s\left\{\frac{1}{a\,c}\left[|\nabla\chi|^2-1\right]A^{(1)} - \frac{2}{c}\nabla\chi\cdot\nabla A^{(0)} - \frac{1}{c}\Delta\chi\,A^{(0)}\right\} \\
& +\left\{\frac{1}{c^2}\left[|\nabla\chi|^2-1\right]\left[-\gamma\frac{c}{a}A^{(1)}+\frac{c^2}{a^2}A^{(2)}\right] - \frac{2}{a}\nabla\chi\cdot\nabla A^{(1)} - \frac{1}{a}\Delta\chi\,A^{(1)} + \Delta A^{(0)}\right\} \\
& +\sum_{k=1}^{\infty}\frac{1}{(s+\gamma)^k}\left\{\frac{1}{c^2}\left[|\nabla\chi|^2-1\right]\left[\gamma^2\frac{c^k}{a^k}A^{(k)} - 2\gamma\frac{c^{k+1}}{a^{k+1}}A^{(k+1)} + \frac{c^{k+2}}{a^{k+2}}A^{(k+2)}\right]\right. \\
& \qquad -\frac{2}{c}\nabla\chi\cdot\left[-\gamma\frac{c^k}{a^k}\nabla A^{(k)} + \frac{c^{k+1}}{a^{k+1}}\nabla A^{(k+1)}\right] \\
& \qquad \left. -\frac{1}{c}\Delta\chi\left[-\gamma\frac{c^k}{a^k}A^{(k)} + \frac{c^{k+1}}{a^{k+1}}A^{(k+1)}\right] + \frac{c^k}{a^k}\Delta A^{(k)}\right\} = 0.
\end{aligned} \tag{10}$$

By setting to zero the coefficients of the various powers of s and of the monomials $(s+\gamma)$ in (10) there result the following differential equations for the functions $\chi(\boldsymbol{x})$ and $A^{(k)}(\boldsymbol{x},t),(k\geq 0)$ with $\boldsymbol{x}\in\bar{\Omega}^+$:

$$|\nabla\chi|^2 = 1, \tag{11a}$$

$$2\nabla\chi\cdot\nabla A^{(0)} + \Delta\chi\,A^{(0)} = 0, \tag{11b}$$

$$2\nabla\chi\cdot\nabla A^{(k)} + \Delta\chi\,A^{(k)} = a\Delta A^{(k-1)} + \frac{\gamma a}{c}\left[2\nabla\chi\cdot\nabla A^{(k-1)} + \Delta\chi\,A^{(k-1)}\right], \quad k\geq 1. \tag{11c}$$

3. Absorbing Boundaries of Arbitrary Shape

The differential equations (11a,b) as well as the recursive set (11c) can be used to determine the unknown functions $\chi(\boldsymbol{x})$ and $A^{(k)}(\boldsymbol{x},t)$ for any k. To this end, we introduce a new coordinate system in $\bar{\Omega}^+$ to aid in subsequent calculations. Let Γ_a be described by the parametric representation $\boldsymbol{X}(u,w)$ where $\boldsymbol{X}$ denotes the position vector on Γ_a and u,w are the surface parameters. Then a new coordinate system is introduced in $\bar{\Omega}^+$ by the description:

$$\boldsymbol{R}(u,w,\xi) = \boldsymbol{X}(u,w) + \xi\,\boldsymbol{\nu}(u,w), \quad \xi\geq 0, \tag{12}$$

where $\boldsymbol{R}$ denotes position vector in Ω^+, ξ is a scalar and $\boldsymbol{\nu}$ is the outward normal to Γ_a (Fig. 1b). Since Γ_a is convex and smooth, the new system is global in Ω^+. Notice that for $\xi=0$, (13) provides the parametric representation of Γ_a. Using the gradient, the Laplacian and the normal derivative in terms of the components of the Euclidean metric tensor g_{ij} of the new coordinate system and their limiting values h_{ij} on Γ_a ($h_{ij} = g_{ij}|_{\xi=0}$), the Gaussian (K) and mean (H) curvatures of Γ_a, along with equations (8b) and (11), one obtains for the normal derivatives of the unknown functions $\chi, A^{(0)}$ and $A^{(1)}$:

$$\chi_\nu(\boldsymbol{x}) = \chi_\xi(\boldsymbol{x}) = 1, \quad \boldsymbol{x} \in \Gamma_a, \tag{13a}$$

$$A_\nu^{(0)}(\boldsymbol{x},t) = A_\xi^{(0)}(\boldsymbol{x},t) = H(\boldsymbol{x})\, A^{(0)}(\boldsymbol{x},t), \quad \boldsymbol{x} \in \Gamma_a, \tag{13b}$$

$$\begin{aligned} A_\nu^{(1)}(\boldsymbol{x},t) = A_\xi^{(1)}(\boldsymbol{x},t) = \frac{a}{2\sqrt{h(\boldsymbol{x})}} &\left\{ \left[\frac{1}{\sqrt{h(\boldsymbol{x})}} \left(h_{22}(\boldsymbol{x})\, A_u^{(0)}(\boldsymbol{x},t) - h_{12}(\boldsymbol{x})\, A_w^{(0)}(\boldsymbol{x},t) \right) \right]_u \right. \\ &+ \left. \left[\frac{1}{\sqrt{h(\boldsymbol{x})}} \left(-h_{12}(\boldsymbol{x})\, A_u^{(0)}(\boldsymbol{x},t) + h_{11}(\boldsymbol{x})\, A_w^{(0)}(\boldsymbol{x},t) \right) \right]_w \right\} \\ &+ \frac{a}{2} \left(H^2(\boldsymbol{x}) - K(\boldsymbol{x}) \right) A^{(0)}(\boldsymbol{x},t), \quad \boldsymbol{x} \in \Gamma_a. \end{aligned} \tag{13c}$$

Letter subscripts above denote partial derivatives. Next, by truncating the series in (9b) one can construct successive approximations to $\hat{p}_\nu$ and hence to the functional $\hat{\mathcal{F}}[\hat{p}]$ in (7b). It can, therefore, be shown that by substituting the Laplace transforms of functions (13) into (9b), while taking into account (8b) and (7b), and keeping none, one or two terms from the series in (9b), the first three approximations for $\hat{p}_\nu$ on Γ_a are given as:

$$\text{0th order}: \ \hat{p}_\nu = \hat{\mathcal{F}}_0[\hat{p}] = -\frac{s}{c}\hat{p}, \tag{14a}$$

$$\text{1st order}: \ \hat{p}_\nu = \hat{\mathcal{F}}_1[\hat{p}] = -\frac{s}{c}\hat{p} + H\,\hat{p}, \tag{14b}$$

$$\begin{aligned} \text{2nd order}: \ \hat{p}_\nu = \hat{\mathcal{F}}_2[\hat{p}] = &-\frac{s}{c}\hat{p} + H\,\hat{p} \\ &+ \frac{c}{2(s+\gamma)} \left\{ \frac{1}{\sqrt{h}} \left\{ \left[\frac{1}{\sqrt{h}} (h_{22}\,\hat{p}_u - h_{12}\,\hat{p}_w) \right]_u + \left[\frac{1}{\sqrt{h}} (-h_{12}\,\hat{p}_u + h_{11}\,\hat{p}_w) \right]_w \right\} \right. \\ &+ \left. (H^2 - K)\hat{p} \right\} \end{aligned} \tag{14c}$$

where $\hat{\mathcal{F}}_k$ ($k = 0,1,2$) denotes the approximate functional of k-th order. Translation of (14) back to the time-domain yields:

$$\text{0th order}: \ p_\nu = -\frac{1}{c}\dot{p}, \tag{15a}$$

$$\text{1st order}: \ p_\nu = -\frac{1}{c}\dot{p} + H\,p, \tag{15b}$$

$$\begin{aligned} \text{2nd order}: \ \dot{p}_\nu + \gamma\, p_\nu = &-\frac{1}{c}\ddot{p} + \left(H - \frac{\gamma}{c} \right) \dot{p} + H\,\gamma\, p \\ &+ \frac{c}{2} \left\{ \frac{1}{\sqrt{h}} \left\{ \left[\frac{1}{\sqrt{h}} (h_{22}\, p_u - h_{12}\, p_w) \right]_u + \left[\frac{1}{\sqrt{h}} (-h_{12}\, p_u + h_{11}\, p_w) \right]_w \right\} \right. \\ &+ \left. (H^2 - K)p \right\}. \end{aligned} \tag{15c}$$

Conditions (15) are the desired approximations of the operator symbol in (2). Notice that, whereas (15a) and (15b) involve only the normal derivative of the pressure p_ν, (15c) introduces the first time derivative of p_ν as well. The latter calls for a special implementation scheme, which is addressed in the next section.

4. Decomposition of the Boundary Operator

The application of Galerkin-type finite element methods to the exterior acoustics problem commonly gives rise to the term:

$$I_a = -\int_{\Gamma_a} \delta p \, p_\nu \, d\Gamma_a. \tag{16}$$

Replacing p_ν in the integral (16) by conditions (15a) or (15b) is trivial; on the other hand, the higher-order condition (15c), while expected to yield improved accuracy and computational economy due to the need for a smaller buffer region (Ω_a), cannot be readily implemented since it involves both the normal derivative of the pressure p_ν and its first time derivative $\dot{p}_\nu$. Our aim is to make the condition amenable to standard integration techniques typically used for interior problems. We show that this can be achieved through the introduction of additional degrees of freedom on the artificial boundary.

Let $q^{(1)}$ and $q^{(2)}$ denote auxiliary variables on Γ_a. Then, one can easily show using Laplace transforms that the following set of three equations is equivalent to (15c):

$$-p_\nu = \frac{1}{c}\dot{p} - H\,p - \frac{c}{2\gamma}\mathcal{J} q^{(1)} - \frac{c}{2\gamma}(H^2 - K) q^{(2)}, \tag{17a}$$

$$\mathcal{J} p - \mathcal{J} q^{(1)} - \frac{1}{\gamma}\mathcal{J}\dot{q}^{(1)} = 0, \tag{17b}$$

$$p - q^{(2)} - \frac{1}{\gamma}\dot{q}^{(2)} = 0. \tag{17c}$$

Here $\mathcal{J}$ is the generalized Beltrami-Mitchell operator defined by:

$$\mathcal{J}\cdot = \frac{1}{\sqrt{h}}\left\{\left[\frac{1}{\sqrt{h}}\left(h_{22}\,(\cdot)_u - h_{12}\,(\cdot)_w\right)\right]_u + \left[\frac{1}{\sqrt{h}}\left(-h_{12}\,(\cdot)_u + h_{11}\,(\cdot)_w\right)\right]_w\right\}. \tag{18}$$

Now multiply (17a) by δp, (17b) by $\delta q^{(1)}$ and (17c) by $\delta q^{(2)}$, where $\delta q^{(1)}$ and $\delta q^{(2)}$ are appropriate test functions and integrate by parts the terms associated with the operator $\mathcal{J}$. There results:

$$-\int_{\Gamma_a} \delta p\, p_\nu\, d\Gamma_a = \frac{1}{c}\int_{\Gamma_a} \delta p\, \dot{p}\, d\Gamma_a - \int_{\Gamma_a} H\,\delta p\, p\, d\Gamma_a + \frac{c}{2\gamma}\int_{\Gamma_a} \nabla^s \delta p \cdot \nabla^s q^{(1)}\, d\Gamma_a \\ -\frac{c}{2\gamma}\int_{\Gamma_a} (H^2 - K)\,\delta p\, q^{(2)}\, d\Gamma_a, \tag{19a}$$

$$\frac{c}{2\gamma}\int_{\Gamma_a} \nabla^s \delta q^{(1)} \cdot \nabla^s p\, d\Gamma_a - \frac{c}{2\gamma}\int_{\Gamma_a} \nabla^s \delta q^{(1)} \cdot \nabla^s q^{(1)}\, d\Gamma_a \\ -\frac{c}{2\gamma^2}\int_{\Gamma_a} \nabla^s \delta q^{(1)} \cdot \nabla^s \dot{q}^{(1)}\, d\Gamma_a = 0, \tag{19b}$$

$$-\frac{c}{2\gamma}\int_{\Gamma_a} (H^2 - K)\,\delta q^{(2)} p\, d\Gamma_a + \frac{c}{2\gamma}\int_{\Gamma_a} (H^2 - K)\,\delta q^{(2)} q^{(2)}\, d\Gamma_a \\ +\frac{c}{2\gamma^2}\int_{\Gamma_a} (H^2 - K)\,\delta q^{(2)} \dot{q}^{(2)}\, d\Gamma_a = 0, \tag{19c}$$

in which ∇^s is the surface gradient.

Equations (19) can then be used to complete the weak-form formulation of the exterior acoustics problem. The right side of (19a) replaces (16), and (19b) and (19c) are added to the resultant functional. It is important to observe that upon spatial discretization, the replacement for (15c) of (19) will lead to a symmetric system of ordinary differential equations. In other words, the contributions from the absorbing boundary maintain both the symmetric structure of the interior problem and the sparsity of the associated system matrices.

5. Absorbing Boundary Element

With the introduction of standard isoparametric approximations ψ_1, ψ_2 and ψ_3, for the functions p, $q^{(1)}$, $q^{(2)}$ and their respective variations appearing in (19), it can be shown that (19) is equivalent to the following pair of element stiffness $\boldsymbol{k}^a$ and damping $\mathbf{c}^a$ matrices:

$$\boldsymbol{k}^a = \frac{c}{2\gamma}\begin{bmatrix} \frac{2\gamma}{c}\boldsymbol{k}_{11} & \boldsymbol{k}_{12} & -\boldsymbol{k}_{13} \\ \boldsymbol{k}_{12}^T & -\boldsymbol{k}_{22} & 0 \\ -\boldsymbol{k}_{13}^T & 0 & \boldsymbol{k}_{33} \end{bmatrix}, \qquad \mathbf{c}^a = \frac{c}{2\gamma^2}\begin{bmatrix} \frac{2\gamma^2}{c^2}\mathbf{c}_{11} & 0 & 0 \\ 0 & -\mathbf{c}_{22} & 0 \\ 0 & 0 & \mathbf{c}_{33} \end{bmatrix}, \tag{20a}$$

with the following definitions for the individual matrices:

$$\boldsymbol{k}_{11} = -\int_{\Gamma_a^e} H\,\boldsymbol{\psi}_1\boldsymbol{\psi}_1^T\,d\Gamma_a^e, \quad \boldsymbol{k}_{22} = \int_{\Gamma_a^e} \overline{\nabla}^s\boldsymbol{\psi}_2\cdot\overline{\nabla}^s\boldsymbol{\psi}_2^T\,d\Gamma_a^e, \quad \boldsymbol{k}_{33} = \int_{\Gamma_a^e}(H^2-K)\,\boldsymbol{\psi}_3\boldsymbol{\psi}_3^T\,d\Gamma_a^e,$$

$$\boldsymbol{k}_{12} = \int_{\Gamma_a^e} \overline{\nabla}^s\boldsymbol{\psi}_1\cdot\overline{\nabla}^s\boldsymbol{\psi}_2^T\,d\Gamma_a^e, \quad \boldsymbol{k}_{13} = \int_{\Gamma_a^e}(H^2-K)\,\boldsymbol{\psi}_1\boldsymbol{\psi}_3^T\,d\Gamma_a^e,$$

$$\mathbf{c}_{11} = \int_{\Gamma_a^e}\boldsymbol{\psi}_1\boldsymbol{\psi}_1^T\,d\Gamma_a^e, \quad \mathbf{c}_{22} = \boldsymbol{k}_{22}, \quad \mathbf{c}_{33} = \boldsymbol{k}_{33}, \tag{20b}$$

where $d\Gamma_a^e$ and the operator $\overline{\nabla}^s$ denote the area differential and the approximation of the surface gradient ∇^s on an element Γ_a^e of Γ_a, respectively.

The element matrices (20a) essentially give rise to a new finite element, the absorbing boundary element (Fig. 2), which is capable of absorbing the waves that reach the artificial boundary while simulating the effect of the truncated infinite domain. We further remark that the element defined by (20) is a surface-only element (Fig. 2); one need only mesh the finite region Ω_a and simply attach the impedance-infinite element on the boundary Γ_a without any further discretization within the infinite exterior region.

6. Conclusions

A systematic procedure that leads to a family of three-dimensional absorbing boundary conditions for the wave equation has been presented. The coupling of the third member of this family –a condition of second-order, to the finite element method gives rise to a new, simple, surface-only element (ABE) that can be used to advantage for the efficient solution of large structural acoustics problems. Numerical examples that are based on the use of the ABE are presented in a related article in these proceedings [Bielak et al, 1997a] and elsewhere [eg in Kallivokas et al, 1997b]. Therein, the accuracy and ease of the use of the new elements is demonstrated. Thus far, conditions (15) have only been implemented for spherical geometries; their strength, however, lies in their ability to accommodate boundaries of arbitrary convex geometry. Results based on implementations for non-spherical geometries

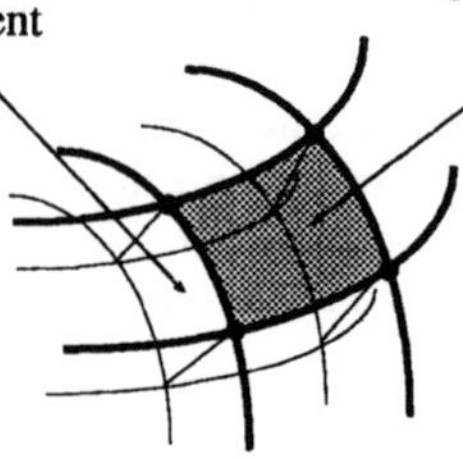

Figure 2: Typical geometry of ABE (Absorbing Boundary Element).

will appear in future communications.

7. References

[1] Barry A., Bielak J., and MacCamy R. C., 1988 "On absorbing boundary conditions for wave propagation," *J. Comput. Phys.*, **79**, pp 449–468

[2] Kallivokas L. F., Bielak J., and MacCamy R. C., 1991, "Symmetric local absorbing boundaries in time and space," *ASCE J. Eng. Mech.*, **117**(9), pp 2027–2048

[3] Bielak J., Kallivokas, L. F., and MacCamy R. C., 1997a, "Absorbing Boundaries Acoustic Wave Propagation Problems," in *Proc. IUTAM Conference on Computational Methods for Unbounded Domains*

[4] Kallivokas, L. F., Bielak J., and MacCamy R. C., 1997b, "A Simple Impedance-Infinite Element for the Finite Element Solution of the Three-Dimensional Wave Equation in Unbounded Domains," *Comput. Methods. Appl. Mech. Engrg.*, **147**, pp 235–262

Acoustic and Electromagnetic Scattering by Large Resonant Structures

G. A. Kriegsmann
Department of Mathematics
Center for Applied Mathematics and Statistics
New Jersey Institute of Technology
Newark, NJ 07102

I. Introduction.

We describe a hybrid method which will efficiently and accurately computes acoustic/electromagnetic waves scattered by a large slowly changing cavity. The cavity is constructed of a waveguide with a flanged opening at one end that couples it to free space, and another flanged opening at the other end. A schematic is shown in Figure 1. The length of this structure $L >> \lambda$ and its walls vary on a scale length H where $L >> H >> \lambda$. This problem crudely models the air intake of a jet engine. Its size and scaling make a direct numerical approach prohibitively expensive.

We use S-matrix theory and adiabatic mode theory to decompose the problem into separate pieces, each one which is solved independently of the others. The first problems require the numerical analyses of a single flanged waveguide in both the receiving and transmitting modes. That is, we must numerically determine how this structure scatters a plane incident wave and how the incident energy is coupled into the propagating modes of the waveguide. And we must also numerically determine how each propagating mode reflects from the aperture and radiates from it. Both of these sub-problems require the accurate numerical approximation of solutions to Helmholtz equation in an infinite domain.

The second sub-problem requires the determination of the propagation of modes in a slowly changing environment. We use adiabatic mode theory to determine how the modes change from one end of the waveguide to the other. Finally, we combine these results to yield an approximate solution of the original field problem. In a sense this procedure is a kind of "Domain Decomposition Scheme" which makes explicit use of the propagating and evanescent mode structure of the underlying waveguide.

The computational burden of this hybrid method is considerably less than a straightforward application of either a finite element or a finite difference method to the entire structure. Only the transmitting and receiving problems require numerical simulations and these do not, with the exception of very high frequency waves, need musch computational overhead. However, these techniques can be used on these sub-problems.

T.L. Geers (ed.), IUTAM Symposium on Computational Methods for Unbounded Domains, 229–234.

II. The Receiving and Transmitting Problem.

For both the receiving and transmitting problems the pressure satisfies the Helmholtz equation

$$\nabla^2 P + k^2 P = 0 \tag{1a}$$

to the left of the structure, as well as inside the guide, where $k = \omega A/c_0$, ω is the frequency of the excitation, c_0 is the speed of sound, and A is the width of the guide to which all lengths have been scaled. On all the bounding surfaces the normal derivative of the field is zero, i.e.,

$$\frac{\partial}{\partial \nu} P = 0. \tag{1b}$$

For the receiving problem the field to the left of the structure is given by

$$P_R = P_0(x, z; \alpha) + P_R^S, \qquad z < 0 \tag{2a}$$

where

$$P_0(x, z; \alpha) = P_{inc}(x, z, \alpha) + P_{inc}(x, -z; \alpha) \tag{2b}$$

$$P_{inc}(x, z; \alpha) = e^{ik(z\cos\alpha + x\sin\alpha)}. \tag{2c}$$

Here P_{inc} is the incident plane wave which is inclined at an angle α with the positive z axis. The second term involving P_{inc} in (2b) is the field caused by a perfectly reflecting sound-hard plane at $z = 0$, i.e. $\frac{\partial}{\partial z} P_0 = 0$, and P_R^S in (2a) represents the field caused by the guide. All these parts each satisfy (1). In addition, the scattered field is given by

$$P_R^S \sim \sigma_R(\theta, \alpha) \frac{e^{ikr}}{\sqrt{r}}, \qquad r \to \infty. \tag{2d}$$

And inside the guide the field is given by the eigenfunction expansion

$$P_R = \sum_{n=0}^{\infty} \tau_n(\alpha) e^{ik_n z} \psi_n(x), \quad z > 0 \tag{3a}$$

where the eigenfunctions are

$$\psi_n = 1, \quad n = 0, \qquad \sqrt{2}\cos(n\pi x), \quad n \geq 1 \tag{3b}$$

$$k_n = \sqrt{k^2 - n^2\pi^2} \tag{3c}$$

and the unknown transmission coefficients are the $\tau_n(\alpha)$.

The transmitting problem is very similar, but now, however, the exciting pressure is the Jth mode propagating from $z = +\infty$. The field in the guide is now given by

$$P_T = e^{-ik_J z} \psi_J(x) + \sum_{n=0}^{\infty} \gamma_n^J e^{ik_n z} \psi_n(x), \quad z > 0 \tag{4a}$$

where the γ_n^J are the unknown reflection coefficients. The field to the left of the structure satisfies (1a) and

$$P_T =\sim \sigma_T^J(\theta)\frac{e^{ikr}}{\sqrt{r}}, \quad r \to \infty. \tag{4b}$$

These two problems can be combined and the concept of an S-Marix can be introduced as follows. Suppose a plane wave of strength a_0 excites the flange waveguide structure and each of the propagating modes also excite it, too. Then by linearity, the cross section σ is given by

$$\sigma = a_0\sigma_R + \sum_{n=0}^{M} \sigma_T^n a_n \tag{5a}$$

where M is the number of propagating modes and a_n is the strength of the nth incident mode. Similarly, if b_i denotes the amplitude of the ith propagating mode which propagates away from the aperture, then

$$\mathbf{b} = a_0\tau + \Gamma\mathbf{a} \tag{5b}$$

where the M vector $\mathbf{b}$ has components b_i, τ components τ_i, and the $M \times M$ matrix Γ is composed of the propagating reflection coefficeints γ_n^i. The linear relationships in (5) forms the S-Matrix.

III. Adiabatic Mode Theory.

We assume here that the waveguide is straight a few wavelengths to the right of the aperture at $z = 0$ and the same is true at the other end of the structure at $z = l$. We also assume for ease of description that the slowly changing guide is symmetric about the line $x = 1/2$ and that modes are cut-off as a result of these changes. Then, if b_i denotes the amplitude of the ith mode which is incident from the left, c_i is the amplitude incident from the right, and a_i is the amplitude of the wave reflected from this part of the structure and propagating to the left, and d_i the analogous amplitude of the reflected wave to the left (see Figure 2), then

$$\mathbf{a} = A\mathbf{c}, \qquad \mathbf{d} = A\mathbf{b} \tag{6a}$$

where A is the diagonal matrix whose entries are

$$A_{nn} = e^{ik\chi_n}. \tag{6b}$$

The phase χ_n is determined from adiabatic mode theory and is given by $\chi_n = \int_0^l w(\lambda z/H)\,dz$, where H is the length scale on which the surface w of the guide is changing.

IV. Solution Reconstruction.

All the necessary ingredients are now in place to construct our approximate solution to the original scattering problem. First we note that the modes with strength d_n that leave the slowly changing portion of the waveguide impinge upon the same aperture discussed in the (II). The modes reflected from it have strength c_n and these are related to the d_n by

$$\mathbf{c} = E\Gamma E\mathbf{d} \tag{7a}$$

where the diagonal matrix E has entries

$$E_{nn} = e^{ik_n l}. \tag{7b}$$

The presence of the matrix E takes into account the location of the aperture at $z = l$.

Next from the adiabatic results (6a) and (7) we have

$$\mathbf{a} = AE\Gamma EA\mathbf{b}. \tag{8}$$

Finally combining this with (5b) we obtain

$$\mathbf{b} = a_0[I - \Gamma AE\Gamma EA]^{-1}\tau \tag{9a}$$

which gives the strengths of the modes propagating to the right in the structure. The strengths of the modes going to the left are given by

$$\mathbf{a} = a_0 AE\Gamma EA[I - \Gamma AE\Gamma EA]^{-1}\tau. \tag{9b}$$

Finally, the cross section is given by combining these results with (5a)

$$\sigma = a_0\sigma_R + \sigma\cdot E\Gamma E\mathbf{b} \tag{9c}$$

where the vector σ has elements σ_T^n.

V. A Numerical Example.

In this example we shall assume for simplicity that the waveguide is straight between the two apertures. Then, $\chi_n = 1$ and the matrix $A = I$. We have used a truncated modal decomposition, or equivalently a Galerkin method using the eigenfunctions as a basis, to obtain approximate solutions to both the receiving and transmitting problems. To keep the results simple and easy to exhibit, we shall only present here the total cross section of the structure as a function of the scaled wave number k, the angle of incidence α and the length l of the structure. The total cross section is defined here by

$$\Sigma = \int_{\pi/2}^{3\pi/2} |\sigma|^2\, d\theta. \tag{10}$$

Figure 3 shows the results for $\alpha = 0$ and $k = .2$ where only the lowest mode propagates and causes a significant resonance at critical lengths. In Figure 4 we have kept the same value of α and have taken $k = 10$, so that 4 modes propagate. Figure 5 has the same value of k, but now $\alpha = \pi/4$. Our code also produces detailed cross sections as function of l, but these will be given elsewhere. Typical running times which include the numerical determination of the impedance matrix take less than a minute or two depending upon the size of k and the number of propagating modes. However, these running times do not significantly depend upon the length of the structure l.

VI. Conclusions.

We have presented a general hybrid method for numerically simulating the scattering of a plane time harmonic wave from a doubly flanged, slowly changing waveguide. Although this formulation assumed no modes were cutoff by these changes, the modifications are straightforward to handle the reflections caused by these effects. In addition, if other obstacles were placed within the guide, then that portion of scattering problem can also be separated out. The S-Matrix for this scattering sub-problem would require, in general, a numerical simulation , too.

The numerical results presented were for a straight waveguide. More complicated structures are presently being studied.

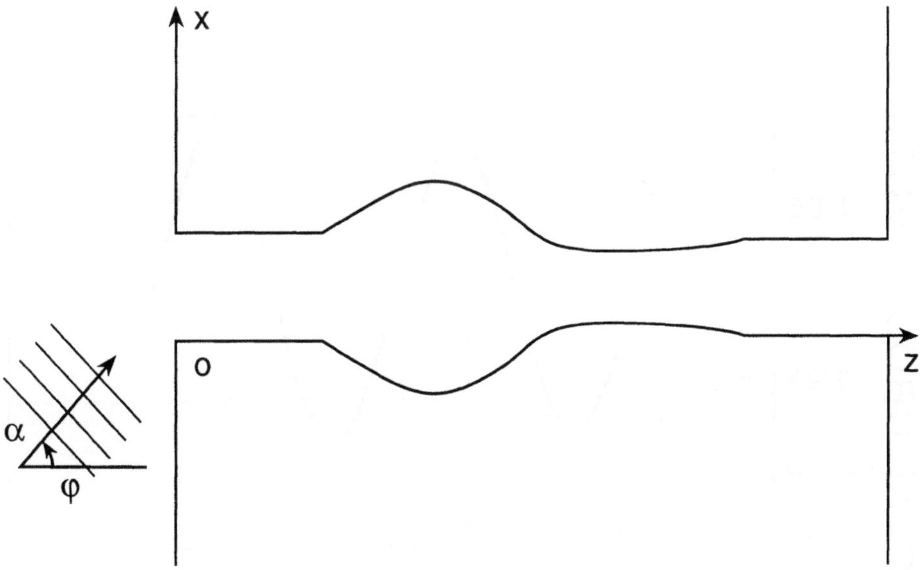

Figure 1

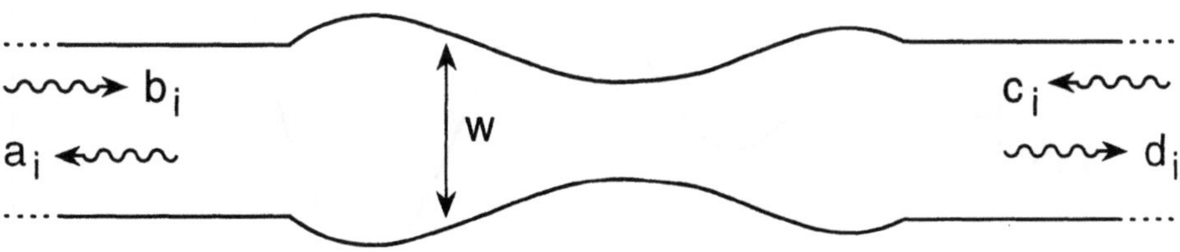

Figure 2

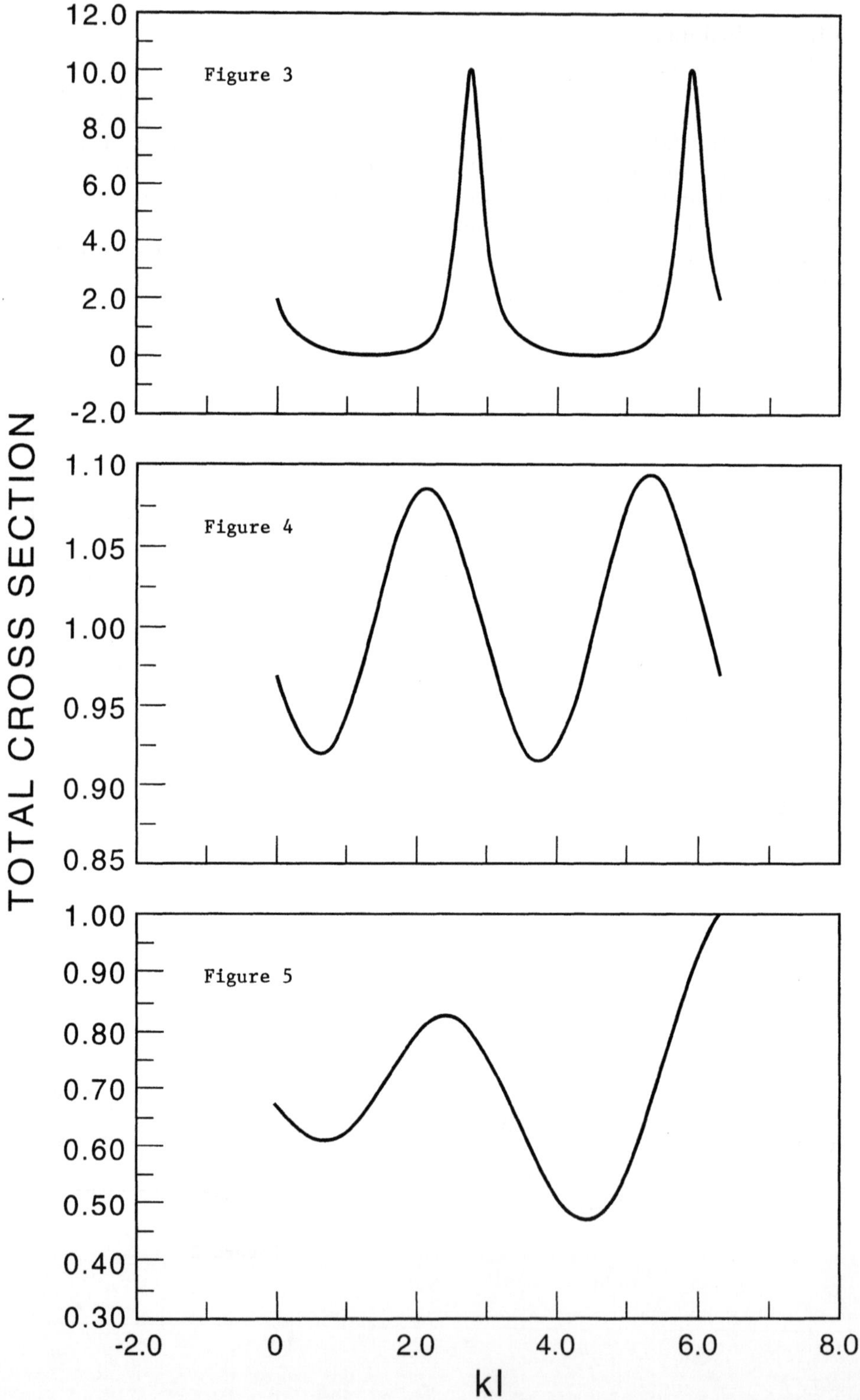
Figure 3
Figure 4
Figure 5
TOTAL CROSS SECTION
kl
12.0
10.0
8.0
6.0
4.0
2.0
0
-2.0
1.10
1.05
1.00
0.95
0.90
0.85
1.00
0.90
0.80
0.70
0.60
0.50
0.40
0.30
-2.0
0
2.0
4.0
6.0
8.0

A Comparison between Time and Frequency Domain Approaches for Rigid Body Scattering Problems

I.C. Mathews and S. Newhouse

Structural Acoustics Group,
Department of Aeronautics,
Imperial College,
Prince Consort Road,
London, UK. SW7 2BY

1 Introduction

All wave problems, such as acoustic or electro-magnetic scattering or underwater shock analysis, have some form of time dependency. The time dependency can be simplified by assuming that the wave has a steady sinusoidal form. This assumption leads to the formulation of the frequency domain method of solving the wave equation. An alternative approach is to explicitly model the progression of the wave through the fluid medium. This is termed the time domain method.

A numerical solution to the differential equations can be found through finite difference, finite elements or the boundary element method. The boundary element method have been used extensively to model the frequency domain over the last 20 years (Mathews 1986, Jeans & Mathews 1992, Newhouse & Mathews 1995). It has the considerable numerical advantage of reducing an analysis over an infinite three dimensional domain to a finite two dimensional surface integration. Boundary elements have also been used in the time domain method through the retarded time integral equation (Antes 1991, Bluck & Walker 1996, Coyette 1995). The main differences between these time domain methods relate to the derivation of the temporal pressure derivative. This variable may be expressed through a finite difference, an isoparametric interpolation of the surface pressure with time, or retained as an independent variable.

T.L. Geers (ed.), IUTAM Symposium on Computational Methods for Unbounded Domains, 235–244.

2 Kirchoff's Retarded Time Integral Equation

The acoustic analysis of a rigid closed object in an infinite exterior fluid domain with speed of sound c and exterior pressure field P is expressed by the wave equation,

$$\nabla^2 P - \frac{1}{c^2}\frac{\partial^2}{\partial t^2}P = 0. \tag{1}$$

The pressure P is assumed to vary across the surface S of the finite body in both space and time. At any instant, the surface pressure can be derived from the wave equation (Morse & Feshbach 1953) and expressed as,

$$2\pi P(r,t) = \int_S \left(\frac{\boldsymbol{n}\cdot\boldsymbol{R}}{R^2}\left[\frac{1}{R}P(x,t_x) + \frac{1}{c}\frac{\partial P}{\partial t}(x,t_x)\right] - \frac{1}{R}\frac{\partial P}{\partial n}(x,t_x)\right) dS(x) \tag{2}$$

where $\boldsymbol{R} = r - x$ and $R = |\boldsymbol{R}|$. The time t_x at point x is expressed by $t_x = t - R/c$. If an incident wave intersects with the surface of the body, a pressure $P_I(r,t)$ exists on the surface. This leads to,

$$\frac{1}{2}P(r,t) = P_I(r,t) + \int_S \left(\frac{\boldsymbol{n}\cdot\boldsymbol{R}}{4\pi R^2}\left[\frac{1}{R}P(x,t_x) + \frac{1}{c}\frac{\partial P}{\partial t}(x,t_x)\right] - \frac{1}{4\pi R}\frac{\partial P}{\partial n}(x,t_x)\right) dS(x) \tag{3}$$

3 Numerical Approach

The surface pressure variation has to be discretised in both space and time. The spatial discretisation of a body or its surface into iso-parametric elements is well established. It has been extensively applied in structural finite elements and acoustic boundary elements in both the time and frequency domains. The variation in pressure across across the body over time is found by advancing from a known state in increments Δt. At each time step the incident wave and boundary conditions are applied, and the current pressure is evaluated by considering the pressure history over the body.

The retarded time integral equation can be expressed as a series of coefficient matrices representing the influences of a pressure disturbance on other parts of the surface as it propagates across the body. These matrices are only dependent on the model's geometry, the surrounding fluid and the analysis time step. They do not need to be re-evaluated for a different excitation frequency which is the case in a frequency domain analysis. The number of disturbance matrices m is related to the maximum distance, R_{MAX} across the surface where m is a rounded up integer

$$m = \frac{R_{MAX}}{c\Delta t}. \tag{4}$$

In general, the discretised retarded time integral equation can be expressed as operations on a set of matrices and vectors. The matrices A_k, B_k and C_k where k varies from 1 to m represent the disturbance coefficient matrices associated with the pressure propagation across the surface. Each coefficient matrix relates the influence of pressure disturbances between a distance $(k-1)c\Delta t$ and $kc\Delta t$ and a point on the surface. The nodal surface pressure vector for each time step i is defined by P_i, $\frac{\partial P}{\partial \boldsymbol{n}_i}$ and $\frac{\partial P}{\partial t_i}$. With this approach the retarded time wave equation is discretised in time by,

$$\sum_{j=1}^{m} A_j P_{(i-j+1)} - \frac{1}{2} P_{(i)} + \sum_{j=1}^{m} B_j \frac{\partial P}{\partial \boldsymbol{n}}_{(i-j+1)} + \sum_{j=1}^{m} C_j \frac{\partial P}{\partial t}_{(i-j+1)} + P_{I(i)} = 0 \tag{5}$$

The disturbance matrices and the pressure history are known and combined into a vector g and gathered on the right hand side of the equation.

$$\left[A_1 - \frac{1}{2} I\right] P_{(i)} + B_1 \frac{\partial P}{\partial \boldsymbol{n}}_{(i)} + C_1 \frac{\partial P}{\partial t_{(i)}} = g \tag{6}$$

The $\frac{\partial P}{\partial \boldsymbol{n}}$ term is defined from the applied boundary conditions, the elastic structural response or in a rigid scattering problem is set to zero. The $\frac{\partial P}{\partial t}$ term is obtained by considering the pressure history. Several different approaches to finding this term will be considered in the next section. The equation therefore reduces to a simple expression which is straight forward to evaluate,

$$\left[A_1 - \frac{1}{2} I\right] P_i = g - C_1 \frac{\partial P}{\partial t}_{(i)} \tag{7}$$

The A_1 matrix can be expressed as a sparse diagonally banded matrix and where g is defined as,

$$g = -P_{I(i)} - \sum_{j=2}^{m} A_j P_{(i-j+1)} - \sum_{j=1}^{m} B_j \frac{\partial P}{\partial \boldsymbol{n}}_{(i-j+1)} - \sum_{j=2}^{m} C_j \frac{\partial P}{\partial t}_{(i-j+1)} \tag{8}$$

4 Derivation of the temporal pressure derivative

4.1 Finite Difference

This approach expresses the temporal pressure derivative as a finite difference of the pressure term (Antes 1991). As it no longer exists as an independent variable it is eliminated from equation (5) and incorporated into matrix A_k. A linear variation of P and a constant variation in $\frac{\partial P}{\partial \boldsymbol{n}}$ is assumed between each time step. The pressure at a time t between time step t_i and t_{i+1} is expressed by,

$$P(x,t) = \frac{1}{\Delta t}\left[(t_{i+1} - t)P_i + (t - t_i)P_{i+1}\right] \tag{9}$$

The disturbance matrices A and B are expressed as,

$$A_i = \int_S \frac{\boldsymbol{n}\cdot\boldsymbol{R}}{4\pi R^2}\left[iH(i-\frac{R}{c\Delta t})H(\frac{R}{c\Delta t}-i+1)-(i-2)H(i-1-\frac{R}{c\Delta t})H(\frac{R}{c\Delta t}-i+2)\right]dS(x) \tag{10}$$

$$B_i = \int_S \frac{1}{4\pi R}H(i-\frac{R}{c\Delta t})H(\frac{R}{c\Delta t}-i+1)dS(x) \tag{11}$$

where H is the Heaviside step function.

4.2 Isoparametric Interpolation

This method allows the temporal variation in pressure to be described through isoparametric interpolating shape functions (Bluck & Walker 1996). These can be defined to be of constant, linear, quadratic or higher interpolating order. As the variation of P over time is described by the shape functions, it is a straight forward to use the derivatives of the shape functions to obtain $\frac{\partial P}{\partial t}$. With this term eliminated as an independent variable the disturbance matrices A and B are expressed by,

$$A_i = \int_S \frac{\boldsymbol{n}\cdot\boldsymbol{R}}{4\pi R^2}H(i-\frac{R}{c\Delta t})H(\frac{R}{c\Delta t}-i+1)\left[\frac{1}{R}-\frac{1}{c}\right]dS(x) \tag{12}$$

$$B_i = \int_S \frac{1}{4\pi R}H(i-\frac{R}{c\Delta t})H(\frac{R}{c\Delta t}-i+1)dS(x) \tag{13}$$

The one dimensional temporal shape functions are identified by T and their derivative by $\dot{T}$ and the temporal parametric coordinate τ. The parametric time coordinate represents the exact position of the quadrature point in retarded time within the current temporal element. The temporal element contains N_{TN} nodes which are Δt apart. The temporal coordinate is defined as,

$$\tau_g = 2.0\left(\frac{R_g}{(N_{TN}-1)c\Delta t} - int\left\{\frac{R_g}{(N_{TN}-1)c\Delta t}\right\}\right) - 1.0 \tag{14}$$

5 Results

5.1 Gaussian Pulse Excitation

A plane wave passes through a point p at time $t = 0$ travelling towards the centre of a sphere. The point p is located where the wave has no initial influence on the sphere. The pressure from the incident wave at point outside the sphere is determined by,

$$P_I(r,t) = exp\left[-f^2(t - D/c)^2\right] \tag{15}$$

where f is the shape factor and D is the perpendicular distance between the point r and wave front at $t = 0$,

$$D = |\boldsymbol{p}| - \frac{\boldsymbol{r} \cdot \boldsymbol{p}}{|\boldsymbol{p}|} \tag{16}$$

No analytical solution exists for this problem in the time domain, but a solution can be found by using the analytical solution in the frequency domain. The Fast Fourier Transform (FFT) algorithm is used to convert the incident wave in the time domain to a representation in the frequency domain (Press, Flannery, Teukolsky & Vetterling 1988). The response to a plain incident wave at the various transformed frequency points is obtained analytically (Junger & Feit 1986). Once the transformed response has been calculated the inverse FFT algorithm is used to convert the frequency domain surface pressure response back into the time domain.

The sphere is discretised by a coarse mesh of 24 elements (98 nodes) and a fine mesh of 96 elements (386 nodes). The convergence of the different numerical methods to the analytical solution at a time step of 0.1s is shown in figure 1 for a 96 element spherical mesh. The variation in surface pressure response is shown for a unit sphere immersed in a unit fluid and excited by a Gaussian Pulse load with a shape factor of 2. Results will be presented in the symposium demonstrating the importance of the mesh and the step size in the solution procedure.

The Courant-Friedrichs-Lewey (CFL) constraint specifies the minimum time step Δt in relation to the minimum mesh spacing d_{min},

$$c\Delta t < \frac{d_{min}}{\sqrt{2}}. \tag{17}$$

This provides an upper limit to Δt in a general finite difference analysis. By using a predictor corrector step as an averaging technique, work with the Electrical Field Integral Equation has shown the $\sqrt{2}$ factor can be reduced to 1.0 (Rynne 1991). This constraint does not seem to present a hard upper limit to the analysis in terms of stability, but rather a point above which the accuracy and stability deteriorates. However, it does provide a useful guide to defining an upper limit to the range of stable time steps. The properties of the spherical discretisations used in this study are shown in in table 1.

5.2 Surface Acceleration

A sphere of radius a which is excited from rest by a normal sinusoidal surface acceleration of amplitude V has an analytical solution for the surface and near field pressure response

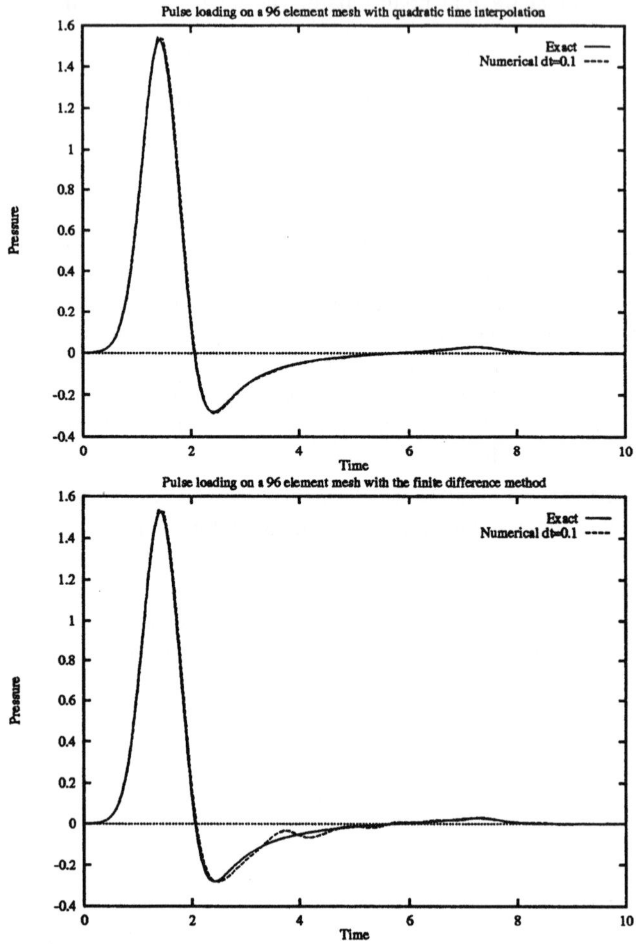

Figure 1: Numerical and analytical surface pressure response from a Gaussian Pulse load at a time step of 0.1s on a 96 element sphere.

Nodes	Quadrilaterals Elements	Minimum Nodal Distance	CFL Limit	Minimum Interior Gauss Distance	Minimum Exterior Gauss Distance
98	24	0.307	0.217	0.249	0.135
386	96	0.152	0.107	0.121	0.065

Table 1: Spacial Properties of two spherical discretisations of a unit sphere

(Astley 1996). The acceleration load is described by,

$$\frac{\partial P}{\partial \boldsymbol{n}}(t) = \begin{cases} 0 & t < 0 \\ -\rho V \sin \omega t & t > 0 \end{cases} \tag{18}$$

The resulting surface pressure at a distance r from the centre of the sphere is obtained through a Laplace transform of the loading to yield,

$$P(r,t) = \frac{\rho a^3 kV}{r(k^2a^2+1)}\left[e^{-d/a} + \frac{1}{ka}\sin kd - \cos kd\right] \tag{19}$$

when $d > 0$, where $d = ct - r + a$ and k is ω/c. If $d \leq 0$ the excitation has not reached the point r in space and the pressure is zero.

Figure 2 shows the variation in surface pressure for a surface acceleration of frequency 0.5Hz applied for 3 complete cycles with an amplitude of 1.0. The initial surface acceleration is then cancelled by applying an additional acceleration that is 180 degrees out of phase with the initial loading. The surface acceleration is neutralised and the surface pressure decays exponentially to zero. The load is applied to a unit sphere which is immersed in a unit fluid. The convergence of the different numerical methods to an analytical solution is shown in figure 2 for a 96 element spherical mesh.

5.3 Comparison with Frequency Domain

Any time domain based loading can be converted to a frequency domain excitation by using the Fast Fourier Transform algorithm. A Gaussian Pulse loading with a shape factor of 2.0 was transformed into the frequency domain and a reduced number of analytical frequency points were used to form the surface pressure response which is shown in figure 3. By using a minimum of 14 analytical frequency domain points, there is no significant differences between the surface pressure response in the time domain for this particular load.

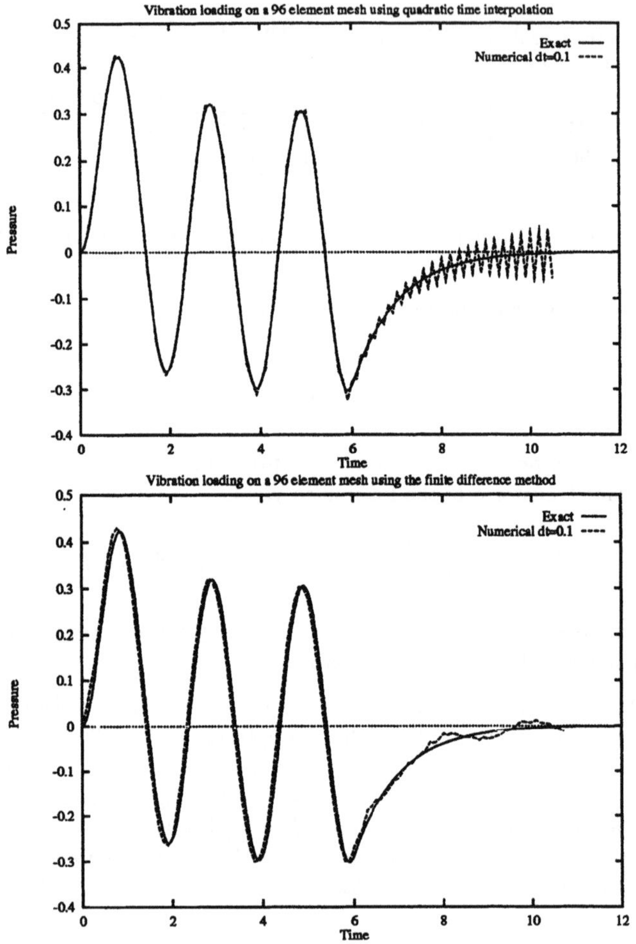

Figure 2: Numerical and analytical surface pressure response from a surface acceleration of 0.5Hz applied for 3 cycles at a time step of 0.1s on a 96 element sphere.

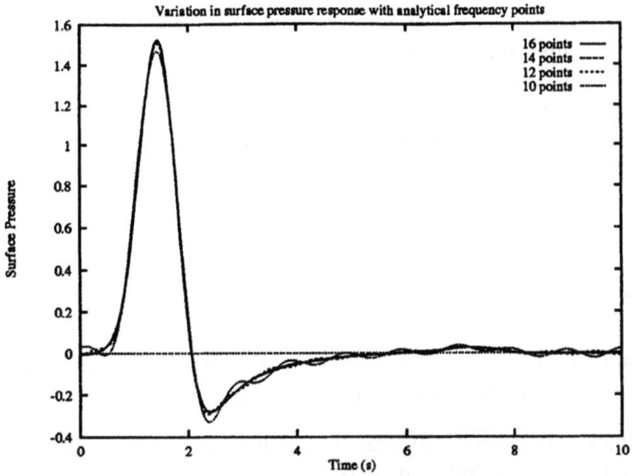

Figure 3: Variation in surface pressure response by using a reduced number of analytical frequency domain points.

To solve this loading problem numerically in the frequency domain would require a solution of a spherical mesh at various frequencies up to a wavenumber of $ka = 8.07$. To accurately solve the problem in the frequency domain requires a significantly finer mesh than the 96 element mesh used in this comparison study.

The equivalent time domain problem was solved using 100 time-steps each of size 0.1 s in 31s using the coarse mesh discretization and 474s using the fine mesh discretization. The frequency domain solutions obtained using a Galerkin type integration procedure required 25s per frequency for the 24 element coarse mesh and 476s per frequency for the 96 element fine mesh. Using a collocation type integration procedure the execution times per frequency were 65s for the coarse mesh and 121s for the fine mesh.

6 Conclusions

A finite difference and an isoparametric based temporal interpolation method have been presented for solving the acoustic pressure using the Kirchoff's Retarded Time Integral Equation. The effectiveness of each method has been shown for both a Gaussian Pulse and a surface acceleration load. Although both of the methods produced converged stable results on the Gaussian Pulse load, their performance with the surface acceleration was influenced by the analysis time step and the mesh spacing. However, stable results seem to be obtained if a time step is selected that is of the order of the minimum gauss point separation. The finite difference method is computationally cheaper than the quadratic temporal pressure interpolation. Although, the execution times for the frequency and time domain methodologies are problem dependent it would appear from our example problems that, if the overall time history response is required, than the time domain method is significantly the computationally faster approach.

References

Antes, H. (1991), Applications in Environmental Noise, *in* R. D. Ciskowski & C. A. Brebbia, eds, 'Boundary Element Methods in Acoustics'.

Astley, R. J. (1996), 'Transient Wave Envelope Elements for Wave Problems', *Journal of Sound and Vibration* **192**(1), 245–261.

Bluck, M. J. & Walker, S. P. (1996), 'Analysis of Three Dimensional Transient Acoustic Wave Propagation using the Boundary Integral Equation', *International Journal for Numerical Methods in Engineering* **39**, 1419–1431.

Coyette, J. P. (1995), Recent advances in transient acoustics using finite element and boundary element methods, *in* 'Euronoise 1995', pp. 515–520.

Jeans, R. & Mathews, I. C. (1992), 'A comparison of numerical collocation and variational procedures to the hypersingular acoustic integral operator', *Computer Methods in Applied Mechanics and Engineering* **101**, 5–26.

Junger, M. C. & Feit, D. (1986), *Sound, structures and their interaction*, MIT Press (Cambridge).

Mathews, I. C. (1986), 'Numerical techniques for three-dimensional steady-state fluid-structure interaction', *Journal of the Acoustic Society of America* **79**, 1317–1325.

Morse, P. M. & Feshbach, H. (1953), *Methods of Theoretical Physics, Parts I & II*, McGraw-Hill.

Newhouse, S. J. & Mathews, I. C. (1995), Adaptivity and three dimensional hierarchical boundary elements for rigid acoustic scattering problems, *in* 'ASME Conference on Mechanical Vibration and Noise', Vol. 3C.

Press, W. H., Flannery, B. P., Teukolsky, S. A. & Vetterling, W. T. (1988), *Numerical Recipes in C*, Cambridge University Press.

Rynne, B. P. (1991), 'Time Domain Scattering from Arbitrary Surfaces Using the Electric Field Integral Equation', *Journal of Electromagnetic Waves and Applications* **5**, 93–112.

OPTIMIZING THE PERFECTLY MATCHED LAYER

P. MONK
Mathematical Sciences
University of Delaware
Newark, DE 19716
USA

F. COLLINO
INRIA
B.P. 105
78153 Le Chesnay
France

1 Introduction

The Bérenger perfectly matched layer (or PML) [4, 3] is used in computational electromagnetism as a "sponge layer" to terminate finite element approximations of scattering problems. An infinite perfectly matched layer creates no reflection for incident waves of any frequency or any incident direction, and the waves decay exponentially in magnitude into the layer. This property is lost when the layer is truncated since the truncation boundary generates a reflected wave. Discretization of the differential equations further changes the reflection coefficient of the layer. In this talk we shall describe the PML in the simple case of a two dimensional electromagnetic problem. The first part of this abstract is expository. We shall show how to derive the layer using a change of variable approach [5, 9, 6]. After this derivation we describe some new analysis to show the effects of truncating and discretizing the PML [9]. We end by showing how to compute an optimal layer, and make some remarks about PML for the full Maxwell system.

Using the change of variables approach, perfectly matched layers have also been developed in acoustics, quantum mechanics [8] and elastodynamics [7]. The literature concerning the PML is very large. We will mention only a small fraction of the available papers here.

2 The PML constructed using change of variables

In this section we shall show how Bérenger's PML can be viewed as a complex change of variables applied to the Maxwell system. The transverse electric wave in free space (with $\epsilon_0 = \mu_0 = 1$) is governed by

$$\left\{\begin{array}{l} \dfrac{\partial H_z}{\partial t} = \dfrac{\partial E_x}{\partial y} - \dfrac{\partial E_y}{\partial x}, \\ \dfrac{\partial E_y}{\partial t} = -\dfrac{\partial H_z}{\partial x}, \quad \dfrac{\partial E_x}{\partial t} = \dfrac{\partial H_z}{\partial y}. \end{array}\right. \tag{1}$$

Suppose we wish to construct an infinite 2D PML in the region $x > 0$. The classical construction of Bérenger is performed in two steps. First, we split the magnetic field:

$$H_z = H_{zx} + H_{zy} \tag{2}$$

T.L. Geers (ed.), IUTAM Symposium on Computational Methods for Unbounded Domains, 245–254.

and rewrite equations (1) as

$$\begin{cases} \dfrac{\partial H_{zy}}{\partial t} = \dfrac{\partial E_x}{\partial y}, & \dfrac{\partial H_{zx}}{\partial t} = -\dfrac{\partial E_y}{\partial x}, \\ \dfrac{\partial E_y}{\partial t} = -\dfrac{\partial H_z}{\partial x}, & \dfrac{\partial E_x}{\partial t} = \dfrac{\partial H_z}{\partial y}. \end{cases} \tag{3}$$

In the second step, a damping factor $\sigma(x)$, ($\sigma(x) = 0$ for $x < 0$) is introduced in each equation in those places where the normal derivative operator ∂_x appears and the other parts of the equations remain unchanged:

$$\begin{cases} \dfrac{\partial H_{zy}}{\partial t} = \dfrac{\partial E_x}{\partial y}, & \dfrac{\partial H_{zx}}{\partial t} + \sigma H_{zx} = -\dfrac{\partial E_y}{\partial x}, \\ \dfrac{\partial E_y}{\partial t} + \sigma E_y = -\dfrac{\partial H_z}{\partial x}, & \dfrac{\partial E_x}{\partial t} = \dfrac{\partial H_z}{\partial y}. \end{cases} \tag{4}$$

System (4) is Bérenger's PML model. We shall refer to this as a split field PML.

If we look for the time harmonic solutions with frequency ω, we get

$$\begin{cases} -i\omega \hat{H}_{zy} = \dfrac{\partial \hat{E}_x}{\partial y}, \quad (-i\omega + \sigma(x))\hat{H}_{zx} = -\dfrac{\partial \hat{E}_y}{\partial x}, \\ (-i\omega + \sigma(x))\hat{E}_y = -\dfrac{\partial \hat{H}_z}{\partial x}, \quad -i\omega \hat{E}_x = \dfrac{\partial \hat{H}_z}{\partial y}, \quad \hat{H}_z = \hat{H}_{zx} + \hat{H}_{zy}. \end{cases} \tag{5}$$

We can rewrite this system of equations with the only unknown being $\hat{H}_z$ to obtain

$$\omega^2 \hat{H}_z + \frac{\partial^2 \hat{H}_z}{\partial y^2} + \frac{i\omega}{i\omega - \sigma(x)} \frac{\partial}{\partial x}\left(\frac{i\omega}{i\omega - \sigma(x)} \frac{\partial \hat{H}_z}{\partial x}\right) = 0. \tag{6}$$

In the region where σ is zero, we recover the well known property that the magnetic field of a time harmonic 2D transverse electric wave satisfies the Helmholtz equation.

Inside the PML layer, we can see that the PML model is derived by making the simple substitution

$$\frac{\partial}{\partial x} \longrightarrow \frac{\partial}{\partial x'} = \frac{i\omega}{i\omega - \sigma(x)} \frac{\partial}{\partial x}, \quad x' = x + \frac{i}{\omega}\int_0^x \sigma(\xi) d\xi, \tag{7}$$

in the initial equation and leaving the tangential (y) derivatives unchanged.

From this observation we can see how to derive the PML by a change of variables. We now derive (5) by this approach. We start by moving to the frequency domain. Let

$$E_1 = \hat{E}_1(x, y)\exp(-i\omega t), \; E_2 = \hat{E}_2(x, y)\exp(-i\omega t), \; H_z = \hat{H}_z(x, y)\exp(-i\omega t).$$

and substituting these expressions into the time dependent system (1) we obtain

$$\begin{cases} -i\omega \hat{H}_z = \dfrac{\partial \hat{E}_x}{\partial y} - \dfrac{\partial \hat{E}_y}{\partial x}, \\ -i\omega \hat{E}_y = -\dfrac{\partial \hat{H}_z}{\partial x}, \quad -i\omega \hat{E}_x = \dfrac{\partial \hat{H}_z}{\partial y}. \end{cases} \tag{8}$$

Then we introduce the stretched variable x' as defined in (7). Using x', we modify the Maxwell system to read (we still use $\hat{E}_1$, $\hat{E}_2$ and $\hat{H}$ as the solution of the modified system)

$$\begin{cases} -i\omega\hat{H}_z = \dfrac{\partial\hat{E}_x}{\partial y} - \dfrac{\partial\hat{E}_y}{\partial x'}, \\ -i\omega\hat{E}_y = -\dfrac{\partial\hat{H}_z}{\partial x'}, \quad -i\omega\hat{E}_x = \dfrac{\partial\hat{H}_z}{\partial y}. \end{cases} \tag{9}$$

and using the chain rule to replace x' by x we obtain

$$\begin{cases} -i\omega\hat{H}_z = \dfrac{\partial\hat{E}_x}{\partial y} - \dfrac{1}{1+i\sigma/\omega}\dfrac{\partial\hat{E}_y}{\partial x}, \\ -i\omega\hat{E}_y = -\dfrac{1}{1+i\sigma/\omega}\dfrac{\partial\hat{H}_z}{\partial x}, \quad -i\omega\hat{E}_x = \dfrac{\partial\hat{H}_z}{\partial y}. \end{cases} \tag{10}$$

This is exactly (5) with the split field equations combined into one equation. We could now split the fields and move back to the time domain to obtain the split field PML.

It is easy to see that this infinite PML produces no reflection at the boundary $x = 0$ and that waves are absorbed exponentially as x increases (see [9]). However, to obtain a practical method, the PML must be truncated, say at the line $x = \delta$. Bérenger uses a simple Dirichlet condition

$$\hat{E}_y(x = \delta, y, t) = 0, \quad \forall y, t > 0.$$

It is also possible to use an absorbing boundary condition on $x = \delta$ (see [10, 9]).

For simplicity, we assume the use of the Dirichlet boundary condition in this abstract. This reflecting boundary at $x = \delta$ generates a reflected wave. In [9] we generalize Bérenger's original estimate of the reflection coefficient to show that the reflection coefficient for the PML layer with Dirichlet condition is

$$R = -\exp\left(-2ik_x \int_0^\delta (1 + i\sigma(\xi)/\omega)\, d\xi\right) \tag{11}$$

where the incident wave is a plane wave $\hat{H}_z = \exp(i(k_x x + k_y y))$ and $k_x^2 + k_y^2 = \omega^2$. Obviously the reflection coefficient depends on the angle of incidence of the wave, but does not depend on the shape of σ (provided the integrated value of σ is fixed). For a given angle of incidence, the reflection coefficient can be made arbitrarily small by taking $\int_0^\delta \sigma(\xi)\, d\xi$ sufficiently large. Unfortunately this conclusion does not hold when the problem is discretized.

3 Discretization of the planar PML

Now we depart from Bérenger original method and follow [14] to derive a non-split field PML. We define the Bérenger variables

$$\tilde{E}_1 = (1 + i\sigma/\omega)\hat{E}_1, \quad \tilde{E}_2 = \hat{E}_2, \quad \tilde{H}_z = \hat{H}_z$$

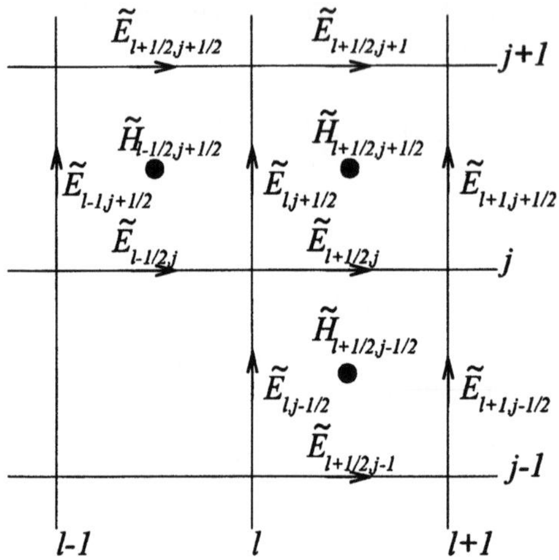

Figure 1: Notation for an interior cell

which satisfy

$$\begin{cases} -i\omega(1+i\sigma/\omega)\tilde{H}_z = \dfrac{\partial \tilde{E}_x}{\partial y} - \dfrac{\partial \tilde{E}_y}{\partial x}, \\ -i\omega(1+i\sigma/\omega)\tilde{E}_y = -\dfrac{\partial \tilde{H}_z}{\partial x}, \quad -i\omega\tilde{E}_x = \dfrac{\partial \tilde{H}_z}{\partial y}. \end{cases} \tag{12}$$

$$\begin{aligned} -i\frac{\omega}{1+i\sigma/\omega}\tilde{E}_1 &= \frac{\partial \tilde{H}_z}{\partial y}, \\ -i\omega\left(1+i\sigma/\omega\right)\tilde{E}_2 &= -\frac{\partial \tilde{H}_z}{\partial x}, \\ -i\omega\left(1+i\sigma/\omega\right)\tilde{H}_z &= \frac{\partial \tilde{E}_1}{\partial y} - \frac{\partial \tilde{E}_2}{\partial x}. \end{aligned}$$

We wish to discretize these equations using mass-lumped edge elements (or equivalently FDTD [13]). First consider a cell with all its edges away from the boundaries, using the notation shown in Figure 1.

We suppose that σ is a piecewise constant function of x and with jumps at $x = lh$, $l = 0, 1, 2, 3, \cdots$. We denote by $\sigma_{l+1/2}$ the value of σ on the interval $(lh, (l+1)h)$. Of course $\sigma_{l+1/2} = 0$ if $l < 0$. Then we define

$$\gamma_{l+1/2} = 1 + i\sigma_{l+1/2}/\omega.$$

Using the mass-lumped finite element method to discretize (12) we obtain:

$$-i\frac{\omega}{\gamma_{l+1/2}}\tilde{E}_{l+1/2,j} = \frac{\tilde{H}_{l+1/2,j+1/2} - H_{l+1/2,j-1/2}}{h} \tag{13}$$

$$-i\omega\left(\frac{\gamma_{l+1/2}+\gamma_{l-1/2}}{2}\right)\tilde{E}_{l,j+1/2} = -\left(\frac{\tilde{H}_{l+1/2,j+1/2}-H_{l-1/2,j+1/2}}{h}\right) \tag{14}$$

$$-i\omega\gamma_{l+1/2}H_{l+1/2,j+1/2} = \frac{\tilde{E}_{l+1/2,j+1}-\tilde{E}_{l+1/2,j}}{h} - \frac{\tilde{E}_{l+1,j+1/2}-\tilde{E}_{l,j+1/2}}{h}. \tag{15}$$

These equations hold for all j and for $l \leq n_l - 1$ where $\delta = n_l h$ is the thickness of the absorbing layer. Next we eliminate $\tilde{E}$ from the above equations to obtain

$$\begin{aligned} -h^2\omega^2\gamma_{l+1/2}\tilde{H}_{l+1/2,j+1/2} &= \gamma_{l+1/2}\left(\tilde{H}_{l+1/2,j+3/2} - 2\tilde{H}_{l+1/2,j+1/2} + \tilde{H}_{l+1/2,j-1/2}\right) \\ &\quad \left(\left(\frac{2}{\gamma_{l+3/2}+\gamma_{l+1/2}}\right)\left(\tilde{H}_{l+3/2,j+1/2} - \tilde{H}_{l+1/2,j+1/2}\right)\right. \\ &\quad \left. + \left(\frac{2}{\gamma_{l+1/2}+\gamma_{l-1/2}}\right)\left(\tilde{H}_{l+1/2,j+1/2} - \tilde{H}_{l-1/2,j+1/2}\right)\right). \end{aligned} \tag{16}$$

This is a discrete form of (6).

Now we assume that

$$\tilde{H}_{l+1/2,j+1/2} = exp(i\kappa_2(j+1/2)h)\tilde{H}_{l+1/2}$$

then using this in (16) and defining

$$\lambda = 1 - \frac{4}{h^2\omega^2}\sin^2(\kappa_2 h/2)$$

we obtain

$$\begin{aligned} -h^2\omega^2\lambda\gamma_{l+1/2}\tilde{H}_{l+1/2} &= \frac{2}{\gamma_{l+3/2}+\gamma_{l+1/2}}\left(\tilde{H}_{l+3/2} - \tilde{H}_{l+1/2}\right) \\ &\quad - \frac{2}{\gamma_{l+1/2}+\gamma_{l-1/2}}\left(\tilde{H}_{l+1/2} - \tilde{H}_{l-1/2}\right), n_l - 1 > l. \end{aligned}$$

We start by asking if the layer is still perfectly matched after discretization. We consider an infinite layer with constant σ and suppose that

$$\gamma_{l+1/2} = \gamma = 1 + i\frac{\sigma}{k} \text{ for } l > 0.$$

If we assume that

$$\begin{aligned} \tilde{H}_{l+1/2} &= \exp(i\kappa_1 h(l+1/2)) + R\exp(-i\kappa_1 h(l+1/2)),\ l \leq -1 \\ \tilde{H}_{l+1/2} &= T\exp(i\kappa_1^\sigma h(l+1/2)),\ l \geq 0 \end{aligned}$$

then κ_1 must satisfy the usual dispersion relation

$$4\sin^2(\kappa_1 h/2) = \lambda\omega^2 h^2. \tag{17}$$

and κ_1^σ must satisfy the corresponding equation for the absorbing layer

$$4\sin^2(\kappa_1^\sigma h/2) = \lambda\omega^2 h^2\gamma^2.$$

Then R and T are found by solving the equations at the interface

$$\begin{aligned}\lambda h^2\omega^2\tilde{H}_{-1/2} + \frac{2}{\gamma+1}\left(\tilde{H}_{1/2} - \tilde{H}_{-1/2}\right) - \left(\tilde{H}_{-1/2} - \tilde{H}_{-3/2}\right) &= 0,\\ \lambda h^2\omega^2\tilde{H}_{1/2} + \frac{1}{\gamma}\left(\tilde{H}_{3/2} - \tilde{H}_{1/2}\right) - \frac{2}{\gamma+1}\left(\tilde{H}_{1/2} - \tilde{H}_{-1/2}\right) &= 0.\end{aligned}$$

Using MAPLE we obtain that the reflection coefficient is given by

$$R = -\frac{1}{16}\frac{\sigma((\omega^2-\kappa_2^2)\sigma + 2\omega(\kappa_2^2-\omega^2)i)}{\omega^2}h^2 + O(h^4)$$

Thus, discretizing the PML has introduced a small reflection from the interface at $x = 0$. Since the magnitude of R is proportional (at least for small h) to $\sigma^2 h^2$, we see that σ cannot be arbitrarily large in the discrete PML. This leads us to the problem of determining an optimal PML that balances the requirement of large σ for good absorption over the need for a small value of σ to ensure a small numerical reflection. We resort to numerical methods for this optimization, but can avoid the complex genetic algorithm approach of [6].

We start by completing the finite difference equations at $x = \delta$. The final equation, at $l = n_l - 1$, depends on the boundary data. In the case of Dirichlet boundary conditions, we simply choose $\tilde{E}_{n_l,j+1/2} = 0$ for all j. Thus given $\tilde{H}_{-3/2}$ we can compute

$$\vec{H} = (\tilde{H}_{-1/2}, \tilde{H}_{1/2}, \cdots, \tilde{H}_{n_l-1/2})^T$$

by solving the matrix equation

$$M\vec{H} = -\tilde{H}_{-3/2}\vec{F}$$

where M is the $(n_l+1)\times(n_l+1)$ matrix defined by

$$M = \begin{pmatrix} c_{-1/2} & d_{1/2} & 0 & \cdots & \\ d_{1/2} & c_{1/2} & d_{3/2} & 0 & \cdots \\ 0 & d_{3/2} & c_{3/2} & d_{5/2} & \cdots \\ & & \ddots & \ddots & \ddots \\ & \cdots & 0 & d_{N_x-3/2} & c_{N_x-1/2}\end{pmatrix}.$$

Here

$$\begin{aligned}c_{j+1/2} &= h^2\omega^2\lambda\gamma_{j+1/2} - 1/(\gamma_{j+3/2}+\gamma_{j+1/2}) - 2/(\gamma_{j+1/2}+\gamma_{j-1/2}),\ -1\le j\le n_l-2\\ d_{j+1/2} &= 2/(\gamma_{j+1/2}+\gamma_{j-1/2}),\quad 0\le j\le n_l-1,\\ c_{n_l-1/2} &= h^2\omega^2\lambda\gamma_{n_l-1/2} - 2/(\gamma_{n_l-1/2}+\gamma_{n_l-3/2}).\end{aligned}$$

We can show that the reflection coefficient for this layer is given by

$$R = -\frac{1+\left(F^T\cdot M^{-1}\cdot F\right)e^{-i\kappa_1 h}}{1+\left(F^T\cdot M^{-1}\cdot F\right)e^{i\kappa_1 h}}.$$

	$N=5$	$N=10$	$N=20$	$N=40$	$N=1000$
$h\sigma_{1/2}$	0.0335	0.178	0.231	0.241	0.259
$h\sigma_{3/2}$	0.147	0.110	0.206	0.253	0.305
$h\sigma_{5/2}$	0.375	0.496	0.230	0.290	0.374
$h\sigma_{7/2}$	0.750	0.516	0.415	0.369	0.482
$h\sigma_{9/2}$	1.32	1.13	0.915	0.549	0.667
$h\sigma_{11/2}$	2.24	1.67	1.30	1.03	1.052
$h\sigma_{13/2}$	4.07	3.22	2.33	2.10	2.13
$h\sigma_{15/2}$	8.35	6.41	4.90	4.74	6.73
$h\sigma_{17/2}$	19.5	16.3	14.5	15.3	11.9
$h\sigma_{19/2}$	64.6	53.2	38.6	18.0	12.4
Average $R(\%)$	0.301%	0.481%	0.753%	1.18%	1.36%

Table 1: Optimized values of σ in each strip of the PML using Dirichlet boundary conditions and a layer of width $n_l = 10$.

where $F = (1, 0, \cdots, 0)^T$. To describe the parametric dependence of R we write

$$\kappa_1 = |\kappa| \cos(\theta) \quad \text{and} \quad \kappa_2 = |\kappa| \sin(\theta)$$

where θ is the angle of incidence of the plane wave. For a given angle θ we can use the dispersion relation for free space (17) to determine $|\kappa|$ provided $\omega h < 2$. Furthermore λ only depends on θ. This reparametrization, together with the notion of number of points per wavelengths defined by $N = 2\pi/(\omega h)$ allows us to write

$$R = R\left(\theta, N, h\sigma_{1/2}, \cdots, h\sigma_{n_l-1/2},\right).$$

To compute an optimal layer $h\sigma_{1/2}, \cdots, h\sigma_{n_l-1/2}$, we minimize

$$\frac{1}{100}\sum_{q=1}^{100} \cos(\theta_q) |R\left(\theta_q, N, h\sigma_{1/2}, \cdots, h\sigma_{n_l-1/2},\right)|^2$$

where $\theta_q = \pi(q-1)/200$. The use of a weight function $\cos(\theta_q)$ avoids local minima. Results of this computation for $n_l = 10$ are shown in Table 1. For the optimal layer, σh generally increases into the layer approximately geometrically. The resulting layers have an average reflection coefficient close to 1% even for long wavelength incident waves. The angular dependence of the optimal reflection coefficient is shown in Figures 2 and 3. Remarkabley, the optimal reflection coefficients differ considerabley from the reflection coefficient in the absence of discretization given by (11).

4 Further Developments

The PML can be generalized to curvilinear coordinate systems using the change of variables approach [9, 12, 14]. In the time harmonic case, using polar coordinates some

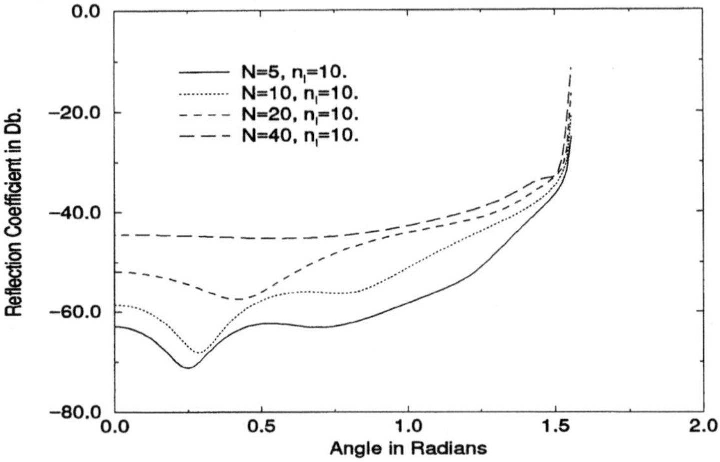

Figure 2: We show the optimal reflection coefficient against angle of incidence of the plane wave for a fixed layer size $n_l = 10$, and changing number of grid points per wavelength ($N = 5$, $N = 10$, $N = 20$, $N = 40$).

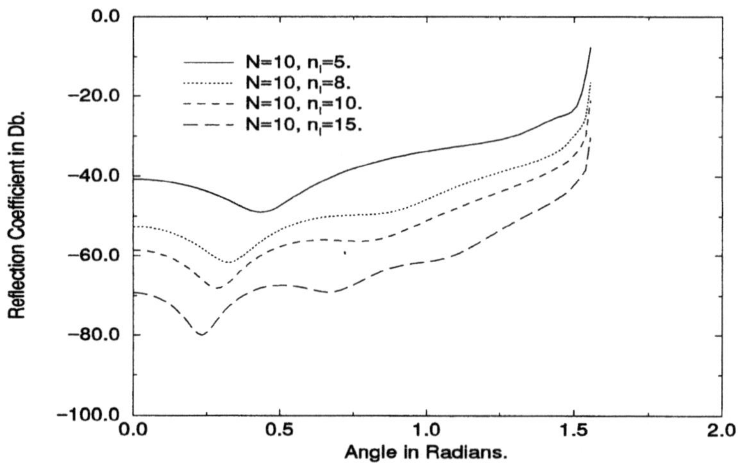

Figure 3: We show the optimized coefficient against angle of incidence of the plane wave for a fixed number of grid points per wavelength $N = 10$, and changing layer size ($n_l = 5$, $n_l = 8$, $n_l = 10$, $n_l = 15$).

theoretical analysis has been performed. In [9] existence was proved except for a possible discrete set of irregular frequencies. In [11], using a very special choice of σ, existence, uniqueness and convergence was proved. Convergence in the general case remains to be analyzed.

For the time domain problem, both split field (as in the original Bérenger PML) and non-split field PML [14, 2] are used. The split field approximation may be unstable for long time computations [1]. A complete analysis of the time domain PML is not currently available.

Both the split field and non-split field can be extended to the full vector 3D Maxwell equations. Either versions can be derived using the change of variables approach either scaling the unknowns (non-split) or without scaling the unknowns (split).

5 Conclusion

The PML can be used in the time or frequency domain to terminate scattering calculations. This technique has proved very successful in practice and is easy to implement. The effect of discretizing the PML leads to some reflection from the layer but this can be controlled by a suitable design of the layer.

A theoretical understanding of the PML is not yet available, and the best implementation in the time domain is still under debate.

Acknowledgement and Disclaimer

Effort of Peter Monk sponsored by the Air Force Office of Scientific Research, Air Force Materials Command, USAF, under grant number F49620-95-1-0067. The US Government is authorized to reproduce and distribute reprints for governmental purposes notwithstanding any copyright notation thereon. The views and conclusions contained herein are those of the authors and should not be interpreted as necessarily representing the official policies or endorsements, either expressed or implied, of the Air Force Office of Scientific Research or the US Government.

References

[1] S. Abarbanel and D. Gottlieb. A mathematical analysis of the PML method. Preprint, 1996.

[2] S. Abarbanel and D. Gottlieb. On the construction and analysis of absorbing layers in C.E.M. Preprint, 1996.

[3] J.-P. Bérenger. Perfectly matched layer for the FDTD solution of wave-structure interaction problems. *IEEE Trans. Antennas Propagat.*, 44:110–117, 1996.

[4] J.P. Berenger. A Perfectly Matched Layer for the Absorption of Electromagnetics Waves. *J. Comput. Phys.*, 114:185–200, 1994.

[5] W. C. Chew and W. H. Weedon. A 3D perfectly matched medium from modified Maxwell's equations with stretched coordinates. *Microwave Opt. Technol. Lett.*, 7(13):599–604, 1994.

[6] W.C. Chew, J.M. Jin, and E. Michielssen. Complex coordinate stretching as a generalized absorbing boundary condition. Technical Report CCEM-20-96, Center for Computational Electromagnetics, University of Illinois at Urbana-Champaign, 1996.

[7] W.C. Chew and Q.H. Liu. Perfectly matched layers for elastodynamics: A new absorbing boundary condition. Technical Report GEO-002-95-20, Schlumberger-Doll Research, 1995.

[8] F. Collino. Perfectly matched layer for the paraxial equation. J. Comp. Phys., submitted in 1995.

[9] F. Collino and P. Monk. The perfectly matched layer in curvilinear coordinates. Technical Report 3049, INRIA, 1996.

[10] J.-M. Jin and W.C. Chew. Combining PML and ABC for the finite-element analysis of scattering problems. *Microwave and Optical Technology Letters*, 12:192–197, 1996.

[11] M. Lassas and E. Somersalo. On the existence and convergence of the solution of the PML equations. Preprint, 1997.

[12] B. Yang, D. Gottlieb, and J.S. Hesthaven. On the use of PML ABC's in spectral time-domain simulations of electromagnetic scattering. Preprint.

[13] K.S. Yee. Numerical solution of initial boundary value problems involving Maxwell's equations in isotropic media. *IEEE Trans. on Antennas and Propagation*, AP–16:302–307, 1966.

[14] L. Zhao and A.C. Cangellaris. A general approach for developing unsplit-field time-domain implementations of perfectly matched layers for FDTD grid truncation. *IEEE Trans. Microwave Theory Tech.*, 44:2555–2563, 1996.

Implementing highly accurate non-reflecting boundary conditions for large scale problems in structural acoustics

Assad A. Oberai, Manish Malhotra, Peter M. Pinsky
Division of Mechanics and Computation, Stanford University,
Stanford, CA, USA.

1 Introduction

Time-harmonic problems of acoustic radiation and scattering in unbounded domains are governed by the Helmholtz equation in the fluid domain, coupled to the structural equations for elastic bodies in contact with the fluid. In addition to the boundary conditions applied to the structure and the fluid-structure interface conditions, a radiation condition is applied at infinity to render the solution to the problem unique. When modeling such problems with domain based methods such as the finite element method, the finite element mesh is truncated at a finite distance from the structure and either an artificial boundary condition that replicates the infinite domain is applied, or special elements such as infinite elements are used.

In this study we discuss the implementation of a highly accurate artificial boundary condition, the so-called DtN map and an infinite element formulation proposed by Burnett [1], based on the multipole expansion. We focus on large-scale problems in structural acoustics implying thereby, that the problem size is large enough to rule out the use of direct solvers in solving the resulting matrix equations. In particular, we study the performance of the DtN map and infinite elements in conjunction with Krylov subspace iterative methods.

We observe that a naive implementation of the DtN boundary condition results in excessive storage and computation costs, whereas the infinite element formulation leads to a matrix system with very poor convergence properties. We propose alternative formulations to overcome these difficulties and demonstrate their efficacy. In section 1 we introduce the problem of interest and describe the DtN and infinite element formulations. In section 2 we focus on the implementation of the DtN map and in section 3 we consider the implementation of the infinite element formulation.

T.L. Geers (ed.), IUTAM Symposium on Computational Methods for Unbounded Domains, 255–264.

2 Helmholtz equation on unbounded domains

We consider the problem of radiation in d dimensions. The strong form of the problem is: Find u such that

$$\mathcal{L}u \equiv -\nabla^2 u - k^2 u = f \qquad \text{in } \Omega \tag{1}$$

$$u = g \qquad \text{on } \Gamma_a \tag{2}$$

$$\lim_{r\to\infty} r^{\frac{d-1}{2}}(\frac{\partial}{\partial r}u - iku) = 0 \tag{3}$$

In equation (1) Ω represents the unbounded domain, k is the wavenumber associated with the problem and f is the prescribed forcing function. In equation (2) g is the prescribed Dirichlet data on the the surface of the radiator denoted by Γ_a. Equation (3) is the Sommerfeld radiation condition which requires all waves at infinity to be outgoing.

In Ω consider a hypersurface denoted by Γ_R that divides Ω into a bounded domain Ω_R and an unbounded domain Ω_∞. Γ_R is chosen such that Γ_a completely encloses Γ_a, and $f|_{\Omega_R} \equiv 0$. In this study Γ_R is a "d" dimensional ball of radius R.

2.1 The DtN boundary condition

The strong form of the problem associated with the non-reflecting boundary condition formulation is given by

$$\mathcal{L}u = f \qquad \text{in } \Omega_R \tag{4}$$

$$u = g \qquad \text{on } \Gamma_a \tag{5}$$

$$\nabla u \cdot n = M(u) \qquad \text{on } \Gamma_R \tag{6}$$

where $M(u)$ is the boundary condition imposed on u at Γ_R. Let $M_{ex}(u)$ be the boundary operator that makes the problems (1) through (3) and (4) through (6) equivalent. It can be shown that in two dimensions (see for *eg.* [3])

$$M_{ex}(u) = \sum_{n=-\infty}^{\infty} z_n(k,R) \int_{\Gamma_R} e^{in\theta} e^{-in\theta'} u(R,\theta')d\Gamma' \tag{7}$$

where

$$z_n(k,R) = \frac{kH_n(kR)'}{H_n(kR)} \tag{8}$$

In d dimensions $M_{ex}(u)$ has the following functional form

$$M_{ex}(u) = \sum_{n=-\infty}^{\infty} \sum_{m=0}^{n_d} z_n(k,R;d)\beta_{nm} \int_{\Gamma_R} \psi_{nm}(x;d)\psi^*_{nm}(x';d)u(R,x')d\Gamma' \tag{9}$$

The DtN map is given by $M_{dtn} = M_{ex}$. The weak form associated with the DtN formulation is obtained from equations (4) through (6). It is given by: Find $u \in \mathcal{S}$ such that $\forall w \in \mathcal{V}$

$$a(w, u) - (w, M_{dtn}(u))_{\Gamma_R} \; = \; (w, f)_{\Omega_R} \tag{10}$$

where

$$a(w, u)_{\Omega_R} \; = \; (\nabla w, \nabla u)_{\Omega_R} - k^2 (w, u)_{\Omega_R} \tag{11}$$

$$(w, M_{dtn}(u))_{\Gamma_R} \; = \; \int_{\Gamma_R} w^* M(u) d\Gamma \tag{12}$$

2.2 Infinite element formulation

The weak form associated with the infinite element formulation is obtained from equations (1) through (3). Find $u \in \mathcal{S}$ such that $\forall w \in \mathcal{V}$

$$a(w, u)_{\Omega_R} + b(w, u)_{\Omega_\infty} = (w, f)_{\Omega_R} \tag{13}$$

In equation (13) $a(\cdot, \cdot)$, $b(\cdot, \cdot)$ and $(\cdot, f)$ may be either bilinear or sesquilinear which leads to the conjugated or the unconjugated form of the infinite element. The analysis in this study applies to both these cases. For the definition of $b(w, u)_{\Omega_\infty}$ and the spaces $\mathcal{V}$ and $\mathcal{S}$ the reader is referred to [6].

3 Implementing the DtN map

Consider the Galerkin method applied to equation (10). Find $u^h \in \mathcal{S}^h \subset \mathcal{S}$ such that $\forall w^h \in \mathcal{V}^h \subset \mathcal{V}$

$$a(w^h, u^h) - (w^h, M^h_{dtn}(u^h))_{\Gamma_R} = (w^h, f)_{\Omega_R} \tag{14}$$

where the superscript is used to denote a finite dimensional subset. The discretized DtN map $M^h_{dtn}(u^h)$ is given by

$$M^h_{dtn}(u^h) = \sum_{n=-N_t}^{N_t} \sum_{m=0}^{n_d} z_n(k, R; d) \beta_{nm} \int_{\Gamma_R} \psi^h_{nm}(x; d) \psi^{h*}_{nm}(x'; d) u^h(R, x') d\Gamma' \tag{15}$$

In [6] we have derived a modal error estimate for $M^h_{dtn}(u^h)$ which indicates that for the harmonics included in its kernel the DtN map produces an extremely accurate boundary condition.

We examine the matrix equations that emanate from using finite element shape functions in equation (14). These are

$$\boldsymbol{K}_{dtn}\boldsymbol{u} = \boldsymbol{f} \tag{16}$$

$$\boldsymbol{K}_{dtn} := \boldsymbol{K} - \boldsymbol{B}_{dtn} \tag{17}$$

Here $\boldsymbol{K}$ is the part of the matrix associated with the interior domain and $\boldsymbol{B}_{dtn}$ is the part of the matrix associated with the DtN boundary condition. Due to the non-local nature of the DtN map $\boldsymbol{B}_{dtn}$ is a dense $N_{\Gamma_R} \times N_{\Gamma_R}$ matrix. If we were to solve equation (16) using an iterative method like QMR there would be the following penalties associated with assembling $\boldsymbol{B}_{dtn}$

1. We would need to store $O(\times N_{\Gamma_R}^2)$ complex numbers.
2. To perform a matrix-vector product of the form $\boldsymbol{u} = \boldsymbol{B}_{dtn}\boldsymbol{p}$ at each iteration would require $O(N_{\Gamma_R}^2)$ flops.

These penalties in storage and computation can be severe if the number of points on the boundary is large. In the remaining part of this section we examine the structure of $\boldsymbol{B}_{dtn}$ and propose a new algorithm to perform matrix-vector products of the form $\boldsymbol{u} = \boldsymbol{B}_{dtn}\boldsymbol{p}$ that addresses the difficulties listed above.

Let $w^h(R,x) = \boldsymbol{w}^T\boldsymbol{N}(R,x)$; $u^h(R,x) = \boldsymbol{N}^T(R,x)\boldsymbol{u}$; $\psi^h_{nm}(x) = \boldsymbol{N}^T(R,x)\boldsymbol{\psi}_{nm}$, where $\boldsymbol{N}(R,x)$ is the restriction of the standard finite element shape function to Γ_R. Then

$$(w^h, M^h_{dtn}(u^h)) = \sum_{n=-N_t}^{N_t}\sum_{m=0}^{n_d}\int_{\Gamma_R}\int_{\Gamma_R} z_n\beta_{nm}w^h(x)\psi^h_{nm}(x)\psi^{h*}_{nm}(x')u^h(x')d\Gamma' d\Gamma \tag{18}$$

$$= \sum_{n=-N_t}^{N_t}\sum_{m=0}^{n_d} z_n\beta_{nm}\boldsymbol{w}^H\boldsymbol{M}\boldsymbol{\psi}_{nm}\boldsymbol{\psi}^H_{nm}\boldsymbol{M}\boldsymbol{u} \tag{19}$$

$$= \boldsymbol{w}^H\boldsymbol{M}(\sum_{n=-N_t}^{N_t}\sum_{m=0}^{n_d} z_n\beta_{nm}\boldsymbol{\psi}_{nm}\boldsymbol{\psi}^H_{nm})\boldsymbol{M}\boldsymbol{u} \tag{20}$$

where $\boldsymbol{M} = \int_{\Gamma_R}\boldsymbol{N}\boldsymbol{N}^T d\Gamma$ Thus we have

$$\boldsymbol{B}_{dtn} = \boldsymbol{M}(\sum_{n=-N_t}^{N_t}\sum_{m=0}^{n_d} z_n\beta_{nm}\boldsymbol{\psi}_{nm}\boldsymbol{\psi}^H_{nm})\boldsymbol{M} \tag{21}$$

In equation (21) the matrix $\boldsymbol{M}$ is sparse and the storage associated with it is $O(N_{\Gamma_R})$. Also the vectors $\boldsymbol{\psi}_{nm}$ contain the values of the harmonics at nodes on Γ_R. We do not need to store these vectors, instead we calculate them when performing the matrix vector product. Thus the overall storage required to store $\boldsymbol{B}_{dtn}$ in the form given by equation (21) is $O(N_{\Gamma_R})$.

To calculate $\boldsymbol{u} = \boldsymbol{B}_{dtn}\boldsymbol{p}$ we propose the following algorithm which makes use of equation (21).

1. $\bar{\boldsymbol{u}} = \boldsymbol{M}\boldsymbol{p}$
2. For each harmonic $\gamma_{nm} = \boldsymbol{\psi}_{nm}^{H}\bar{\boldsymbol{u}}$
3. And then $\epsilon_{nm} = z_n\beta_{nm}\gamma_{nm}$
4. $\tilde{\boldsymbol{u}} = \sum_{n=-N_t}^{N_t}\sum_{m=0}^{n_d}\epsilon_{nm}\boldsymbol{\psi}_{nm}$
5. $\boldsymbol{u} = \boldsymbol{M}\tilde{\boldsymbol{u}}$

Flops required for the above algorithm are $O(N_{\Gamma_R}N_T)$, where N_T is the total number of modes included in the DtN kernel. Usually $N_T << N_{\Gamma_R}$ and we can get significant computational gains from this algorithm. The above algorithm also allows us to implement the DtN boundary condition in the matrix free form since the matrix $\boldsymbol{M}$ has a local structure. See [5] for details.

For the case of equiangular meshes on Γ_R even when $N_T \approx N_{\Gamma_R}$ we can specialize the above algorithm to effect further reduction in flops per matrix-vector product. In two dimensions $\{\boldsymbol{\psi}_n\}_j = \exp(inj\Delta\theta)$ and we can use the Fast Fourier Transform algorithm in steps (2) and (4). The modified algorithm is:

1. $\bar{\boldsymbol{u}} = \boldsymbol{M}\boldsymbol{p}$
2. $\gamma = FFT(\bar{\boldsymbol{u}})$
3. $\epsilon_n = z_n\beta_n\gamma_n$
4. $\tilde{\boldsymbol{u}} = InverseFFT(\boldsymbol{\epsilon})$
5. $\boldsymbol{u} = \boldsymbol{M}\tilde{\boldsymbol{u}}$.

Using this algorithm flops per matrix-vector product reduce from $O(N_{\Gamma_R}^2)$ to $O(N_{\Gamma_R}log(N_{\Gamma_R}))$. The extension of this idea to three dimensions is contingent on a Fast spherical convolution algorithm. The stumbling block in this case is the absence of a fast Legendre transform. Recent advances made in this area by Healy and co-workers [2] appear promising. However, even using a naive Legendre transform algorithm we can reduce the flops in three dimensions from $O(N_{\Gamma_R}^2)$ to $O(N_{\Gamma_R}^{3/2})$.

Finally we present a numerical example where we use the algorithm written above. The problem that we wish to solve is that of an infinite cylindrical radiator of radius $a = 1.0$. The DtN boundary condition is applied at $R = 1.25$ and the wavenumber associated with the problem is $k = 10$. We use an equiangular mesh on Γ_R with $N_{\Gamma_R} = 512$. The total degrees of freedom in the matrix problem are $N_{dof} = 4608$. The matrix equations are solved using QMR with no preconditioner. We require 212 iterations to converge to an accuracy of $1e-06$. The table below is a comparison of storage and computation costs associated with the naive DtN algorithm and the DtN-FFT algorithm.

Method	Storage (nnz $\times 10^3$)	CPU time (secs)
DtN	152	366
DtN-FFT	21	101

We note a 7-fold reduction in storage and at least a 3-fold reduction computing time.

4 Implementing infinite elements

We begin with the Galerkin approximation to equation (13) *viz.* Find $u^h \in \mathcal{S}^h \subset \mathcal{S}$ such that $\forall w^h \in \mathcal{V}^h \subset \mathcal{V}$

$$a(w^h, u^h)_{\Omega_R} + b(w^h, u^h)_{\Omega_\infty} = (w^h, f)_{\Omega_R} \tag{22}$$

Here u^h and w^h are constructed as tensor products of finite element shape functions in the angular direction and the terms from the multipole expansion in the radial direction. For example,

$$u^h = \frac{e^{ikr}}{r^{(d-1)/2}} \sum_{n=1}^{N_{\Gamma_R}} \sum_{m=1}^{N_r} a_{nm} N_n(\theta, \phi) \hat{p}_m(\frac{1}{r^n}) \tag{23}$$

where $\hat{p}_m(x)$ is a polynomial in x of order $m-1$. In particular, we work with the following polynomials suggested by Shirron.

$$p_m(R) := \frac{e^{ikr}}{r^{(d-1)/2}} \hat{p}_m(\frac{1}{r^n}) \tag{24}$$

$$p_1(R) = 1 \quad ; \quad p_m(R) = 0 \; m \neq 1 \tag{25}$$

In [4] we have proposed an equivalent formulation for the infinite elements based on the non-reflecting boundary condition formulation. This reads, find $u^h \in \mathcal{S}^h$ such that $\forall w^h \in \mathcal{V}^h$

$$a(w^h, u^h) - (w^h, M_{ie}^h(u^h))_{\Gamma_R} = (w^h, f)_{\Omega_R} \tag{26}$$

M_{ie}^h is determined by equating equations (22) and (26). To examine the matrix equations associated with these equations, we note we can write u^h as

$$u^h = u_1^h + u_2^h \qquad ; \quad u_2^h|_{\Omega_R} = 0 \tag{27}$$

Then the matrix equations associated with (22) are

$$\begin{bmatrix} \boldsymbol{K}_{11} & \boldsymbol{K}_{12} \\ \boldsymbol{K}_{21} & \boldsymbol{K}_{22} \end{bmatrix} \begin{bmatrix} \boldsymbol{u}_1 \\ \boldsymbol{u}_2 \end{bmatrix} = \begin{bmatrix} \boldsymbol{f} \\ \boldsymbol{0} \end{bmatrix} \tag{28}$$

and the matrix equations associated with (26) are

$$\boldsymbol{K}_{sc}\boldsymbol{u}_1 = \boldsymbol{f} \tag{29}$$

$$\boldsymbol{K}_{sc} := \boldsymbol{K}_{11} - \boldsymbol{B}_{sc} \tag{30}$$

$$\boldsymbol{B}_{sc} := \boldsymbol{K}_{12}\boldsymbol{K}_{22}^{-1}\boldsymbol{K}_{21} \tag{31}$$

Note that $\boldsymbol{K}_{sc}$ is obtained from (28) by taking the Schur complement of the variables with subscript 2 with respect to $\boldsymbol{K}_{ie}$. We refer to this formulation as the Schur complement formulation. The modal accuracy of M_{ie}^h for both conjugated and unconjugated formulation has been examined in [6].

We examine the convergence characteristics associated with solving equation (28) in conjunction with an iterative solver (QMR in this case) and compare it to the convergence for the DtN map. The following observations and analysis hold for both the conjugated and the unconjugated versions of infinite elements. We solve a radiation problem in 2 dimensions, with $N_r = 3$, and $N_{dof} = 210$. From figure 1. we observe that the convergence of the infinite element formulation is poor. This can be understood by comparing the spectra of $\boldsymbol{K}_{ie}$ and $\boldsymbol{K}_{dtn}$ (see figure 3.). We observe that in comparison to the eigenvalues of $\boldsymbol{K}_{dtn}$, the eigenvalues of $\boldsymbol{K}_{ie}$ are

1. Clustered closely about the origin.
2. Spread out farther in the complex plane.

It is well known that convergence of linear systems described by matrices with such spectral behavior is typically very poor (see [7]).

We now consider the Schur complement formulation given by equation (29). In [6] we prove the following:

1. For small n the quantity $|\lambda_n(\boldsymbol{B}_{dtn}) - \lambda_n(\boldsymbol{B}_{sc})|$ is small.
2. $\lambda_{max}(\boldsymbol{K}_{sc})$ grows with h at the same rate as $\lambda_{max}(\boldsymbol{K}_{dtn})$.

The observations are borne out in figure 4. We observe that the spectra of $\boldsymbol{K}_{sc}$ and $\boldsymbol{K}_{dtn}$ are very similar and are motivated to test the convergence of solving the equation (29). We plot the result in figure 1. We observe that the convergence of the Schur complement formulation is remarkably better than the standard infinite element formulation. To implement the Schur complement version of the infinite element formulation we propose the following algorithm to calculate matrix-vector products efficiently. Let

$$\begin{aligned} \boldsymbol{u} = &= \boldsymbol{K}_{sc}\boldsymbol{p} \\ &= (\boldsymbol{K}_{11} - \boldsymbol{K}_{12}\boldsymbol{K}_{22}^{-1}\boldsymbol{K}_{21})\boldsymbol{p} \\ &= \boldsymbol{u}_1 - \boldsymbol{u}_2 \end{aligned}$$

$\boldsymbol{u}_1$ is easy to calculate but to calculate but the equation for $\boldsymbol{u}_2$ involves $\boldsymbol{K}_{22}^{-1}$. We do not compute $\boldsymbol{K}_{22}^{-1}$ explicitly, instead at the beginning of the iterative process, we compute and store $\boldsymbol{K}_{22} = \boldsymbol{L}\boldsymbol{U}$. Then to compute $\boldsymbol{u}_2$ at each iteration:

1. $\bar{\boldsymbol{p}} = \boldsymbol{K}_{21}\boldsymbol{u}$
2. Solve $\boldsymbol{L}\tilde{\boldsymbol{p}} = \bar{\boldsymbol{p}}$
3. Solve $\boldsymbol{U}\hat{\boldsymbol{p}} = \tilde{\boldsymbol{p}}$

4. $\boldsymbol{u}_2 = \boldsymbol{K}_{12}\hat{\boldsymbol{p}}$

We now solve a numerical problem using this algorithm for the Schur complement formulation. We model radiation from an infinite cylindrical radiator, with the following problem parameters: $kR = 20$, $a = 1$, $N_{dof} = 5200$ and $N_{\Gamma_R} = 200$. We solve this problem using the standard IE formulation, the Schur complement formulation and a preconditioned (SSOR) Schur complement formulation. The convergence results are shown in figure 2. and also summarized in the table below. We observe that for this problem the standard IE formulation does not converge, whereas the Schur complement formulation does converge. Further the preconditioner which is a variant of the SSOR preconditioner is effective in reducing the total number of flops. This development is discussed in [4].

Method	n	M flops/iteration	total M flops
IE	-	0.81	-
IE-Schur	361	1.00	364
IE-Schur (SSOR)	193	1.31	261

5 Conclusions

In this work we have

1. Developed an algorithm to implement the discrete DtN map on general unstructured meshes that significantly reduces storage and computational overheads.
2. Proposed and implemented a computationally efficient scheme with improved convergence characteristics for the infinite element formulation.

References

[1] D.S. Burnett. A three-dimensional acoustic infinite element based on a prolate spheroidal multipole expansion. *J. Acoust. Soc. Am.*, 96(5):2798–2816, 1994.

[2] D. M. Healy Jr., D. Rockmore, and S. S. B. Moore. Ffts for the 2-sphere- improvements and variations. *Technical report PCS-TR96-292, Dartmouth college, Department of Computer Science*, 1996.

[3] J. B. Keller and D. Givoli. Numerical methods for problems in infinite domains. *J. Comput. Physics*, 82(1):172–192, 1989.

[4] M. Malhotra, A. A. Oberai, and P. M. Pinsky. Implementing infinite elements for large-scale problems in structural acoustics. *In preparation*, 1997.

[5] M. Malhotra and P. M. Pinsky. A matrix-free interpretation of the nonlocal dirichlet-to-neumann radiation boundary condition. *Int. J. Numer. Methods Engrg*, 39:3705–3713, 1996.

[6] A. A. Oberai, P. M. Pinsky, and M. Malhotra. Analysis of infinite elements based on multipole expansions as non reflecting boundary conditions. *In preparation*, 1997.

[7] Y. Saad. *Iterative methods for sparse linear systems.* PWS Publishing Company, Boston, MA, 1st edition, 1995.

Figures

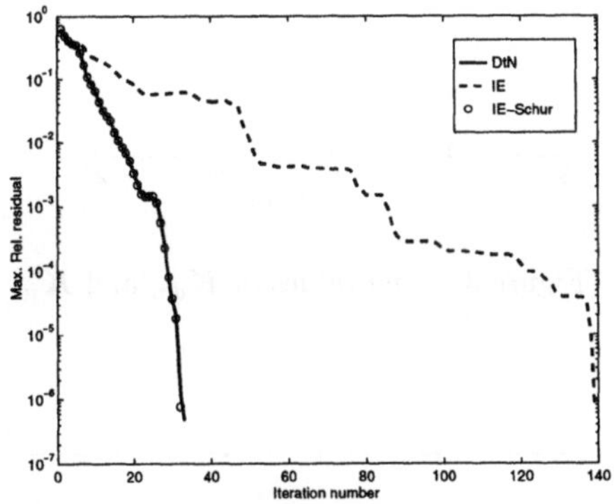

Figure 1: Convergence of QMR iterations for conjugated IE

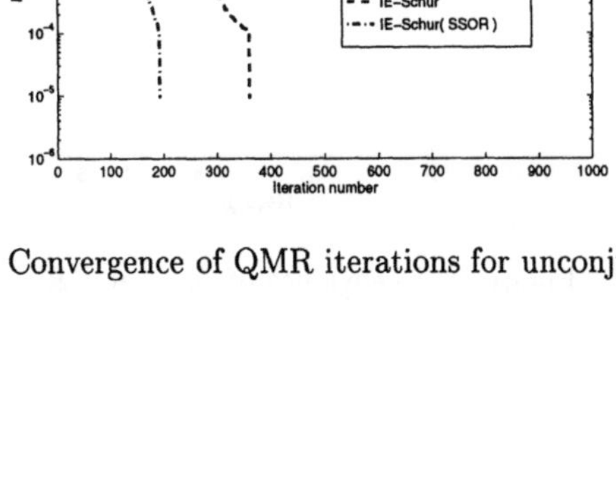

Figure 2: Convergence of QMR iterations for unconjugated IE

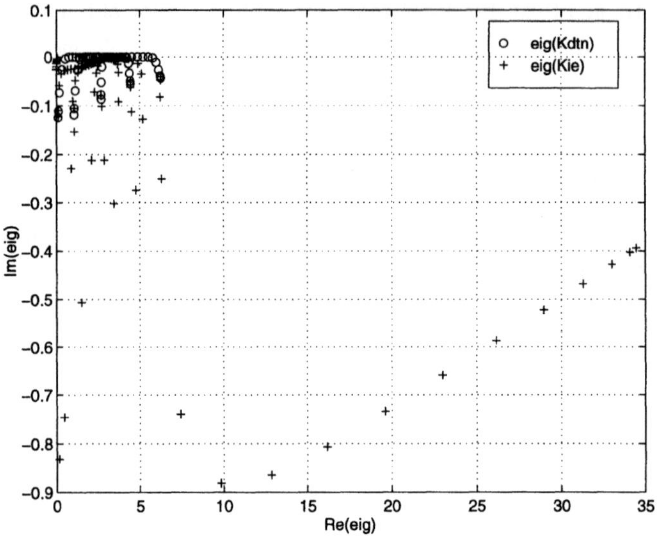

Figure 3: Eigenvalues of $\boldsymbol{K}_{dtn}$ and $\boldsymbol{K}_{ie}$

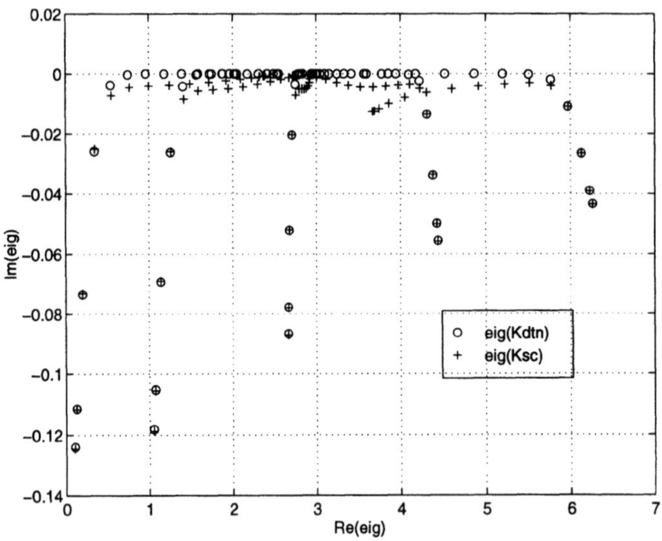

Figure 4: Eigenvalues of $\boldsymbol{K}_{dtn}$ and $\boldsymbol{K}_{sc}$

Progress on Highly Accurate Non-Reflecting Boundary Conditions for Finite Element Simulations of Transient Acoustics Problems

Andrzej J. Safjan
Department of Engineering Mechanics
University of Nebraska-Lincoln
212 Bancroft Hall, Lincoln, NE 68588, U.S.A.

1. Introduction

The focus of this work is on developing a family of highly accurate boundary operators for solving wave-type equations in unbounded domains. In particular, we consider wave propagation problems characterized by the classical wave equation, and we are interested in radiating boundary conditions which can be easily fit into the framework of a typical finite element code.

Various approaches have been proposed for treating computational boundaries (cf., e.g., [1], [2], [3], [4], and the literature therein). Usually, a sequence of boundary operators $\{\mathcal{B}_n\}_{n=1}^{\infty}$ is derived, which approximate an operator $\mathcal{B}$ which provides no reflections at the boundary of computational domain. On the other hand, operators $\mathcal{B}_n$ may involve high derivatives of the solution, and, therefore, may be difficult to implement in a standard finite element code. Moreover, how many terms in the sequence are needed to produce satisfactory results is often unknown.

Recently, new families of boundary operators have been introduced which overcome some of those difficulties [6]. These operators may achieve an arbitrary accuracy (in the sense of an arbitrary small reflection coefficient), and have the form $\mathcal{B}u = g$, where $\mathcal{B}$ is the Dirichlet, Neumann or Robin operator. This particular form guarantees the well-posedness of the ensuing truncated domain problem,

T.L. Geers (ed.), IUTAM Symposium on Computational Methods for Unbounded Domains, 265–273.

and makes the operators suitable for implementation in a standard finite element code.

2. A Model Problem

Consider the wave equation in Cartesian coordinates

$$u_{,tt} - \nabla^2 u = 0 \tag{2.1}$$

for $x > 0, t > 0, y \in \mathbb{R}$. Here u is a function of $\mathbf{x} = (x, y)$ and t, and the units have been normalized so that the wave propagation speed is equal to unity. The goal is to construct boundary conditions that cause waves propagating from the interior of $\Omega = \{\mathbf{x} : x > 0,\ y \in \mathbb{R}\}$, to pass through the boundary $\Gamma_c = \{\mathbf{x} : x = 0,\ y \in \mathbb{R}\}$, without being reflected.

More specifically, we seek a boundary operator $\mathcal{B}$, such that if w is the solution of the initial-value problem in $\mathbb{R}^2$

$$\begin{cases} w_{,tt} - \nabla^2 w = 0\ , & \mathbf{x} \in \mathbb{R}^2\ ,\ t \in (0, T\,]\ , \\ w\big|_{t=0} = u_0\ , & w_{,t}\big|_{t=0} = u_1\ ; \end{cases} \tag{2.2}$$

and u is the solution of the initial-boundary value problem in the truncated domain $\Omega = \mathbb{R}_+ \times \mathbb{R}$

$$\begin{cases} u_{,tt} - \nabla^2 u = 0\ , & \mathbf{x} \in \Omega\ ,\ t \in (0, T]\ , \\ u\big|_{t=0} = u_0\ , & u_{,t}\big|_{t=0} = u_1\ , \\ \mathcal{B}u\big|_{x=0} = 0\ ; \end{cases} \tag{2.3}$$

then the difference $u - w$, measured in an appropriate norm, $\|\cdot\|_{\Omega,T}$, is less than a prescribed tolerance ϵ

$$\|u - w\|_{\Omega,T} < \epsilon\ . \tag{2.4}$$

Our notion of a solution is classical; we also assume that the Cauchy data u_0 and u_1 have compact support in Ω.

3. Split Operators [6]

Consider the following sequence of initial-boundary value problems for the wave equation:

$$\begin{cases} u^{(j)}_{,tt} - \nabla^2 u^{(j)} = 0 \,, & \mathbf{x} \in \Omega,\ t \in (0,T] \,, \\ u^{(j)}\big|_{t=0} = u_0 \,, & \\ u^{(j)}_{,t}\big|_{t=0} = u_1 \,, & \\ \mathcal{B}_j u^{(j)} = \begin{cases} 0 \,, & j = 1 \,, \\ \mathcal{A}_{j-1} u^{(j-1)}, & j > 1 \,, \end{cases} & \text{on } \Gamma_c \\ j = 1, 2, \ldots, N \,; & \end{cases} \tag{3.1}$$

where $\Omega = \{\mathbf{x} : x > 0,\ y \in \mathbb{R}\}$ and $\Gamma_c = \{\mathbf{x} : x = 0,\ y \in \mathbb{R}\}$. The goal is to construct boundary operators $\mathcal{B}_j$ and $\mathcal{A}_{j-1}$ such that the magnitude of the resulting reflection coefficient $|R_N|$,

$$u^{(N)} = f(t + xc - ys) + R_N\, f(t - xc - ys) \,, \tag{3.2}$$

is as small as possible for a wide range of angles of incidence θ. Here $f(t{+}xc{-}ys)$ is an arbitrary plane wave propagating in the negative x direction and striking the artificial boundary Γ_c at the angle θ, $R_N\, f(t - xc - ys)$ is the associated reflected wave, $c \equiv \cos\theta$, $s \equiv \sin\theta$, and f is some function.

The choice of the left-hand side operators $\mathcal{B}_j$ is of utmost importance. Since the governing wave equation is of the second order, $\mathcal{B}_j$ should contain derivatives of at most order one. Consequently, the choice of $\mathcal{B}_j$ is limited to $\mathcal{B}_D$, the Dirichlet operator,

$$\mathcal{B}_D\, u \overset{\text{def}}{=} u\big|_{x=0} \,, \tag{3.3}$$

$\mathcal{B}_N$, the Neumann operator,

$$\mathcal{B}_N\, u \overset{\text{def}}{=} u_{,n}\big|_{x=0} \,, \tag{3.4}$$

and $\mathcal{B}_R$, the Robin operator

$$\mathcal{B}_R\, u \overset{\text{def}}{=} (\lambda\, u_{,t} + u_{,n})\big|_{x=0} \quad , \quad \lambda > 0 \,. \tag{3.5}$$

Here $u_{,n}$ denotes the normal derivative of u at Γ_c .

Consider now a typical one-stage problem. Assume that $u^{(j-1)}$

$$u^{(j-1)} = f(t + xc - ys) + R_{j-1}\, f(t - xc - ys) \;, \tag{3.6}$$

is already known, and that its reflection coefficient, $|R_{j-1}|$, is of the order I,

$$|R_{j-1}| = O\left(|1-c|^{I}\right). \tag{3.7}$$

The goal of a multistage process is to find operators $\mathcal{B}_j$ and $\mathcal{A}_{j-1}$ such that the resulting j-th stage solution $u^{(j)}$ is better in the sense that R_j, the reflection coefficient associated with $u^{(j)}$, is of higher order,

$$|R_j| = O\left(|1-c|^{J}\right), \quad J > I. \tag{3.8}$$

The reflection coefficient R_j, associated with the j-th stage problem

$$\mathcal{B}_j\, u^{(j)} = \mathcal{A}_{j-1} u^{(j-1)} \;, \tag{3.9}$$

is of the following general form

$$R_j = g(c) R_{j-1} + h(c), \tag{3.10}$$

where $g(c)$ and $h(c)$ are rational functions of c. In order that R_j be of order J, $J > I$, $g(c)$ and $h(c)$ have to satisfy the following conditions:

$$h(c) = O\left(|1-c|^{J}\right), \tag{3.11}$$

and

$$g(c) = O\left(|1-c|^{J-I}\right), \quad J > I. \tag{3.12}$$

Conditions (3.11) and (3.12) are referred to as the *order conditions.*

Finally, in order to assess various boundary operators, it is important to consider not only the order of R, (i.e., the power of $(1-c)$), but also K, the associated constant:

$$K = \lim_{c \to 1}\left(\left|(1-c)^{-J} R\right|\right), \tag{3.13}$$

where $|R| = O\left(|1-c|^{J}\right)$.

Now, let us introduce the following family of boundary operators

$$\begin{cases} \left(u_{,t}^{(1)} - u_{,x}^{(1)}\right)\Big|_{x=0} = 0 \ , \\ \partial_t^{j-1}\left(\partial_t - \partial_x\right) u^{(j)}\big|_{x=0} = \\ = \left(-\frac{1}{2^{j-1}}\left(\partial_t - \partial_x\right)^j + \partial_t^{j-1}\left(\partial_t - \partial_x\right)\right) u^{(j-1)}\Big|_{x=0} \ , \\ j = 2, 3, \ldots, N \ . \end{cases} \tag{3.14}$$

These operators are based on the Robin operator (3.5), with $\lambda = 1$. We claim that the associated reflection coefficient is of order N. The proof is by induction. For $j = 1$, we have $R_1 = -\frac{1-c}{1+c}$. Next, let us assume that $|R_{j-1}| = O\left(|1-c|^{j-1}\right)$. A direct computation of R_j yields

$$R_j = g(c) R_{j-1} - \frac{1}{2^{j-1}} \frac{(1-c)^j}{1+c} \quad , \tag{3.15}$$

where

$$g(c) = \left(-\frac{1}{2^{j-1}}\left(1+c\right)^{j-1} + 1\right) \quad . \tag{3.16}$$

This, together with the observation that $g(1) = 0$, implies that $|R_j| = O\left(|1-c|^j\right)$, which completes the proof.

For boundary operators (3.14), constant K may be suboptimal. To get a better K, sequence (3.14) has to be appended by one extra stage:

$$\begin{aligned} &\partial_t^{N-1}\left(\partial_t - \partial_x\right) u^{(N+1)}\big|_{x=0} = \\ &= \left(-\frac{1}{2^{N-1}}\left(\partial_t - \partial_x\right)^N + \partial_t^{N-1}\left(\partial_t - \partial_x\right)\right) u^{(N)}\Big|_{x=0} \end{aligned} \tag{3.17}$$

Since R_N is of order N, the term $g(c)R_N$ is of order $N+1$, and the constant K associated with R_{N+1} reaches the optimal value $\frac{1}{2^N}$.

4. Temporal and Spatial Approximations

Further steps include temporal and spatial approximations of initial-boundary value problems (3.1). These approximations are achieved by using high-order, unconditionally stable TG (Taylor-Galerkin) schemes for temporal approximation combined with *hp*- adaptive Finite Element approximations in spatial variables. Details are given in [7].

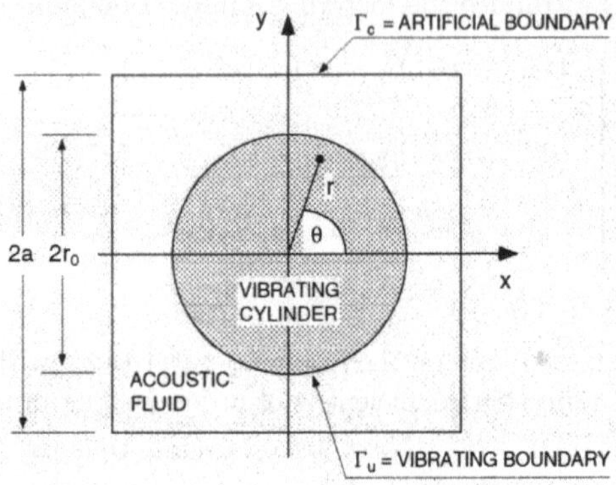

Figure 5.1: Vibrating cylinder problem.

5. Numerical Examples

The test problem is defined as follows (see Fig. 5.1).

- We seek the solution for the classical vibrating cylinder problem which is defined below:

  - governing wave equation is to be solved in $\Omega_\infty \times (0, T]$, where

$$\Omega_\infty = \mathbb{R}^2 \setminus \{(r, \theta) : r \leq r_0\} \ , \tag{5.1}$$

    $r_0 = 0.5$ and $T = 0.5$;

  - the boundary conditions are defined below

$$u_n|_{r=r_0} = A\left[\sin\left(\omega t\right)\right]^5 \ \left[\mathrm{H}(t) - \mathrm{H}(t-1)\right] \ , \tag{5.2}$$

    where $A = 0.05$, $\omega = 2\pi$, and $\mathrm{H}(\cdot)$ is the Heaviside step function; and,

  - the initial conditions are assumed to be zero.

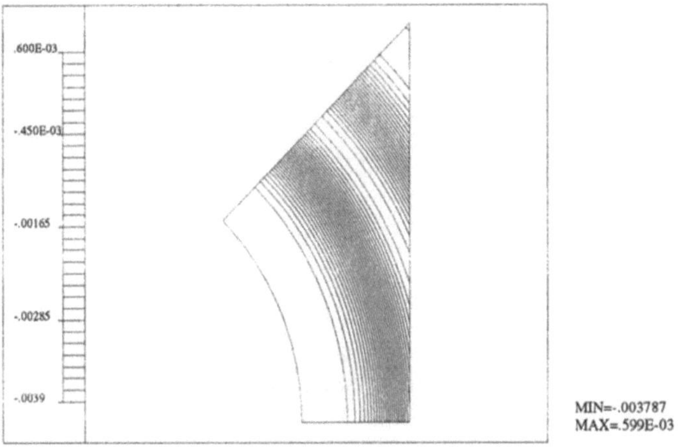

Figure 5.2: Contour plot of the solution $u^{(N)}$ at time $t = 0.5$. Boundary operator B–A with $N = 4$.

- We introduce the computational domain Ω

$$\Omega = \{(x, y) = (-a, a) \times (-a, a)\} \ , \tag{5.3}$$

with $a = 0.7$, and we compare the results for various boundary operators with the exact solution. (Note that the exact solution for the vibrating cylinder problem can be found by means of Fourier-Bessel series.) The following notation is used:

  - $\mathcal{B}$–A = operators defined by (3.14) with one extra stage (3.17).

The resulting bounded domain problem was solved by using a 2-stage 4th order Taylor-Galerkin scheme ([7]), with a time-step $\Delta t = \frac{1}{240}$, and a fixed mesh of elements of order $p = 5$. Note that, since the error introduced by the boundary truncation can be observed at the continuous level only (i.e., when no temporal and spatial approximations are introduced), it was necessary to select a rich enough finite element space, and a highly accurate approximation in time.

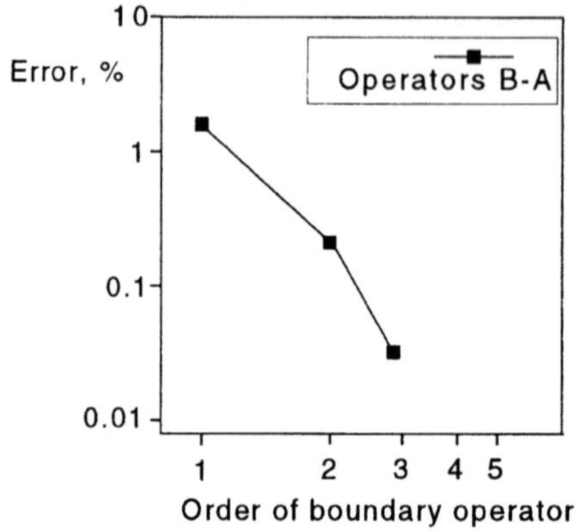

Figure 5.3: Experimental convergence rates for boundary operators.

Figure 5.2 shows a contour plot of pressure $u^{(N)}$ obtained by using the multi-stage operator $\mathcal{B}$–A with $N = 4$, and Fig. 5.3 shows the graph of the relative error

$$Error = \frac{\|u - u^{(N)}\|}{\|u\|} \times 100\% \tag{5.4}$$

as a function of operator order. Here u is the exact solution of the unbounded domain problem, $u^{(N)}$ is the solution obtained by using an N -th order boundary operator, $\|\cdot\|$ is the energy norm, and the graph is plotted in a log-log scale. As can be seen, boundary operators produce results convergent toward the exact solution.

6. Conclusions

The following comments are in order:

1. split operators have the form

$$\left(u^{(j)}_{,t} - u^{(j)}_{,x}\right)\Big|_{x=0} = \mathcal{A}_j u^{(j-1)}\big|_{x=0} \; , \tag{6.1}$$

where $\mathcal{A}_j$ is an operator involving high derivatives of the solution, and $u^{(j-1)}$ is assumed to be known; this particular form can be fit into a typical finite element code inasmuch as $\mathcal{A}_j u^{(j-1)}$ can be obtained by post-processing (in fact, $\mathcal{A}_j$ can be factorized into operators involving first or second derivatives only);

2. the difference between successive approximations $u^j - u^{(j-1)}$, measured in a suitable norm, provides a built-in *a posteriori* indicator of the error introduced by the domain truncation; and,

3. split boundary operators have the same left-hand side which results in their economical implementation.

Acknowledgments. The support of this work by the Office of Naval Research (ONR) under the contract #N00014-89-J-1451 is gratefully acknowledged.

References

[1] A. Bayliss and E. Turkel, "Radiation Boundary Conditions for Wave-like Equations," *Comm. Pure Appl. Math.* **33,** 707 (1980).

[2] B. Engquist and A. Majda, "Radiation Boundary Conditions for the Numerical Simulation of Waves," *Math. Comp.* **31,** 629 (1977).

[3] B. Engquist and A. Majda, "Radiation Boundary Conditions for Acoustic and Elastic Wave Calculations," *Comm. Pure Appl. Math.* **32,** 313 (1979).

[4] R. L. Higdon, "Absorbing Boundary Conditions for Difference Approximations to the Multi-Dimensional Wave Equation", *Math. Comp.* **47,** 437 (1986).

[5] H. O. Kreiss, "Initial Boundary Value Problems for Hyperbolic Systems", Comm. Pure Appl. Math. **23,** 277 (1970).

[6] Safjan, A., Highly Accurate Non-Reflecting Boundary Conditions for Finite Element Simulations of Transient Acoustics Problems, accepted for publication in *Computational Methods in Applied Mechanics and Engineering.*

[7] A. Safjan and J. T. Oden, "High-Order Taylor-Galerkin Methods for Linear Hyperbolic Systems," *J. Comput. Phys.* **120,** 206 (1995).

THE SCALED BOUNDARY FINITE-ELEMENT METHOD: FUTURE DEVELOPMENTS

Chongmin Song and John P. Wolf

Institute of Hydraulics and Energy
Department of Civil Engineering
Swiss Federal Institute of Technology Lausanne
CH-1015 Lausanne
Switzerland

1 Introduction

In the accompanying paper "The scaled boundary finite-element method: state of the art", published in the same proceedings [Wolf and Song, 1998], the present state of the method for elastodynamics is summarized. After *discretizing the boundary with finite elements* and *applying the scaled boundary transformation* (Fig. 1) and *Galerkin's weighted residual*

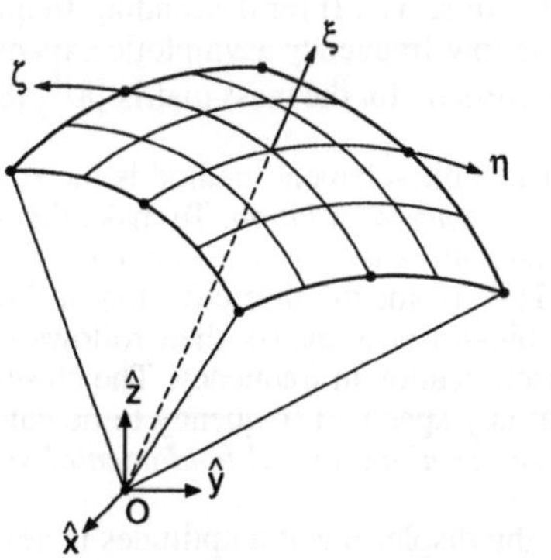

Fig. 1. Scaled boundary transformation of geometry of surface finite element.

method, the governing partial differential equations of linear elastodynamics are reformulated as the scaled boundary finite-element equation in displacements, a system of linear second-order ordinary differential equations, with the dimensionless radial coordinate ξ as independent variable

$$[E^0]\xi^2\{u(\xi)\}_{,\xi\xi}+((s-1)[E^0]-[E^1]+[E^1]^T)\xi\{u(\xi)\}_{,\xi}$$
$$+((s-2)[E^1]^T-[E^2])\{u(\xi)\}+\omega^2[M^0]\xi^2\{u(\xi)\}=0 \quad (1)$$

$\{u(\xi)\}$ are the displacement amplitudes at ξ corresponding to the degrees of freedom in the nodes on the boundary, and s is the spatial dimension ($= 2$ or $= 3$). The coefficient matrices $[E^0]$, $[E^1]$, $[E^2]$ and $[M^0]$ are calculated and assembled similarly as the static-stiffness and mass matrices of finite elements on the boundary [Song and Wolf, 1997].

Introducing the definition of the dynamic stiffness, a system of nonlinear first-order ordinary differential equations in the dynamic-stiffness matrix $[S^\infty(\omega)]$ (superscript ∞ for

T.L. Geers (ed.), IUTAM Symposium on Computational Methods for Unbounded Domains, 275–284.

unbounded) with the frequency as the independent variable is obtained.

$$([S^{\infty}(\omega)]+[E^1])[E^0]^{-1}([S^{\infty}(\omega)]+[E^1]^T)-(s-2)[S^{\infty}(\omega)]$$
$$-\omega[S^{\infty}(\omega)],_{\omega}-[E^2]+\omega^2[M^0]=0 \quad (2)$$

For a bounded medium with the dynamic-stiffness matrix $[S^b(\omega)]$ (superscript b for bounded) the equation is written as

$$([S^b(\omega)]-[E^1])[E^0]^{-1}([S^b(\omega)]-[E^1]^T)+(s-2)[S^b(\omega)]$$
$$+\omega[S^b(\omega)],_{\omega}-[E^2]+\omega^2[M^0]=0 \quad (3)$$

A detailed description of the scaled boundary finite-element method which is also called the consistent infinitesimal finite-element cell method can be found in Part I of [Wolf and Song, 1996a].

For $\omega = 0$, (2) and (3) are solved analytically for the static-stiffness matrices of an unbounded medium $[K^{\infty}]$ and of a bounded medium $[K^b]$ [Section 6 of Wolf and Song, 1998]. The high-frequency asymptotic expansion of $[S^{\infty}(\omega)]$ for an unbounded medium is calculated analytically from (2) which provides the starting value to solve (2) for descending frequencies numerically [Section 4 of Wolf and Song, 1998]. The low-frequency asymptotic expansion of $[S^b(\omega)]$ for a bounded medium yields an analytical expression for the mass matrix $[M^b]$ [Section 7 of Wolf and Song, 1998].

The most striking feature of the scaled boundary finite-element method is the *reduction of the spatial dimension by one without requiring a fundamental solution.* To make the method even more appealing, the *scaled boundary finite-element equation in displacements* $\{u(\xi)\}$ *(Eq. (1)) is solved analytically for any frequency.* This avoids the discretization in the radial coordinate ξ. The analytical solution of the dynamic-stiffness matrix then follows directly from that of the displacements without numerical discretization in frequency. The closed-form solution also permits the dynamic-stiffness matrix at any specified frequency to be calculated directly without starting from a high frequency in this *semi-analytical fundamental-solution-less* boundary-element method.

In the following sections a series solution for the displacement amplitudes is developed from which the analytical expression of the dynamic stiffness follows. Only a summary of the key equations is provided. For details [Song and Wolf, 1998] must be consulted. This closed-form solution extends the high efficiency and accuracy to unbounded and bounded media for any frequency, which is demonstrated with examples. In addition, suggestions for further research are discussed.

2 Analytical Solution of Displacements in Frequency Domain

For the special case when the coefficient matrices $[E^0]$, $[E^1]$, $[E^2]$ and $[M^0]$ in (1) are scalars, (1) can be transformed to the Bessel differential equation, for which an analytical solution exists. Similar to the series solution of the Bessel function, the analytical solution of (1) is formulated as a series in the radial coordinate ξ. After determining the coefficient matrices of the series solution, two independent sets of solutions are obtained.

To solve the second-order ordinary differential equations (1) of order n analytically, a transformation to first-order ordinary differential equations of order $N = 2n$ is performed. The dependent variables $\{X(\xi)\}$ of the first-order ordinary differential equations are defined as

$$\{X(\xi)\} = \left\{ \begin{array}{c} \xi^{\frac{s-2}{2}}\{u(\xi)\} \\ \xi^{-\frac{s-2}{2}}\{R(\xi)\} \end{array} \right\} \quad (4)$$

with the resultant force amplitude

$$\{R(\xi)\} = \xi^{s-2}([E^0]\xi\{u(\xi)\},_\xi + [E^1]^T\{u(\xi)\}) \tag{5}$$

An independent set of particular solutions is grouped in the matrix $[X(\bar{\xi})]$. Eq. (1) is rewritten as

$$\bar{\xi}[X(\bar{\xi})],_{\bar{\xi}} = -[Z][X(\bar{\xi})] - \bar{\xi}^2 \begin{bmatrix} 0 & 0 \\ [M^0] & 0 \end{bmatrix} [X(\bar{\xi})] \tag{6}$$

with the coefficient matrix

$$[Z] = \begin{bmatrix} [E^0]^{-1}[E^1]^T - \frac{s-2}{2}[I] & -[E^0]^{-1} \\ -[E^2] + [E^1][E^0]^{-1}[E^1]^T & -[E^1][E^0]^{-1} + \frac{s-2}{2}[I] \end{bmatrix} \tag{7}$$

and the independent variable

$$\bar{\xi} = \omega\xi \tag{8}$$

To transform the coefficient matrix $[Z]$, the eigenvalue problem is formulated

$$[Z][\Phi] = -[\Phi]\lceil\Lambda\rfloor \tag{9}$$

with the diagonal eigenvalue matrix $-\lceil\Lambda\rfloor$ and the eigenvector matrix $[\Phi]$. For the Hamiltonian matrix $[Z]$ in (7) the eigenvalues consist of two groups with opposite signs

$$-\lceil\Lambda\rfloor = \begin{bmatrix} \lceil\lambda\rfloor & \\ & -\lceil\lambda\rfloor \end{bmatrix} \tag{10}$$

with $\mathrm{Re}(\lambda_i) < 0$. The eigenvalues $-\Lambda_i$ are arranged so that their real parts increase sequentially, i.e.

$$\mathrm{Re}(-\Lambda_1) \leq \mathrm{Re}(-\Lambda_2) \leq \ldots \leq \mathrm{Re}(-\Lambda_N) \tag{11}$$

The solution of (6) is written as

$$[X(\bar{\xi})] = [\Phi][A(\bar{\xi})][Y(\bar{\xi})] \tag{12}$$

The matrix $[A(\bar{\xi})]$ is formulated as a matrix polynomial in $\bar{\xi}$ as

$$[A(\bar{\xi})] = [I] + \bar{\xi}^2[A_1] + \bar{\xi}^4[A_2] + \ldots + \bar{\xi}^{2k}[A_k] + \ldots \tag{13}$$

with the coefficient matrices $[A_k]$ $k = 1, 2, \ldots$ to be determined. The matrix $[Y(\bar{\xi})]$ in (12) satisfies the ordinary differential equations

$$\bar{\xi}[Y(\bar{\xi})],_{\bar{\xi}} = [P^*(\bar{\xi})][Y(\bar{\xi})] \tag{14}$$

where $[P^*(\bar{\xi})]$ is again a matrix polynomial in $\bar{\xi}$

$$[P^*(\bar{\xi})] = [P_0^*] + \bar{\xi}^2[P_1^*] + \bar{\xi}^4[P_2^*] + \ldots + \bar{\xi}^{2k}[P_k^*] + \ldots \tag{15}$$

The coefficient matrices $[P_k^*]$ $k = 0, 1, \ldots$ will be chosen in the simplest form possible.

Substituting (12) in (6), premultiplying by $[\Phi]^{-1}$ and postmultiplying by $[Y(\bar{\xi})]^{-1}$ result in

$$\bar{\xi}[A(\bar{\xi})]_{,\bar{\xi}} - \lceil\Lambda\rfloor\,[A(\bar{\xi})] + [A(\bar{\xi})][P^*(\bar{\xi})] + \bar{\xi}^2[M][A(\bar{\xi})] = 0 \tag{16}$$

with

$$[M] = [\Phi]^{-1}\begin{bmatrix} 0 & 0 \\ [M^0] & 0 \end{bmatrix}[\Phi] \tag{17}$$

The unknown coefficient matrices $[A_k]$, $[P_k^*]$ will be determined from this equation.

Substituting (13) and (15) in (16) leads to a matrix polynomial in $\bar{\xi}$. Setting their coefficients equal to zero yields

$$[P_0^*] = \lceil\Lambda\rfloor \tag{18}$$

$$\lceil\Lambda\rfloor\,[A_k] - [A_k]\,\lceil\Lambda\rfloor - 2k[A_k] = [P_k^*] + [C_k] \qquad k = 1, 2, \ldots \tag{19}$$

with $[A_0] = [I]$ and

$$[C_k] = [M][A_{k-1}] + \sum_{i=1}^{k-1}[A_i][P_{k-i}^*] \tag{20}$$

After determining $[A_i]$ and $[P_i^*]$ up to $i = k-1$, $[C_k]$ is calculated from (20). To determine $[A_k]$ and $[P_k^*]$, (19) is formulated as the corresponding $(N \times N)$ scalar equations

$$(\Lambda_i - \Lambda_j - 2k)[A_k]_{ij} = [P_k^*]_{ij} + [C_k]_{ij} \tag{21}$$

For $\Lambda_i - \Lambda_j - 2k \neq 0$

$$[P_k^*]_{ij} = 0 \tag{22}$$

$$[A_k]_{ij} = \frac{[C_k]_{ij}}{\Lambda_i - \Lambda_j - 2k} \tag{23}$$

and for $\Lambda_i - \Lambda_j - 2k = 0$

$$[P_k^*]_{ij} = -[C_k]_{ij} \tag{24}$$

$$[A_k]_{ij} = \text{arbitrary} \tag{25}$$

are obtained.

As the eigenvalues Λ_i $(i = 1, 2, \ldots N)$ are finite, a maximum k which satisfies $\Lambda_i - \Lambda_j - 2k = 0$ exists. Thus, the number of $[P_k^*]$ different from zero is finite. As $\Lambda_i - \Lambda_j$ is a fixed value for a specified pair i, j, $\Lambda_i - \Lambda_j - 2k = 0$ can only be satisfied for one k, if at all. It follows from (15) that the power of $\bar{\xi}$ equals $2k$ in the corresponding term $\bar{\xi}^{2k}[P_k^*]$. $[P^*(\bar{\xi})]$ in (15) can be written as one matrix

$$[P^*(\bar{\xi})] = \lceil\Lambda\rfloor + \bar{\xi}^{\lceil\Lambda\rfloor}[U]\bar{\xi}^{-\lceil\Lambda\rfloor} \tag{26}$$

with $\bar{\xi}^{\lceil\Lambda\rfloor} = \begin{bmatrix} \bar{\xi}^{\Lambda_1} & \bar{\xi}^{\Lambda_2} & \ldots & \bar{\xi}^{\Lambda_N} \end{bmatrix}$ and the upper-triangular matrix with zeros on the diagonal

$$[U] = \begin{bmatrix} 0 & [P^*_{\frac{\Lambda_1-\Lambda_2}{2}}]_{12} & [P^*_{\frac{\Lambda_1-\Lambda_3}{2}}]_{13} & \cdots & [P^*_{\frac{\Lambda_1-\Lambda_N}{2}}]_{1N} \\ & 0 & [P^*_{\frac{\Lambda_2-\Lambda_3}{2}}]_{23} & \cdots & [P^*_{\frac{\Lambda_2-\Lambda_N}{2}}]_{2N} \\ & & 0 & \cdots & [P^*_{\frac{\Lambda_3-\Lambda_N}{2}}]_{3N} \\ & & & \ddots & \vdots \\ & & & & 0 \end{bmatrix} \tag{27}$$

as the eigenvalues are arranged in the sequence determined by (11). The element $[P^*_{\frac{\Lambda_i-\Lambda_j}{2}}]_{ij}$ only differs from zero when $(\Lambda_i - \Lambda_j)/2$ is a positive integer (Eqs. (22) and (24)). Substituting (26) in (14) and solving the resultant equations yields

$$[Y(\bar{\xi})] = \bar{\xi}^{\lceil\Lambda\rfloor}\bar{\xi}^{[U]} \tag{28}$$

where $\bar{\xi}^{[U]}$ can be calculated with a Taylor expansion with a finite number of terms

$$\bar{\xi}^{[U]} = e^{[U]\ln\bar{\xi}} = [I] + [U]\ln\bar{\xi} + \frac{1}{2!}[U]^2(\ln\bar{\xi})^2 + \ldots + \frac{1}{m!}[U]^m(\ln\bar{\xi})^m \quad m < N \tag{29}$$

as $[U]^i = 0$ for $i \geq m+1$. Substituting (28) in (12) yields

$$[X(\bar{\xi})] = [\Phi][A(\bar{\xi})]\bar{\xi}^{\lceil\Lambda\rfloor}\bar{\xi}^{[U]} \tag{30}$$

Partitioning (30) and applying (4) leads to the general solution of (1)

$$\begin{Bmatrix} \{u(\xi)\} \\ \{R(\xi)\} \end{Bmatrix} = \begin{bmatrix} \xi^{-\frac{s-2}{2}}[X_{11}(\bar{\xi})] & \xi^{-\frac{s-2}{2}}[X_{12}(\bar{\xi})] \\ \xi^{\frac{s-2}{2}}[X_{21}(\bar{\xi})] & \xi^{\frac{s-2}{2}}[X_{22}(\bar{\xi})] \end{bmatrix} \begin{Bmatrix} \{c_1\} \\ \{c_2\} \end{Bmatrix} \tag{31}$$

with the conformably partitioned integration constants $\{c_1\}$, $\{c_2\}$ determined from the boundary conditions at two values of ξ.

3 Dynamic-Stiffness Matrix

From the general solution for the amplitudes of the displacements and resultant forces, the dynamic-stiffness matrix for a specified medium can be determined. The final result is listed for three cases.

A) Domain between two similar boundaries $\xi_i \leq \xi \leq \xi_e$ (Fig. 2)
For the domain bounded by the interior boundary $\xi = \xi_i$ (i for interior) and the exterior boundary $\xi = \xi_e$ (e for exterior) the dynamic-stiffness matrix equals

$$\begin{bmatrix} [S_{ii}(\omega)] & [S_{ie}(\omega)] \\ [S_{ei}(\omega)] & [S_{ee}(\omega)] \end{bmatrix} = \begin{bmatrix} -\xi_i^{\frac{s-2}{2}}[X_{21}(\bar{\xi}_i)] & -\xi_i^{\frac{s-2}{2}}[X_{22}(\bar{\xi}_i)] \\ \xi_e^{\frac{s-2}{2}}[X_{21}(\bar{\xi}_e)] & \xi_e^{\frac{s-2}{2}}[X_{22}(\bar{\xi}_e)] \end{bmatrix} \times \begin{bmatrix} \xi_i^{-\frac{s-2}{2}}[X_{11}(\bar{\xi}_i)] & \xi_i^{-\frac{s-2}{2}}[X_{12}(\bar{\xi}_i)] \\ \xi_e^{-\frac{s-2}{2}}[X_{11}(\bar{\xi}_e)] & \xi_e^{-\frac{s-2}{2}}[X_{12}(\bar{\xi}_e)] \end{bmatrix}^{-1} \tag{32}$$

with $\bar{\xi}_i = \omega\xi_i$, $\bar{\xi}_e = \omega\xi_e$.

B) Bounded domain $0 \leq \xi \leq \xi_e = 1$ (Fig. 3)

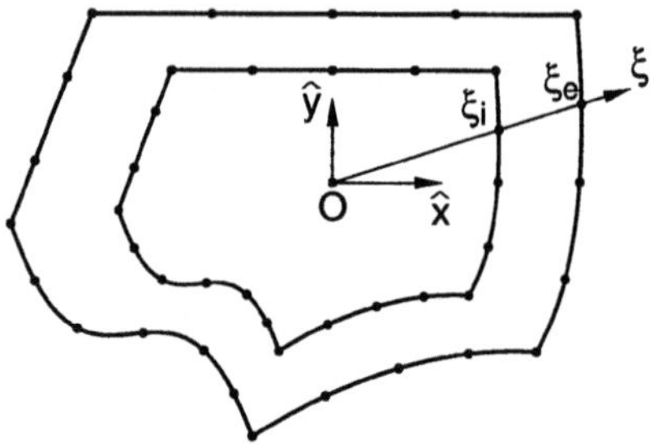

Fig. 2. Domain between interior and exterior boundaries corresponding to $\xi_i \leq \xi \leq \xi_e$.

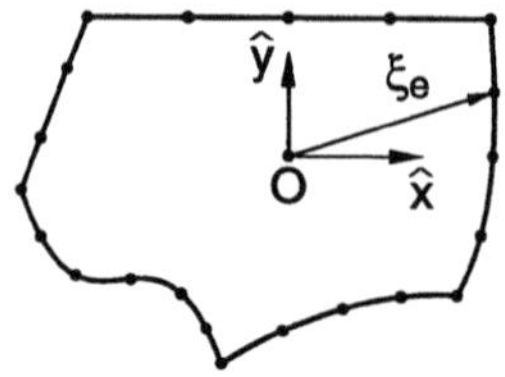

Fig. 3. Domain on interior of boundary corresponding to $0 \leq \xi \leq \xi_e = 1$.

To achieve a finite solution at $\xi = 0$ in the bounded domain, $\{c_2\}$ in (31) vanishes. The dynamic-stiffness matrix with respect to the degrees of freedom on the boundary equals

$$[S(\bar{\xi}_e)] = \xi_e^{s-2}([\Phi_{21}][A_{11}(\bar{\xi}_e)] + [\Phi_{22}][A_{21}(\bar{\xi}_e)])([\Phi_{11}][A_{11}(\bar{\xi}_e)] + [\Phi_{12}][A_{21}(\bar{\xi}_e)])^{-1} \tag{33}$$

with $\xi_e = 1$, $\bar{\xi}_e = \omega$.

C) Unbounded domain $\xi_i = 1 \leq \xi < \infty$ (Fig. 4)

The unbounded domain on the exterior of the boundary corresponding to $\xi_i = 1 \leq \xi < \infty$ is examined. The radiation condition formulated at $\xi = \infty$ must be enforced in (31) when determining the integration constants $\{c_1\}$, $\{c_2\}$. No obvious choice of $\{c_1\}$ and $\{c_2\}$ in (31) is identified which satisfies the radiation condition. This is in contrast to the Hankel functions which are readily defined from the first and second kind of Bessel functions for the scalar case. As an alternative, the high-frequency asymptotic expansion for the dynamic-stiffness matrix $[S^\infty(\omega)]$ of the unbounded domain (superscript ∞ for unbounded) is applied, which satisfies the radiation condition at infinity. The unbounded domain of Fig. 4 is replaced by the bounded domain of Fig. 2. On its exterior boundary ξ_e the nodal force-displacement relationship determined by $[S_e^\infty(\omega)]$ (calculated for a large but finite $\bar{\xi}_e = \omega\xi_e$) serves as the boundary condition.

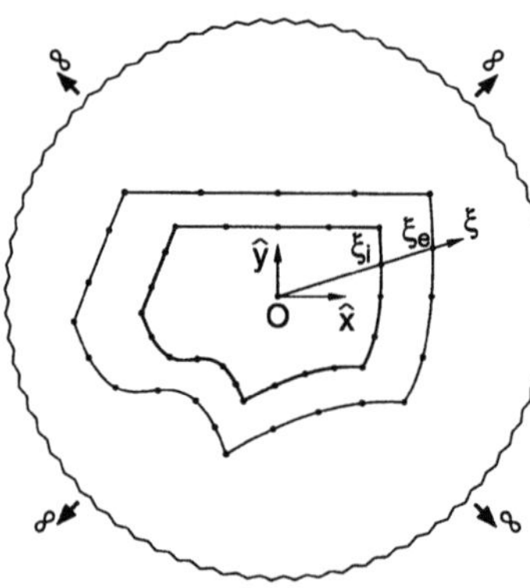

Fig. 4. Domain on exterior of boundary corresponding to $\xi_i = 1 \leq \xi < \infty$.

The high-frequency asymptotic expansion of the dynamic-stiffness matrix of an unbounded domain is formulated for ξ_e as

$$[S_e^\infty(\omega)] = \xi_e^{s-2}\left(\mathrm{i}\bar{\xi}_e[C_\infty] + [K_\infty] + \sum_{j=1}^{m} \frac{1}{(\mathrm{i}\bar{\xi}_e)^j}[A_j]\right) \tag{34}$$

The derivation and calculation of the coefficient matrices $[C_\infty]$, $[K_\infty]$ and $[A_j]$ are the same as those in [Wolf and Song, 1996b].

The dynamic-stiffness matrix of the unbounded medium with respect to the degrees of freedom on the boundary $\xi_i = 1$ is calculated as

$$[S_i^\infty(\omega)] = [S_{ii}(\omega)] - [S_{ie}(\omega)]([S_e^\infty(\omega)] + [S_{ee}(\omega)])^{-1}[S_{ei}(\omega)] \tag{35}$$

with the submatrices $[S_{ii}(\omega)]$, $[S_{ie}(\omega)]$, $[S_{ei}(\omega)]$ and $[S_{ee}(\omega)]$ of the bounded domain $\xi_i \leq \xi \leq \xi_e$ (32). To calculate the dynamic-stiffness matrix $[S_i^\infty(\omega)]$ of the unbounded medium at the boundary $\xi_i = 1$ for a specific ω, a sufficiently large $\bar{\xi}_e$ is selected for which the high-frequency asymptotic expansion of (34) yields an accurate $[S_e^\infty(\omega)]$. The ξ_e then follows as $\xi_e = \bar{\xi}_e/\omega$ and $\bar{\xi}_i = \omega$.

4 Out-Of-Plane Motion of Circular Cavity Embedded in Full-Plane

As an example, a circular cavity of radius r embedded in a full-plane of shear modulus G and density ρ is examined for out-of-plane motion (anti-plane). The out-of-plane displacement u and the shear stresses τ_{zr} and $\tau_{z\theta}$ in the polar coordinate system r, θ are shown in Fig. 5.

The analytical solution of the equation of motion is derived by applying the technique of separation of variables. For a prescribed displacement amplitude in the circumferential direction on the boundary proportional to the n-th eigenfunction $\cos n\theta$ ($n = 0, 1, 2, \ldots$), the displacement amplitude is expressed as

$$u_n = \left(c_{n1} H_n^{(1)}\left(\frac{\omega \hat{r}}{c_s}\right) + c_{n2} H_n^{(2)}\left(\frac{\omega \hat{r}}{c_s}\right)\right) \cos n\theta \tag{36}$$

with the first and second-kind Hankel functions of order n, the shear-wave velocity $c_s = \sqrt{G/\rho}$ and the integration constants c_{n1} and c_{n2}.

First, the bounded domain between the interior boundary with radius r ($\xi_i = 1$) and the exterior boundary with $2r$ ($\xi_e = 2$), an annular ring, is addressed.

The analytical solution of the dynamic-stiffness matrix derived from the displacement amplitude in (36) equals

$$\begin{bmatrix} S_{ii}(a_0) & S_{ie}(a_0) \\ S_{ei}(a_0) & S_{ee}(a_0) \end{bmatrix} = A_n \begin{bmatrix} -nJ_n(\xi_i a_0) + \xi_i a_0 J_{n+1}(\xi_i a_0) & -nY_n(\xi_i a_0) + \xi_i a_0 Y_{n+1}(\xi_i a_0) \\ nJ_n(\xi_e a_0) + \xi_e a_0 J_{n+1}(\xi_e a_0) & nY_n(\xi_e a_0) - \xi_e a_0 Y_{n+1}(\xi_e a_0) \end{bmatrix} \begin{bmatrix} J_n(\xi_i a_0) & Y_n(\xi_i a_0) \\ J_n(\xi_e a_0) & Y_n(\xi_e a_0) \end{bmatrix}^{-1} \tag{37}$$

with $A_n = 2\pi$ for $n = 0$, $A_n = \pi$ for $n = 1, 2, \ldots$ and the dimensionless frequency $a_0 = \omega r/c_s$. J_n and Y_n are the first and second-kind Bessel functions.

In the scaled boundary finite-element method the boundary is discretized with 8 curved 3-node line elements. The dynamic-stiffness matrix $[S(a_0)]$ follows from (32) with 17 terms ($k = 16$) in the matrix polynomial of $[A(\bar{\xi})]$ in (13). After enforcing the out-of-plane motion corresponding to the second ($n = 1$) eigenfunction $\cos\theta$, the four dynamic-stiffness coefficients corresponding to those in (37) result. Excellent agreement with the analytical solution (37) e.g. for $S_{ii}(a_0)$ is achieved up to $a_0 = 6$ as shown in Fig. 6.

Second, the unbounded medium is addressed. Enforcing the radiation condition yields $c_{n1} = 0$ in the analytical solution (36). The dynamic-stiffness coefficient corresponding to the

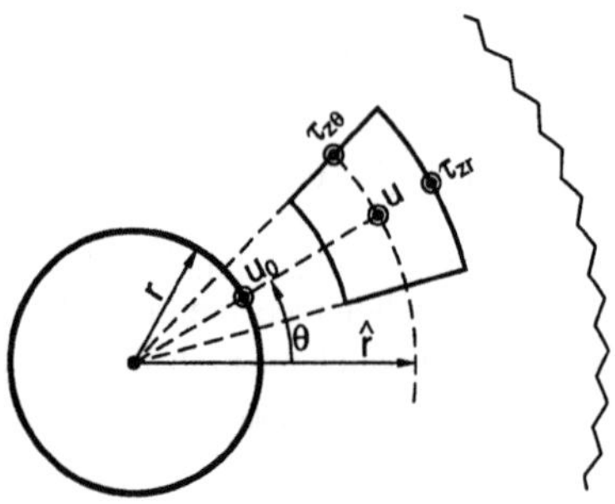

Fig. 5. Circular cavity embedded in full-plane in out-of-plane motion.

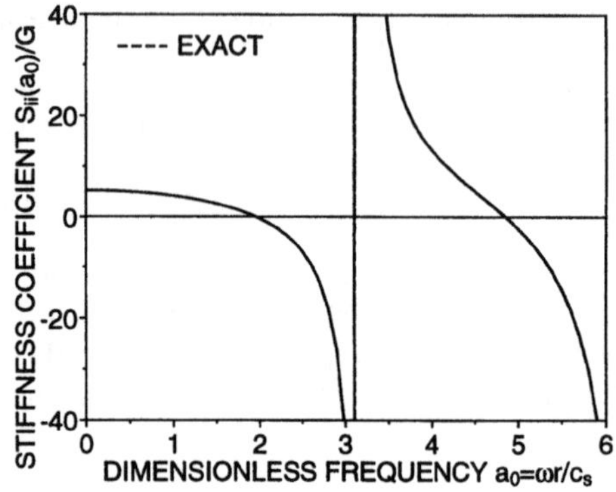

Fig. 6. Dynamic-stiffness coefficient of annular ring (diagonal coefficient for interior boundary).

n-th eigenfunction $\cos n\theta$ is written as

$$S_n^{\infty}(a_0) = A_n \left(a_0 \frac{H_{n+1}^{(2)}(a_0)}{H_n^{(2)}(a_0)} - n \right) \tag{38}$$

with $A_n = 2\pi$ for $n = 0$ and $A_n = \pi$ for $n = 1, 2, \ldots$.

In the scaled boundary finite-element method the same mesh is used as described above. The high-frequency asymptotic expansion for $[S_e^{\infty}(\omega)]$ of order 16×16 is evaluated at $\bar{\xi}_e = 3c_s/r$ with $m = 1$ (34). The corresponding ξ_e follows as $\bar{\xi}_e/\omega$. The dynamic-stiffness matrix $[S_{ii}(a_0)]$, $[S_{ie}(a_0)]$, $[S_{ei}(a_0)]$ and $[S_{ee}(a_0)]$ of the corresponding annular ring (with $\xi_i = 1$) is calculated from (32) with 7 terms ($k = 6$) in the matrix polynomial of $[A(\bar{\xi})]$ in (13). After determining the dynamic-stiffness matrix $[S^{\infty}(\omega)]$ of order 16×16, the equivalent dynamic-stiffness coefficients of the circular cavity corresponding to $n = 0$ and $= 1$ are calculated. The spring coefficient $k_n(a_0)$ and damping coefficient $c_n(a_0)$ are defined as

$$S_n^{\infty}(a_0) = \pi G(k_n(a_0) + \mathrm{i} a_0 c_n(a_0)) \tag{39}$$

Excellent agreement with the analytical solution in (38) is observed for $n = 0$ and $= 1$ in Figs. 7 and 8.

5 Further Research

The *semi-analytical* scaled boundary finite-element method achieves a *closed-form solution* in the domain for bounded and unbounded media with a spatial discretization on the boundary only. Past studies demonstrate its high accuracy and efficiency. The analytical solution permits further developments which otherwise would not be straightforward. Such topics of further research are briefly outlined, whereby no verification has been performed.

1. The high-frequency asymptotic expansion for the dynamic-stiffness matrix of an unbounded medium is assumed as a matrix polynomial and calculated from the scaled

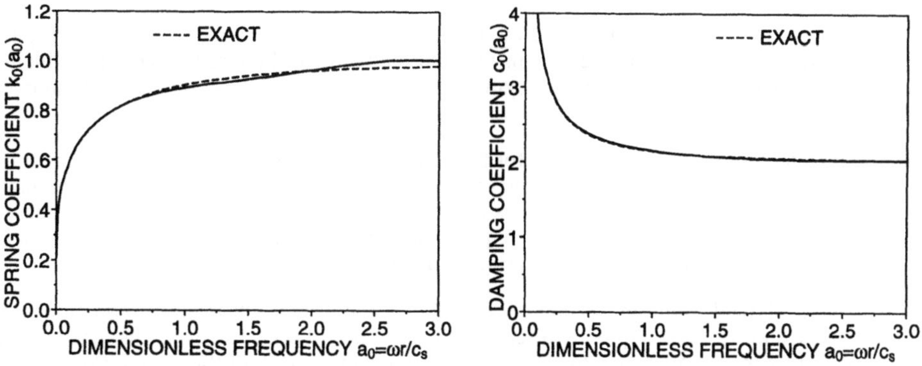

Fig. 7. Dynamic-stiffness coefficient of circular cavity ($n = 0$).

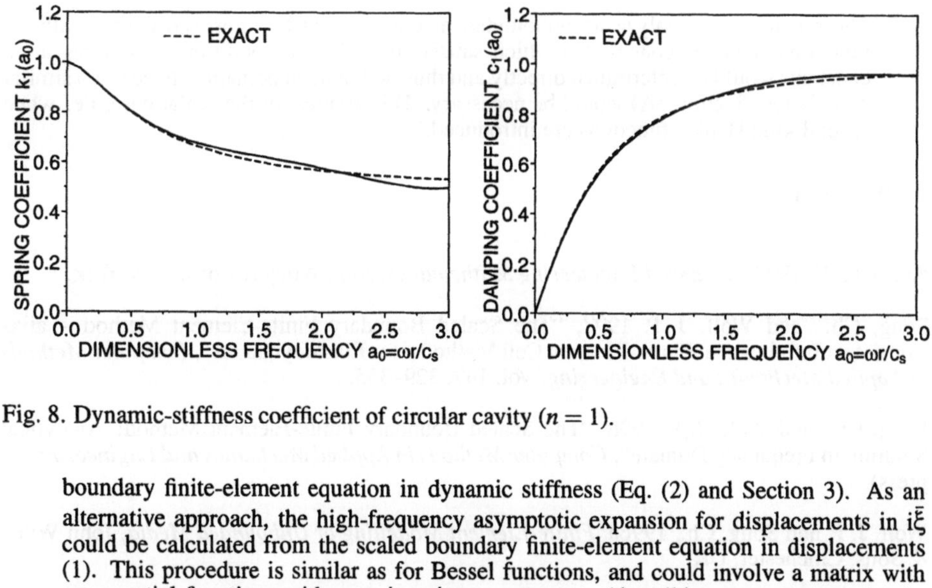

Fig. 8. Dynamic-stiffness coefficient of circular cavity ($n = 1$).

boundary finite-element equation in dynamic stiffness (Eq. (2) and Section 3). As an alternative approach, the high-frequency asymptotic expansion for displacements in $i\bar{\xi}$ could be calculated from the scaled boundary finite-element equation in displacements (1). This procedure is similar as for Bessel functions, and could involve a matrix with exponential functions with pure imaginary arguments (describing wave propagation in the radial direction), a matrix power function and a matrix polynomial. The dynamic-stiffness matrix would follow as a ratio of two matrix polynomials in $i\bar{\xi}$. The applicable range would be much larger than that of assuming the dynamic-stiffness matrix directly as a matrix polynomial. When applied to the analysis of an unbounded medium (Section 3C), the $\bar{\xi}_e$ could be selected as a much smaller value which would permit fewer terms to be chosen in (13) when calculating the dynamic-stiffness matrix of the bounded domain between the two similar boundaries.

2. When a transfer function in the frequency domain is a ratio of two matrix polynomials, many realizations in the time domain exist which do not introduce errors (system theory). For the case where the high-frequency asymptotic expansion of the dynamic-

stiffness matrix of an unbounded medium is a ratio of two matrix polynomials, highly accurate transmitting boundaries in the time domain can be constructed. Introducing state variables (internal variables) symmetric mass, damping and static-stiffness matrices can be obtained.

3. The matrix polynomial expansion for the displacement amplitudes in a bounded medium (Section 2) leads to a dynamic-stiffness matrix which is a ratio of two matrix polynomials (Eq. (33), Section 3B). Thus, various realizations in the time domain exist, as discussed under item 2. This procedure thus requires only a spatial discretization on the boundary and avoids convolution integrals in time in a dynamic analysis of a bounded medium.

4. The analytical solution (Section 2) of the homogeneous differential equations (1) corresponds to vanishing body forces. In the presence of body forces, terms independent of displacements are added to (1), as demonstrated in Eq. (52) of [Song and Wolf, 1997]. A solution based on the method of variation of parameters using the solution of the corresponding homogeneous equation (1) is possible [Kreyszig, 1993]. Thus, for body forces, no discretization inside the domain is required.

5. The efficiency to analyze an unbounded medium would be improved dramatically, if functions could be constructed which satisfy the radiation condition. The integration constants would be determined directly and thus no bounded domain between two similar boundaries (Section 3A) would be necessary. This applies for the scalar case, i.e. when second-kind Hankel functions are introduced.

6 References

Kreyszig, E. 1993, *Advanced Engineering Mathematics*, John Wiley & Sons, New York.

Song, Ch. and Wolf, J. P. 1997, "The Scaled Boundary Finite-Element Method – alias Consistent Infinitesimal Finite-Element Cell Method – for Elastodynamics", *Computer Methods in Applied Mechanics and Engineering*, Vol. 147, 329–355.

Song, Ch. and Wolf, J. P. 1998, "The Scaled Boundary Finite-Element Method: Analytical Solution in Frequency Domain", *Computer Methods in Applied Mechanics and Engineering* (in press).

Wolf, J. P. and Song, Ch., 1996a, *Finite-Element Modelling of Unbounded Media*, John Wiley & Sons, Chichester, UK.

Wolf, J. P. and Song, Ch., 1996b, "Consistent Infinitesimal Finite-Element Cell Method in Frequency Domain", *Earthquake Engineering and Structural Dynamics*, Vol. 25, 1307–1327.

Wolf, J. P. and Song, Ch., 1998, "The Scaled Boundary Finite-Element Method: State of the Art", *Proceedings of IUTAM Symposium on Computational Methods for Unbounded Domains*, Boulder, CO., July 27-31, 1997, edited by Geers, T. L., Kluwer Academic Publishers.

ON THE COMBINED IMPLEMENTATION OF GLOBAL BOUNDARY CONDITIONS WITH CENTRAL DIFFERENCE MULTIGRID FLOW SOLVERS[1]

Semyon V. Tsynkov[2]

Aerodynamic and Acoustic Methods Branch
NASA Langley Research Center
Hampton, VA 23681–2199

1 Introduction

In modern scientific computing, multigrid methods have proven to provide for one of the most efficient ways for the iterative solution of large algebraic systems that arise from discretization of the original continuous problems. These methods have been widely and successfully implemented over the last fifteen years in different applied areas, in particular, computational fluid dynamics (CFD). The key idea behind the multigrid methodology (that the short-wave components of the error expansion on the grid decay faster than the long-wave ones and therefore, the problem should be considered on a sequence of grids so that for any wave there be a grid, on which this wave can be regarded as short) appears quite universal, although every particular case may require working out some specific issues relevant to the particular type of implementation considered.

A big class of problems that present both theoretical and practical interest in scientific computing involves global boundary conditions as an element of the mathematical formulation. An important example is given by the problems originally formulated on unbounded domains. In this case, the domain is typically truncated prior to the numerical solution. Consequently, the so-called artificial boundary conditions (ABC's) should be set at the external boundary of the finite computational domain, so that the truncated problem not be subdefinite. At the same time, the ABC's should ensure that the resulting finite-domain solution be close to the corresponding fragment of the original infinite-domain solution. Generally, the proper treatment of artificial boundaries has a profound

[1]This talk was prepared while the author held a National Research Council Resident Research Associateship at NASA Langley Research Center.

[2]Permanent address: Department of Applied Mathematics, School of Mathematical Sciences, Tel Aviv University, Ramat Aviv, Tel Aviv 69978, Israel. Fax: (972-3)640-9357; E-mail: `tsynkov@math.tau.ac.il`; URL: `http://fmad-www.larc.nasa.gov:80/~tsynkov/`.

T.L. Geers (ed.), IUTAM Symposium on Computational Methods for Unbounded Domains, 285–294.

impact on the overall quality and performance of numerical algorithms and interpretation of the results.

In most cases the exact ABC's, i.e., those that drive the error associated with the domain truncation to zero, are global. The exceptions are rare and, as a rule, restricted to one-dimensional model examples. Other ABC's, i.e., those that do introduce some error because of the domain truncation, basically follow the trend that higher accuracy of the boundary procedure requires more of the nonlocal nature of the exact ABC's to be somehow taken into account.

Recent computational studies by different authors have convincingly demonstrated that highly accurate nonlocal ABC's may be particularly effective if implemented along with the multigrid interior solver. The efficacy amounts not only to the improvement of accuracy of the resulting numerical solution, which is basically expected from the more advanced treatment of the external boundary, but also to the drastic increase, by about a factor of three, of the convergence rate of multigrid iterations. (This behavior is apparently relevant just to the combined implementation of global boundary conditions with multigrid; is other cases, when a more conventional iterative algorithm is used in the interior domain, the nonlocal ABC's can reportedly even slow down the convergence.) The aforementioned studies mostly relate to the area of external flows and duct flows in both inviscid and viscous framework. The comparison of the convergence rates has been conducted against the standard algorithms, in which the approximate local ABC's based on the extrapolation of physical and/or characteristic variables are typically used.

Although the scope of the previous studies has been quite substantial, the results obtained up to date are still of mostly experimental (observational) kind. On one hand, they lack completeness as, for example, some important combinations of flow regime and multigrid strategy have not been investigated. On the other hand, these results also lack thorough theoretical justification.

The primary purpose of this talk is to present/describe the numerical evidence of improved multigrid performance due to the global boundary conditions (available up to date) and to draw the attention of the scientific computing research community to this important phenomenon. In our opinion, the issue of interaction of global ABC's with multigrid certainly deserves a further systematic study. In aerodynamics, in particular, external flow problems represent a wide class of important formulations, and as CFD becomes a widely used tool for the configuration analysis and design and often a substitute for the wind-tunnel tests, the advanced interior solvers and advanced ABC's will eventually lead to improvement of the research and industrial capability in airframe systems. Understanding of the mechanism responsible for the increase of multigrid performance as it caused by the nonlocal highly accurate ABC's will undoubtedly be most beneficial from the standpoint of full-scale flow applications, as well as many other problems.

2 Artificial Boundary Conditions Based on the Difference Potentials Method

A survey of methods for setting the ABC's in different areas of scientific computing can be found in our work [Tsynkov, 1996a; Tsynkov, 1997c].

Previously, we have developed and implemented in practice the highly accurate nonlocal ABC's for computation of steady-state external viscous flows. The methodology is based on the far-field linearization and usage of the Calderon generalized potentials and pseudodifferential boundary projections [Calderon, 1963] (see also [Seeley, 1966]). The potentials and projections are actually employed in the modified form proposed by Ryaben'kii; the corresponding numerical technique is known as the difference potentials method (DPM), see [Ryaben'kii, 1985; Ryaben'kii, 1987; Ryaben'kii, 1996] and also [Mikhlin *et al.*, 1995]. The DPM-based boundary conditions are global; the method allows us to simultaneously meet the high accuracy standards and the requirements of geometric universality and easiness in implementation.

In [Ryaben'kii and Tsynkov, 1995a], we describe the foundations of the two-dimensional DPM-based ABC's (Navier-Stokes equations). In [Tsynkov, 1995], we implement these ABC's along with the code `FLOMG`, see [Swanson and Turkel, 1985; Swanson and Turkel, 1987; Swanson and Turkel, 1997], and present some numerical results for flows over airfoils. In [Tsynkov *et al.*, 1996], we show the subsequent numerical results and propose an approximate treatment of turbulence in the far field. Our work [Ryaben'kii and Tsynkov, 1995b] delineates the algorithm for solving one-dimensional systems that arise when calculating the generalized potentials. In [Tsynkov, 1996b], we extend the DPM-based approach to the case of flows that oscillate in time; we also provide some solvability results for the linearized thin-layer equations used for constructing the ABC's. In [Ryaben'kii and Tsynkov, 1997], we present a general survey of the DPM-based methodology as applied to solving external problems in CFD (both theoretical and computational aspects), including parallel implementation of the algorithm, some observations on the combined implementation of nonlocal ABC's with multigrid, and entry-wise interpolation of the matrices of boundary operators.

The use of two-dimensional DPM-based ABC's along with the multigrid Navier-Stokes code `FLOMG` by Swanson and Turkel has shown that these boundary conditions readily apply to the boundaries of irregular shape and appear very easy to incorporate into the existing solver. In our computations, the DPM-based ABC's have consistently outperformed the standard local methods from the standpoints of accuracy, convergence rate, and robustness. The investigated regimes range from the very low (incompressible limit) to transonic Mach numbers and encompass both laminar and turbulent flows. In Table 1, we show a particular example of how the DPM-based ABC's can improve the accuracy of the solution against the standard two-dimensional point-vortex (p.-v.) methodology of [Thomas and Salas, 1985] for a transonic turbulent flow around the airfoil RAE2822.

The next natural objective after constructing the two-dimensional algorithm was the analysis of three-dimensional flows. This case is undoubtedly the one most demanded by the current computational practice in aerodynamics. In [Tsynkov, 1996c; Tsynkov, 1997a], we outline the basic elements of the DPM-based ABC's for external viscous flows around the wing-shaped configurations. The corresponding numerical results (subsonic) (see [Tsynkov, 1997a]) were obtained with the NASA-developed production code `TLNS3D`, see [Vatsa *et al.*, 1993]. In [Tsynkov and Vatsa, 1997; Tsynkov, 1997b], we further develop the three-dimensional DPM-based algorithm and present the computational results for a variety of flow regimes including low speed flows, transonic flows, and separated turbulent flows. In all cases (see [Tsynkov, 1997a; Tsynkov and Vatsa, 1997; Tsynkov, 1997b]), the DPM-based ABC's provide for a very substantial improvement of accuracy/economy of the computer resources; a particular example for the transonic turbulent flow around

TABLE 1. RAE2822: $M_0 = 0.73$; $Re_0 = 6.5 \cdot 10^6$; $\alpha = 2.79°$.

Domain size	3 chords		8 chords		50 chords	
Dimension of the grid	600 × 104		608 × 112		640 × 128	
Type of ABC'	p.-v.	DPM	p.- v.	DPM	p.-v.	DPM
Wave lift, C_l	0.8653	0.8591	0.8624	0.8589	0.8603	0.8593
Relative error	0.58%	0.02%	0.24%	0.04%	0%	0%
Wave drag, $C_d \times 10$	0.1203	0.1263	0.1209	0.1261	0.1255	0.1260
Relative error	4.14%	0.24%	3.67%	0.08%	0%	0%
Full drag, $C_D \times 10$	0.1755	0.1816	0.1762	0.1815	0.1810	0.1815
Relative error	3.04%	0.05%	2.65%	0%	0%	0%

the ONERA M6 wing is shown in Table 2. Moreover, the new ABC's also provide for a noticeable speedup (up to a factor of three) of the convergence of multigrid iterations.

TABLE 2. ONERA M6: $M_0 = 0.84$; $Re_0 = 11.7 \cdot 10^6$; $\alpha = 3.06°$.

Domain size	3 root chords		10 root chords	
Dimension of the grid	193 × 49 × 33		209 × 57 × 33	
Type of ABC's	characteristics	DPM	characteristics	DPM
Full lift, C_L	0.298±0.004	0.2798	0.2805	0.2786
Relative error	6.24%±1.43%	0.43%	0%	0%
Full drag, $C_D \times 10$	0.168±0.008	0.1537	0.1542	0.1531
Relative error	8.95%±5.19%	0.39%	0%	0%

3 The Speedup of Multigrid Convergence

Both codes that we have used for integrating the flow equations (`FLOMG` in 2D and `TLNS3D` in3D) are based on the central-difference finite-volume discretization in space with the first- and third-order artificial dissipation. The steady-state solution is obtained by means of the multigrid iterations with five-stage Runge-Kutta relaxation, residual smoothing, and local Courant step. The DPM-based ABC's are implemented only on the finest level of multigrid, and for the 3D case, also only on the first and the last Runge-Kutta stages. The standard treatment of the external boundary for `FLOMG` and `TLNS3D` is based on the locally one-dimensional analysis of characteristics and extrapolation, which may or may not be supplemented by the point-vortex correction in 2D.

In Figure 1, we show convergence history for the multigrid calculation of a three-dimensional attached transonic turbulent flow around the ONERA M6 wing; the multigrid strategy involves three subsequent grid levels with V-cycles.

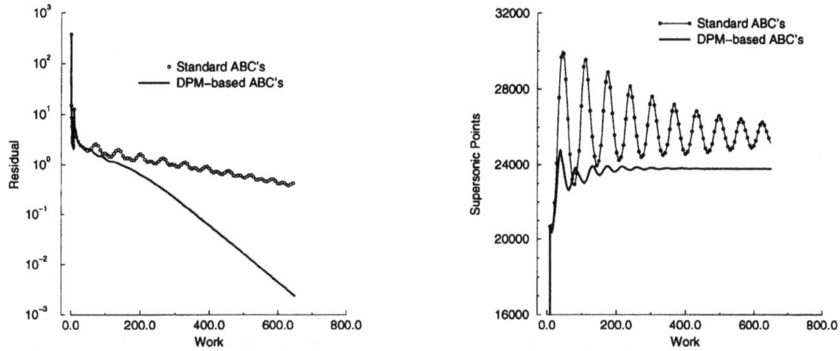

Figure 1. ONERA M6: $M_0 = 0.84$, $Re_0 = 11.7 \cdot 10^6$, angle of attack $\alpha = 3.06°$. Convergence for the residual of the continuity equation (left) and the number of supersonic nodes in the domain (right). Average domain "radius" is 3 root chords of the wing; one-block C-O grid $193 \times 49 \times 33$.

From Figure 1 we conclude that the mere replacement of the standard local boundary conditions by the DPM-based nonlocal ABC's may produce as drastic change in the algorithm performance as a three-fold increase in the multigrid convergence rate. (Slow convergence for the standard ABC's is the reason why the corresponding entries in Table 2 are given with the error bars indicated.) This effect is still visible but becomes less pronounced as we enlarge the size of the computational domain, see Figure 2.

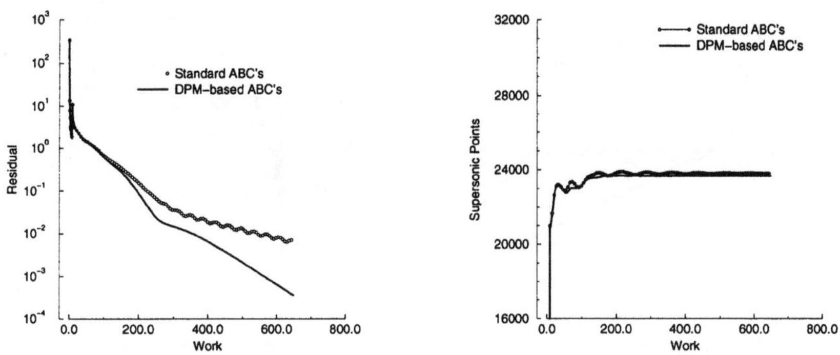

Figure 2. ONERA M6: $M_0 = 0.84$, $Re_0 = 11.7 \cdot 10^6$, angle of attack $\alpha = 3.06°$. Convergence for the residual of the continuity equation (left) and the number of supersonic nodes in the domain (right). Average domain "radius" is 10 root chords of the wing; one-block C-O grid $209 \times 57 \times 33$.

On the other hand, the effect becomes even more significant when we consider a separated turbulent flow around the ONERA M6 wing. This is a more difficult case, for which the multigrid algorithm with standard boundary conditions on the small domain simply fails to converge, whereas the nonlocal DPM-based ABC's still provide for a good convergence;

for the large domain, the convergence is satisfactory for both types of the ABC's. The corresponding results are shown in Figure 3.

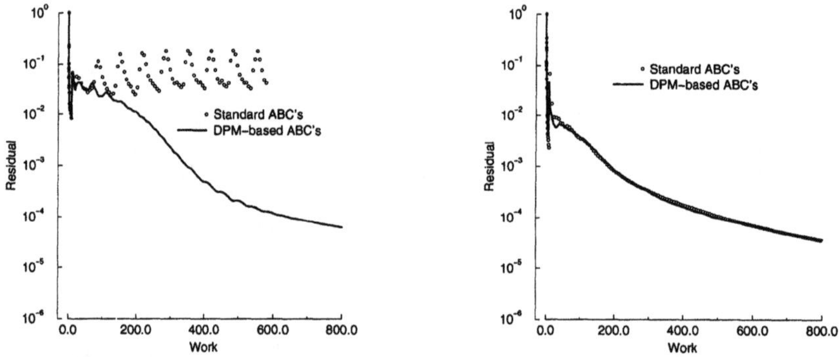

Figure 3. ONERA M6: $M_0 = 0.84$, $Re_0 = 11.7 \cdot 10^6$, angle of attack $\alpha = 5.06°$. Convergence for the residual of the continuity equation. Average domain "radius" is 3 root chords of the wing (left) and 10 root chords of the wing (right); one-block C-O grid 193 × 49 × 33.

At the same time, as opposed to the DPM-based boundary conditions, standard ABC's do not guarantee the convergence even for the big domain if a coarser grid is used, see Figure 4. We should note, however, that the cases of Figures 3 and 4 differ not only by the dimension of the grid; the calculations of Figure 3 have been performed with Menter's two-equation turbulence model and the case of Figure 4 has been run with the Baldwin-Lomax algebraic model, which typically tends to reduce the separation.

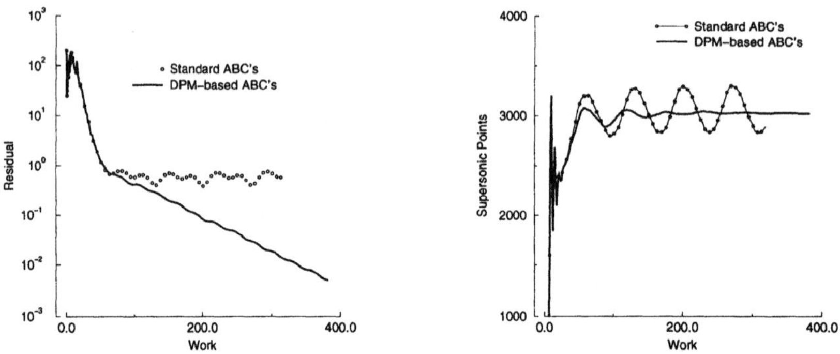

Figure 4. ONERA M6: $M_0 = 0.84$, $Re_0 = 11.7 \cdot 10^6$, angle of attack $\alpha = 5.06°$. Convergence for the residual of the continuity equation (left) and the number of supersonic nodes in the domain (right). Average domain "radius" is 10 root chords of the wing; one-block C-O grid 97 × 25 × 17.

We have also been able to obtain similar results for the two-dimensional flows. Two subsonic laminar examples for the NACA0012 airfoil are shown in Figure 5; W-type multigrid cycles have been used for these computations.

Convergence histories for some turbulent transonic calculations in two space dimensions (RAE2822 airfoil) are shown in Figure 6, the speedup effect, however, happens to

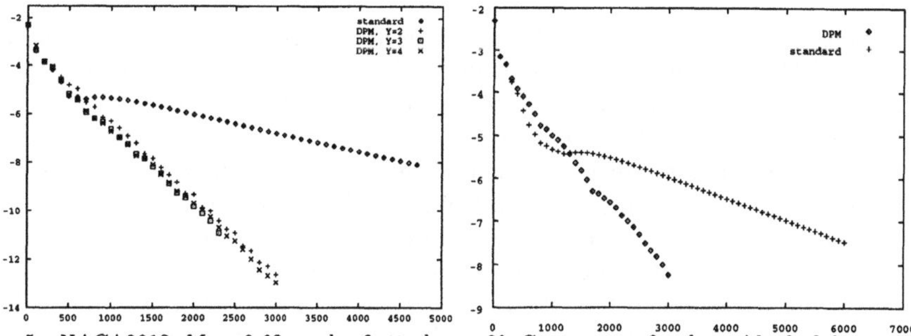

Figure 5. NACA0012: $M_0 = 0.63$, angle of attack $\alpha = 2°$. Convergence for the residual of the continuity equation (vs. number of cycles); one-block C grid 256×64. Left: $Re = 4000$, 5 chords domain; right: $Re = 5000$, 10 chords domain.

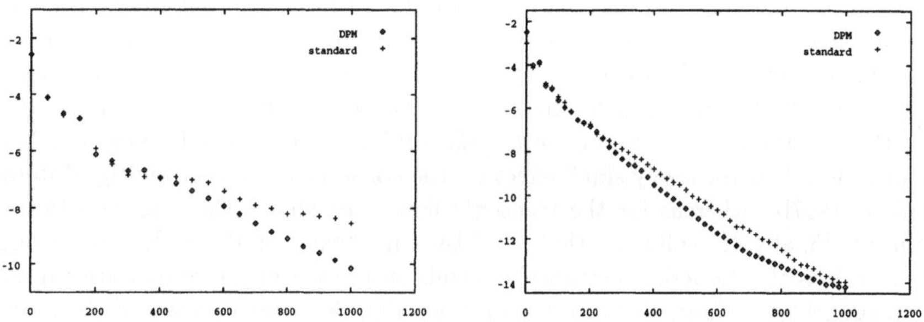

Figure 6. RAE2822: $M_0 = 0.73$, $Re = 6.5 \cdot 10^6$, angle of attack $\alpha = 2.79°$. Convergence for the residual of the continuity equation (vs. number of cycles); average domain "radius" is 11 chords; one-block C grid 256×64 with the near-surface spacing 10^{-4} (left) and 10^{-5} (right).

be less visible in these cases. Finally, similarly to the three-dimensional case (see Figure 3 (left) and Figure 4), we have observed in two space dimensions (for both laminar and turbulent transonic flows) the situations when the algorithm with standard boundary conditions failed to converge whereas the algorithm with the DPM-based ABC's still converged rapidly, the corresponding examples are presented in Figure 7.

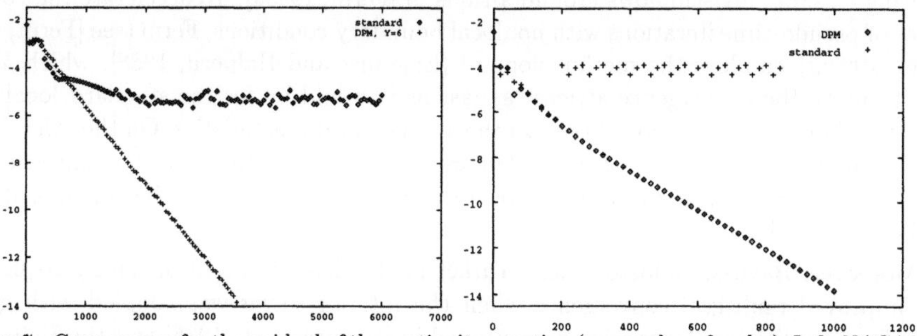

Figure 7. Convergence for the residual of the continuity equation (vs. number of cycles). Left: NACA0012, $M_0 = 0.85$, $Re_0 = 4000$, angle of attack $\alpha = 1°$; one-block C grid 256×64; average domain "radius" is 5.5 chords. Right: RAE2822, $M_0 = 0.73$, $Re = 6.5 \cdot 10^6$, angle of attack $\alpha = 2.79°$; one-block C grid 256×64 with $0.33 \cdot 10^{-4}$ near-surface spacing; average domain "radius" is 6 chords.

As mentioned above, although the acceleration of multigrid convergence due to the improved ABC's is extremely important for applications, neither rigorous mathematical justification nor a conclusive experimental explanation of why and when the speedup occurs is available as of yet. For our two- and three-dimensional computations we have used different multigrid strategies (W- and V-cycles, respectively, with different number of levels). Moreover, the laminar airfoil flows, see Figures 5 and 7 (left), for which the most noticeable convergence acceleration occured in two dimensions, were computed on relatively coarse grids with the Courant numbers determined globally, whereas the transonic turbulent flows around the ONERA M6 wing, see Figures 1, 2, 3 (left), and 4, for which the most noticeable convergence acceleration occured in three dimensions, were calculated on strongly stretched grids with the local time step. On the other hand, for our two-dimensional turbulent transonic computations, which have been conducted using local Courant numbers, see Figure 6, we have not been able to obtain the convergence acceleration as drastic as the one that corresponds to the similar three-dimensional cases. As for the results of laminar computations (both subsonic and transonic), they are available for 2D only. In three dimensions, we have run only turbulent cases and in doing so found that for the purely subsonic flows ($M_0 = 0.5$ and $M_0 = 0.01$ for ONERA M6) the type of ABC's had relatively small effect on the convergence rate of multigrid iterations [Tsynkov, 1997b], whereas for the transonic flows (see above) this effect could be very significant. Finally, the influence that the ABC's may exert on the multigrid convergence rate, as well as the numerical accuracy, depends on the size of the computational domain and apparently, the dimension of the grid. Basically, the overall impact of the ABC's on the computational algorithm decreases as the domain enlarges (compare, e.g., Figures 1 and 2).

Besides our own work, there is other evidence of the improved performance of highly accurate nonlocal ABC's when combined with a multigrid iteration procedure. For example, Ferm, see [Ferm, 1992], notes that for a simple pseudo-time interior solver without multigrid the nonlocal highly accurate ABC's of [Ferm and Gustafsson, 1982; Ferm, 1992] constructed for the Euler flows in ducts would slow down the convergence to steady state. The same phenomenon occurs for the boundary conditions of [Ferm, 1990] as applied to calculating external inviscid flows around airfoils in [Ferm, 1993a]. To accelerate the convergence of pseudo-time iterations with nonlocal boundary conditions, Ferm (see [Ferm, 1992; Ferm, 1993a]) employs the methodology of [Engquist and Halpern, 1988], which allows him to make the convergence at least as fast as it would be for the standard local nonreflecting boundary conditions based on the analysis of characteristics. On the other hand, when boundary conditions of [Ferm, 1990] are implemented along with the multigrid interior Euler solver, they no longer slow down the convergence. This has been demonstrated in [Ferm, 1993b].

Moreover, Bayliss, Goldstein, and Turkel in [Bayliss *et al.*, 1985] have also noticed the improved multigrid convergence when the interior solver was coupled with global boundary conditions; the specific case studied in [Bayliss *et al.*, 1985] was the Helmholtz equation in a waveguide. However, similarly to the aforementioned work by other authors, the convergence speedup of [Bayliss *et al.*, 1985] has not been justified by analysis.

In spite of the existing inconclusiveness as to what causes the foregoing improvement in performance and for what combinations of parameters one can expect to obtain the maximal gain, the results available up to date still prove that the combination of nonlocal boundary conditions with multigrid has a very substantial promise for reducing the overall cost of computations for different applied problems. Therefore, a detailed study of interaction between the nonlocal boundary conditions and multigrid becomes an important issue.

It is natural to start the study of this kind with analyzing some scalar model examples in simple geometric configurations. Then, as it comes to the flow problems, a natural sequence of the consecutively more complex formulations is: full-potential flows, Euler flows, and Navier-Stokes flows. As concerns the full-scale flow applications, many different factors, see Table 3, may influence the overall algorithm performance and more extensive numerical experiments may be required for determining those that could provide the maximal positive effect.

TABLE 3. Factors that may influence the overall algorithm performance for flow applications.

Geometry	Flow regime	Solver	Implementation of ABC's
2D vs. 3D	Subsonic vs. transonic	Type of scheme	Consistency with scheme
Domain size	Laminar vs. turbulent	Multigrid strategy	Consistency with multigrid
Grid dimension	Attached vs. separated	Relaxation process	Consistency with relaxation
Grid stretching		Courant step	
Cell aspect ratio			

References

Bayliss, A., Goldstein, C. I., and Turkel, E., 1985, "The Numerical Solution of the Helmholtz Equation for Wave Propagation Problems in Underwater Acoustics," *Computers and Mathematics with Applications*, Vol. 11, pp. 655-665.

Calderon, A. P., 1963, "Boundary-Value Problems for Elliptic Equations," in *Proceedings of the Soviet-American Conference on Partial Differential Equations at Novosibirsk*, Fizmatgiz, Moscow, USSR, pp. 303–304.

Engquist, B. and Halpern, L., 1988, "Far Field Boundary Conditions for Computation over Long Time," *Appl. Numer. Math.*, Vol. 4, pp. 21–45.

Ferm, L., 1990, "Open Boundary Conditions for External Flow Problems," *J. Comput. Phys.*, Vol. 91, pp. 55–70.

Ferm, L., 1992, *Non-Reflecting Accurate Open Boundary Conditions for the Steady Euler Equations*, Technical Report No. 143, Department of Scientific Computing, Uppsala University, Uppsala, Sweden.

Ferm, L., 1993a, *Modified External Boundary Conditions for the Steady Euler Equations*, Technical Report No. 153, Department of Scientific Computing, Uppsala University, Uppsala, Sweden.

Ferm, L., 1993b, *Multigrid for External Flow Problems*, Technical Report, Department of Scientific Computing, Uppsala University, Uppsala, Sweden.

Ferm, L. and Gustafsson, B., 1982, "A Downstream Boundary Procedure for the Euler Equations," *Computers and Fluids*, Vol. 10, pp. 261–276.

Mikhlin, S. G., Morozov, N. F., and Paukshto, M. V., 1995, *The Integral Equations of the Theory of Elasticity*, B. G. Teubner Verlagsgesellschaft, Stuttgart, Germany.

Ryaben'kii, V. S., 1985, "Boundary Equations with Projections," *Russian Math. Surveys*, Vol. 40, pp. 147–183.

Ryaben'kii, V. S., 1987, *Difference Potentials Method for Some Problems of Continuous Media Mechanics*, Nauka, Moscow, USSR. (in Russian)

Ryaben'kii, V. S., 1996, "Difference Potentials Method and its Applications," *Math. Nachr.*, Vol. 177, pp. 251–264.

Ryaben'kii, V. S. and Tsynkov, S. V., 1995a, "Artificial Boundary Conditions for the Numerical Solution of External Viscous Flow Problems," *SIAM J. Numer. Anal.*, Vol. 32, pp. 1355–1389.

Ryaben'kii, V. S. and Tsynkov, S. V., 1995b, "An Effective Numerical Technique for Solving a Special Class of Ordinary Difference Equations," *Appl. Numer. Math.*, Vol. 18, pp. 489–501.

Ryaben'kii, V. S. and Tsynkov, S. V., 1997, *An Application of the Difference Potentials Method to Solving External Problems in CFD*, NASA Technical Memorandum No. 110338, Langley Research Center, Hampton, Virginia, USA; also to appear in *CFD Review 1997*, M. Hafez and K. Oshima, eds.

Seeley, R. T., 1966, "Singular Integrals and Boundary Value Problems," *Amer. J. Math.*, Vol. 88, pp. 781–809.

Swanson, R. C. and Turkel, E., 1985, *A Multistage Time-Stepping Scheme for the Navier-Stokes Equations*, AIAA Paper No. 85-0035, Reno, Nevada, USA.

Swanson, R. C. and Turkel, E., 1987, *Artificial Dissipation and Central Difference Schemes for the Euler and Navier-Stokes Equations*, AIAA paper No. 87-1107-CP, Honolulu, Hawaii, USA.

Swanson, R. C. and Turkel, E., 1997, *Multistage Schemes with Multigrid for the Euler and Navier-Stokes Equations. Volume I: Components and Analysis*, NASA Technical Paper No. 3631, Langley Research Center, Hampton, Virginia, USA.

Thomas, J. L. and Salas, M. D., 1985, *Far-Field Boundary Conditions for Transonic Lifting Solutions to the Euler Equations*, AIAA Paper No. 85–0020, Reno, Nevada, USA.

Tsynkov, S. V., 1995, "An Application of Nonlocal External Conditions to Viscous Flow Computations," *J. Comput. Phys.*, Vol. 116, pp. 212–225.

Tsynkov, S. V., 1996a, *Artificial Boundary Conditions Based on the Difference Potentials Method*, NASA Technical Memorandum No. 110265, Langley Research Center, Hampton, Virginia, USA.

Tsynkov, S. V., 1996b, *Artificial Boundary Conditions for Computation of Oscillating External Flows*, NASA Technical Memorandum No. 4714, Langley Research Center, Hampton, Virginia, USA; also to appear in *SIAM J. Sci. Comp.*

Tsynkov, S. V., 1996c, "Nonlocal Artificial Boundary Conditions for Computation of External Viscous Flows," in: *Computational Fluid Dynamics'96*, Proceedings of the Third ECCOMAS CFD Conference, September 9–13, 1996, Paris, France, J.-A. Desideri, C. Hirsch, P. Le Tallec, M. Pandolfi, and J. Périaux, eds., John Wiley & Sons, pp. 512–518.

Tsynkov, S. V., 1997a, "Artificial Boundary Conditions for Infinite-Domain Problems," to appear in *Proceedings of the ICASE/LaRC Workshop on Barriers and Challenges in Computational Fluid Dynamics*, Hampton, August 5–7, 1996, M. D. Salas et al., eds., Kluwer Academic Publishers.

Tsynkov, S. V., 1997b, *External Boundary Conditions for Three-Dimensional Problems of Computational Aerodynamics*, NASA Technical Memorandum No. 110337, Langley Research Center, Hampton, Virginia, USA; also submitted to *SIAM J. Sci. Comput.*

Tsynkov, S. V., 1997c, "Numerical Solution of Problems on Unbounded Domains and Application of the Difference Potentials Method," submitted to *Appl. Numer. Math.*

Tsynkov, S. V., Turkel, E., and Abarbanel, S., 1996, "External Flow Computations Using Global Boundary Conditions," *AIAA J.*, Vol. 34, pp. 700–706; also: AIAA Paper No. 95-0562, Reno, Nevada, USA, 1995.

Tsynkov, S. V., and Vatsa V. N., 1997, "An Improved Treatment of External Boundary for Three-Dimensional Flow Computations," AIAA Paper No. 97-2074, in *Proceedings of the 13th AIAA CFD Conference, Part 2*, June 29 – July 2, Snowmass Village, Colorado, USA, AIAA, Reston, VA, USA, pp. 1139–1149.

Vatsa, V. N., Sanetrik, M. D., and Parlette, E. B., 1993, *Development of a Flexible and Efficient Multigrid-Based Multiblock Flow Solver*, AIAA Paper No. 93-0677, Reno, Nevada, USA.

A RESPONSE ANALYSIS OF VERY LARGE FLOATING STRUCTURE UNDER AIRPLANE LANDING BY FEM AND A SPONGE LAYER FOR THE UNBOUNDED DOMAIN

Eiichi Watanabe and Tomoaki Utsunomiya

Department of Civil Engineering
Kyoto University
Kyoto, Japan

1. INTRODUCTION

In small floating structures, the elastic response is smaller than the rigid body motion; in case of very large floating structures (VLFS), however, the depth is much smaller than the horizontal dimensions and thus the floating structures tend to be very flexible [Wu, et al., 1996, Yamamoto, 1994].

The present authors have already presented a transient response analysis of axisymmetric floating plate with closed fluid domain by Finite Element Method (F.E.M.) [Watanabe, et al., 1996, 1997]. In the analysis with the closed fluid domain, the radiation waves generated by the plate vibration are reflected at the boundary wall; thus large distance between the boundary wall and the structure is required in order to obtain an equivalent solution for open sea. When the same method is applied to a general 3-dimensional analysis, requirement of large storage limits the size of the structure. Therefore, in order to analyze a large structure such as a floating airport, open boundary condition should be satisfied at the finite (and desirably small) distance from the structure, which will reduce the size of F.E.M. model.

In case of a single wave frequency, the analysis could be extended to non-reflecting open boundaries by use of the Sommerfeld boundary condition. However, in such a transient problem as the airplane landing, many frequencies will be included in the radiation waves, and thus the Sommerfeld condition alone is insufficient to disperse the wave energy and another appropriate method for unbounded boundary is required.

The main purpose of this study is to analyze the dynamic deflectional response of a very large floating structure at the airplane landing in the time domain. Considered herein in particular is the application of an wave absorption filter which consists of sponge layer and Sommerfeld boundary developed by Ohyama and Nadaoka [Ohyama and Nadaoka, 1991]. The finite element method is adopted for the axisymmetric problem at first, with an open boundary problem. Then, an attempt is made to extend it to the 3-dimensional problem and the time domain analysis is performed by simplifying the landing load of a Jumbo jet plane.

T.L. Geers (ed.), IUTAM Symposium on Computational Methods for Unbounded Domains, 295–304.

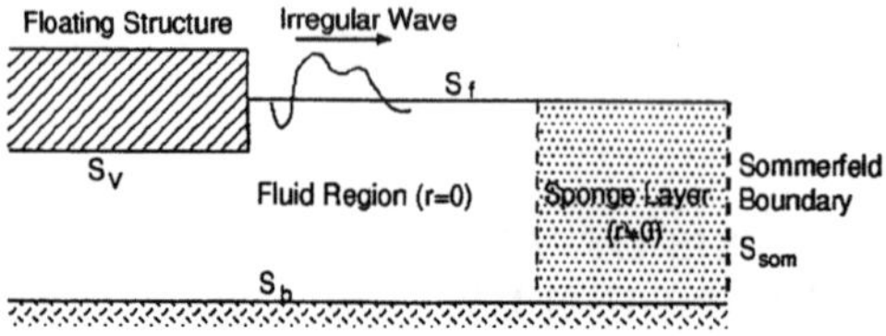

Fig. 1. Numerical Wave Absorption Filter

2. METHOD OF ANALYSIS

2.1. Wave Absorption Filter

The Sommerfeld boundary has been used almost exclusively for the absorption of waves beyond certain domain. However, it assumes a single wave velocity and thus, for irregular waves, the wave energy is only insufficiently reduced. In stead of this type of boundary, an attempt is made using the numerical wave filter as shown in Fig. 1 [Ohyama and Nadaoka, 1991].

This wave absorption filter consists of the sponge layer which is characterized by the damper to absorb the wave energy and a Sommerfeld boundary surrounding it. The wave absorption mechanism of the sponge filter is considered to be very effective even against the irregular waves since it is based on the energy absorption due to the friction of the passing fluid and does not have significant frequency dependence. Only one caution may be such that when the length of an incidental wave component happens to be larger than the radius of the sponge layer, then the energy absorption may not be sufficient. Thus, for this long wave length, the Sommerfeld boundary located outside of the sponge layer may work effectively to disperse the energy by taking into account the fact that the long wave velocity, c, can be approximated by velocity $\sqrt{gh}$ where g and h refers to the gravity and the depth of the fluid, respectively.

2.2. Formulation by Finite Element

The equilibrium equation for small motions of a compressible and inviscid fluid flowing through a resisting matrix material can be written as follows:

$$\nabla p + r\dot{\mathbf{u}} + \rho\ddot{\mathbf{u}} = \mathbf{O} \tag{1}$$

$$p = -B\nabla \cdot \mathbf{u} \tag{2}$$

where p refers to the dynamic pressure; $\mathbf{u}$, $\dot{\mathbf{u}}$, and $\ddot{\mathbf{u}}$ the fluid particle displacement, velocity, and acceleration, respectively; ρ the density of the fluid; r the volumetric drag coefficient (force per unit volume per velocity); and B the bulk modulus of the fluid. In this study, constant value of r is applied at the sponge layer surrounding the fluid region. In the fluid region, no drag resistant is applied, i.e., $r = 0$.

Based on the linear wave theory, appropriate boundary conditions can be represented as follows:

$$\mathbf{n} \cdot \ddot{\mathbf{u}} = 0 \quad \text{on } S_b \tag{3}$$

$$\mathbf{n} \cdot \ddot{\mathbf{u}} = \mathbf{n} \cdot \ddot{\mathbf{u}}^m \quad \text{on } S_V \tag{4}$$

$$\mathbf{n}\cdot\dot{\mathbf{u}} = \frac{1}{\rho g}\dot{p} \quad \text{on } S_f \tag{5}$$

$$\mathbf{n}\cdot\dot{\mathbf{u}} = \frac{1}{\rho\sqrt{gh}}\dot{p} \quad \text{on } S_{som} \tag{6}$$

where $\mathbf{n}$ represents the outward unit normal vector out of the fluid region at the boundaries, and $\ddot{\mathbf{u}}^m$ represents the acceleration of the structural media at the fluid-structure boundary.

When $\nabla\cdot$ is applied to Eq. (1), the following equation is obtained:

$$\nabla\cdot\frac{1}{\rho}\nabla p + \nabla\cdot(\mu\dot{\mathbf{u}}) + \nabla\cdot\ddot{\mathbf{u}} = 0 \tag{7}$$

where

$$\mu = \frac{r}{\rho}. \tag{8}$$

If $\mu =$ constant in an element, then $\nabla\cdot\mu = 0$. Besides, substitution of Eq. (2) into Eq. (7) yields

$$\frac{1}{B}\ddot{p} + \frac{\mu}{B}\dot{p} - \nabla\cdot\frac{1}{\rho}\nabla p = 0 \tag{9}$$

Next, let us consider the integration of the weighted integrand of Eq. (9) with the weight, w:

$$\begin{aligned}&\int_V w\left\{\frac{1}{B}\ddot{p} + \frac{\mu}{B}\dot{p} - \nabla\cdot\frac{1}{\rho}\nabla p\right\} dV \\ &= \int_V w\frac{1}{B}\ddot{p}\,dV + \int_V w\frac{\mu}{B}\dot{p}\,dV + \int_V \nabla w\cdot\frac{1}{\rho}\nabla p\,dV - \int_S w\frac{1}{\rho}\nabla p\cdot d\mathbf{s} = 0\end{aligned} \tag{10}$$

where $\mathbf{s}$ refers to the outward normal vector at the boundary, and $d\mathbf{s} = \mathbf{n}ds$.

For the formulation by Finite Element Method, the fluid, sponge parts should be divided into many regions considering the pressure, p, as unknown being designated in terms of the joint pressure vector, $\mathbf{p}$, and the shape function, $\mathbf{N}$, so that

$$p = \mathbf{N}^T\mathbf{p} \tag{11}$$

Moreover, adopting $\mathbf{N}$ for the weighting function, w, the following discretization can be made:

$$\int_V \mathbf{N}\frac{1}{B}\mathbf{N}^T dV\ddot{\mathbf{p}} + \int_V \mathbf{N}\frac{\mu}{B}\mathbf{N}^T dV\dot{\mathbf{p}} + \int_V \nabla\mathbf{N}\cdot\frac{1}{\rho}\nabla\mathbf{N}^T dV\mathbf{p} - \int_S \mathbf{N}\frac{1}{\rho}\nabla p\cdot d\mathbf{s} = \mathbf{O} \tag{12}$$

where from the 1st to the 3rd terms of Eq. (12) designate the volume integrals of the interior of the element; the 4th term, on the other hand, designates the boundary integrals. Upon substitution of appropriate boundary conditions Eqs. (3)-(6) into Eq. (12), the following equation can be obtained:

$$(\mathbf{M}+\mathbf{M}_f)\ddot{\mathbf{p}} + (\mathbf{C}+\mathbf{C}_f+\mathbf{C}_{som})\dot{\mathbf{p}} + (\mathbf{K}+\mathbf{K}_{som})\mathbf{p} + \mathbf{M}_V\ddot{\mathbf{u}}^m = \mathbf{O} \tag{13}$$

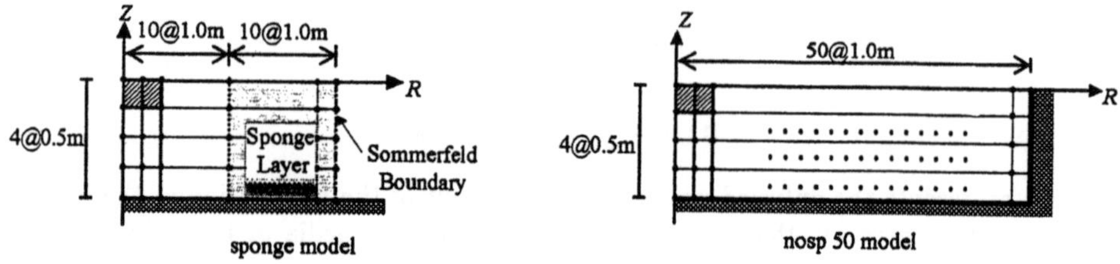

Fig. 2. Axi-symmetric Sponge and Non-sponge Elements

where

$$\mathbf{M} = \int_V \mathbf{N}\frac{1}{B}\mathbf{N}^T\, dV \qquad \mathbf{M}_f = \int_{S_f} \mathbf{N}\frac{1}{\rho g}\mathbf{N}^T\, ds$$

$$\mathbf{C} = \int_V \mathbf{N}\frac{\mu}{B}\mathbf{N}^T\, dV \qquad \mathbf{C}_f = \int_{S_f} \mathbf{N}\frac{\mu}{\rho g}\mathbf{N}^T\, ds$$

$$\mathbf{C}_{som} = \int_{S_{som}} \mathbf{N}\frac{1}{\rho\sqrt{gh}}\mathbf{N}^T\, ds \qquad \mathbf{K} = \int_V \nabla\mathbf{N}\cdot\frac{1}{\rho}\nabla\mathbf{N}^T\, dV$$

$$\mathbf{K}_{som} = \int_{S_{som}} \mathbf{N}\frac{\mu}{\rho\sqrt{gh}}\mathbf{N}^T\, ds \qquad \mathbf{M}_V = \int_{S_V} \mathbf{N}\mathbf{n}\cdot\mathbf{N}_m^T\, ds$$

where **M**, **C**, and **K** refer to the mass, damping and stiffness matrix, respectively. Wherever there are no suffices, they imply the interior of the fluid and the suffices f, V and som refer to the free surface, boundary between the fluid and the floating body and the Sommerfeld boundary, respectively. Moreover, $\mathbf{N}_m$ refers to the shape function of displacement of the structure.

Structural media is also discretized by the usual finite element procedure, and the coupled system equations are solved for the unknown nodal vectors combining the pressure in the fluid and the displacement of the structural media.

The computation is performed by ABAQUS which is a general purpose finite element program. It is equipped with all of the elements in Eq. (13), thus theoretically enables acoustic and interface elements in ABAQUS to be applied to the fluid and sponge elements [Hibbitt, Karlsson and Sorensen, Inc., 1995].

3. RESULTS OF ANALYSIS AND DISCUSSIONS

3.1. Verification of Sponge Elements

At first, in axisymmetric problems, the applicability of the sponge elements will be studied. The model used is shown in Fig. 2. In this figure, "nosp L" refers to non sponge elements but with the rigid wall placed at L m from the axis of axisymmetry; while "sponge" refers to the model with the sponge layer extending from 10 m from the axis over the length of 10 m and with the volumetric drag coefficient, r, and with a Sommerfeld boundary at the exterior boundary. The other features of the models are provided in Table 1.

Now let us consider a floating structural model of radius 2 m to which a sinusoidal heaving motion with the amplitude of 0.5 m is forced. The structural elements are 4-node

Table 1. Characteristics of Verification Model

Young's Modulus	6.617×10^{11} kN/m^2
Density of Fluid	1.025 ton/m^3
Bulk Modulus of Fluid	2.22 GPa
Draft	0.5 m
Depth of Water	2 m

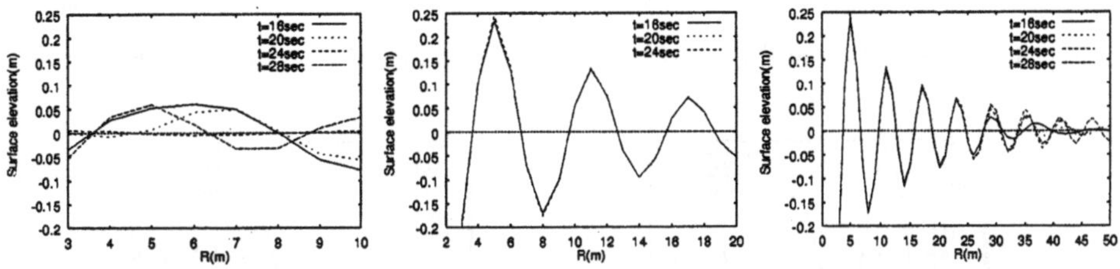

(a) nosp10 model (b) sponge model (c) nosp50 model

Fig. 3. Change of Free Surface Elevation at $f = 0.5$Hz

linear hybrid nonconforming axisymmetric elements; while elements for the fluid and the boundaries other than the fluid-structure boundary are 4-node axisymmetric acoustic elements; moreover, interface elements are used at the fluid-structure boundaries. The length of each of the elements is 1.0 m long in the radial direction and 0.5 m deep in the vertical direction, respectively, and the time interval is taken to be 0.1 sec.

Shown in Fig. 3(a)$\sim$(c) are the free surface elevation when the forced heaving displacement is assigned at the frequency of $f = 0.5$Hz. The volumetric drag coefficient, r, is prescribed as $r = 0.1$kN·sec/m^4 in the sponge model.

If no reflection of waves were expected, the ordinate should coincide. The results show that the waves of nosp50 and sponge coincide very well for different time steps; however the wave of nosp10 varies significantly showing that its surface elevation is significantly influenced by the reflection of waves. nosp50 has not yet received the reflected waves and in sponge the wave energy seems to be well absorbed.

Considered next is the change of free surface when the length of the sponge layer and the coefficient of volumetric drag are changed. First, the sponge length is prescribed to be 20m while changing the coefficient, r, at 3 distinct levels from $r = 0.1$ to 10 kN·sec/m^4. The results are shown in Fig. 4(a)$\sim$(c).

When $r = 10$ kN·sec/m^4, certain variation of the water elevation is observed in the fluid domain; however, no significant change of water elevation in the sponge layer can be observed. This implies the fact that for higher value of r, the resistance is so strong that the waves are reflected at the interior boundary of the sponge layer. In case the drag coefficient, r, is smaller, the reflection does not occur and the energy seems to be absorbed in the sponge layer.

Moreover, analysis is made to find the effect of the length of the sponge layer by changing it from 10 m to 20 m by keeping the distance between the central axis and the interior

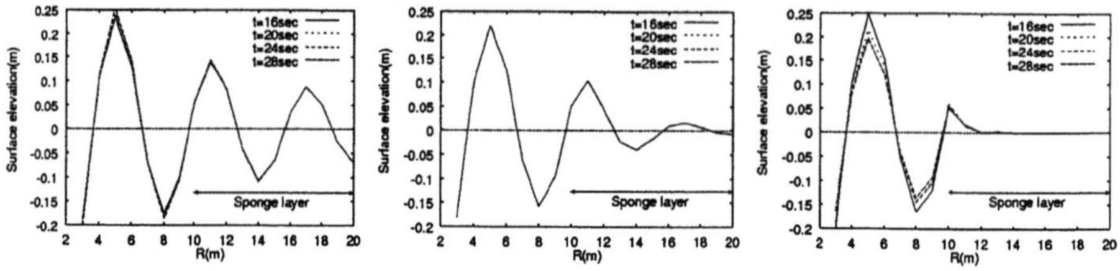

(a) $r = 0.1$ kN sec/m^4 (b) $r = 1.0$ kN sec/m^4 (c) $r = 10$ kN sec/m^4

Fig. 4. Configuration of Free Surface for different Value of Resistance Coefficient, r

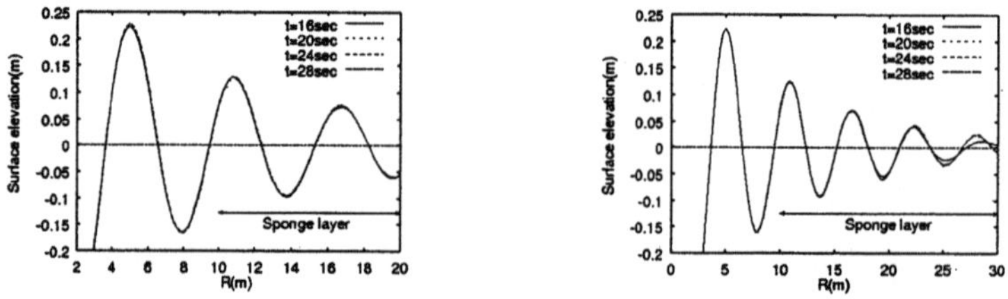

(a) length of sponge layer = 10m (b) length of sponge layer = 20m

Fig. 5. Effect of Sponge Layer Length on the Surface Elevation

boundary of the sponge layer to be 10 m. Shown in Fig. 5 also are change of free surface assuming the volumetric drag coefficient, r, to be $r = 0.1$ kN·sec/m^4 and the element length in the radial direction is kept to be as small as 0.25 m.

When the length of the sponge layer is 20 m, the amplitude of the response at the exterior boundary tends to be zero. From this, it may be seen that even if the volumetric drag coefficient is small, waves can be well absorbed if the length of the sponge filter is taken sufficiently long enough.

3.2. Analysis for Airplane Landing in the Axisymmetric Problems

An attempt is made to evaluate the applicability of the sponge layer for axisymmetric airplane landing problem. Two analytical disk models of radius 200 m are considered. The first model has a rigid wall at 200 m from the boundary of the disk as shown in Fig. 6; while the second model as shown in Fig. 7 has the sponge layer of the length 40 m at the distance 20 m from the boundary of the disk with the coefficient of volumetric drag, $r = 0.1$ kN· sec/m^4. The other characteristics remain the same as shown in Table 1 and the time interval is taken to be 0.1 sec.

The load pattern is as shown in Fig. 8 and the maximum load is assumed to be 675 tonf including the impact force of a jumbo jet plane [Takarada, 1982], and Fig. 9 shows the time history of the vertical deflection at the point of the landing.

It can be seen from this that the deflection of the model without the sponge layer becomes significantly large after 45 seconds from the landing. This implies that once propagated wave is reflected at the rigid wall located at 200 m away from the boundary of

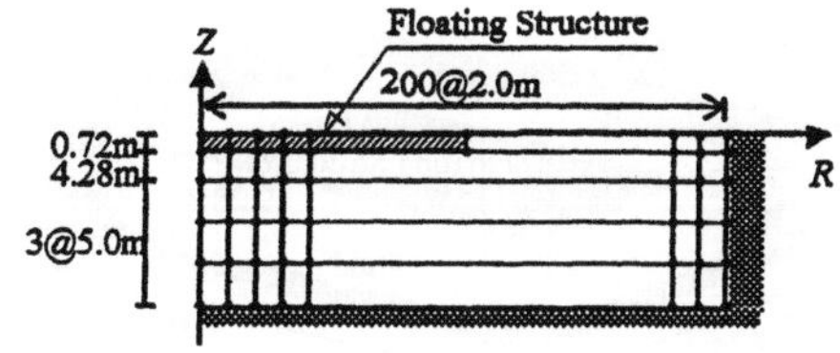

Fig. 6. Model: Rigid wall at 200 m from Disc

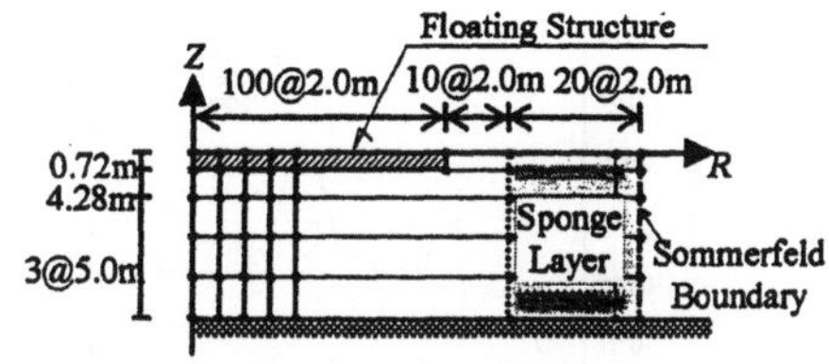

Fig. 7. Model: Sponge Layer at 20 m from Disc

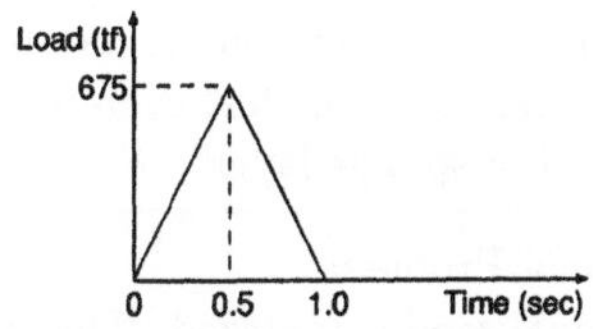

Fig. 8. Loading Pattern for Airplane Landing

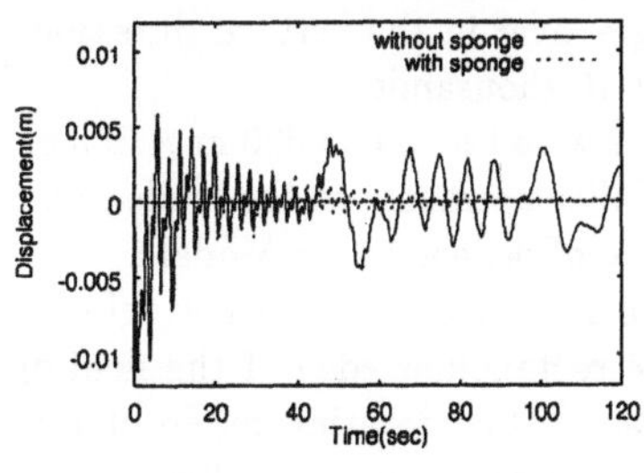

Fig. 9. Time History of Vertical Deflection at the Point of Landing

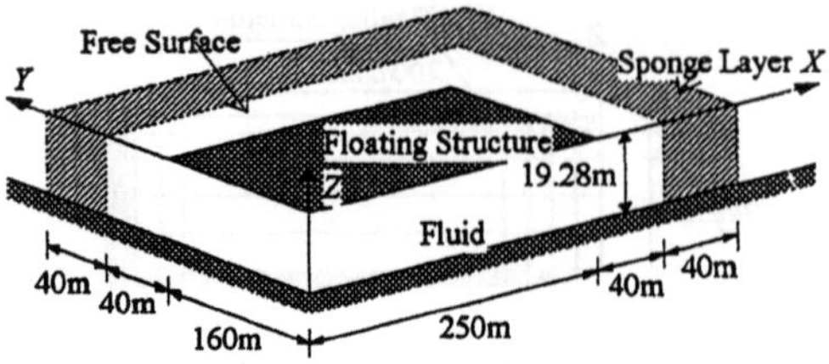

Fig. 10. Model for 3-dimensional Analysis(Model 1)

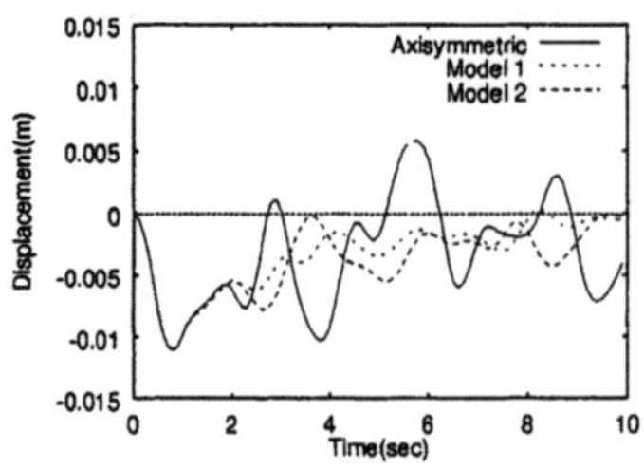

Fig. 11. Comparison of Time History of Vertical Displacement at the Point of Landing

the disk and comes back to the center about 45 seconds later. On the other hand, in the model with the sponge layer the displacement is damped out gradually to become zero finally. From these results, it can be seen that even for irregular waves, the wave energy can be absorbed very well by the sponge layer.

3.3. Analysis of 3-dimensional Problem

Fig. 10 shows a 3-dimensional model analyzed. The size of the floating structure is 500 m $\times$ 320 m (Model 1). The landing load is applied at the center of the structure, and thus the analysis can be made assuming double symmetry about x- and y-axis. The load pattern is as shown in Fig. 8.

For the floating structure, 8-node elements with second order polynomial are used; whereas, for the fluid and sponge layer, 20-node elements with second order polynomial are used with only a single layer in the vertical direction. Even in such an analysis, the D.O.F. amounts to about 70 thousands.

For comparison, Model 2 with the size of 320 m $\times$ 320 m has also been analyzed. Shown in Fig. 11 are the vertical displacements at the loaded point. Also shown in Fig. 12 is the time variation of the shape of deflection for Model 1.

It can be seen that up to 2 sec, the axisymmetric deformation is conspicuous; after 2 sec, however, the effect of the rectangular edge of the floating structure becomes significant reflecting the basic characteristic vibration mode of the floating structures. This will explain why the displacement responses are different to each other after 2 sec. It must be born in mind, however, that for the response analysis immediately after the landing, the axisymmetric model will be thought to be sufficient for practical purposes to find the

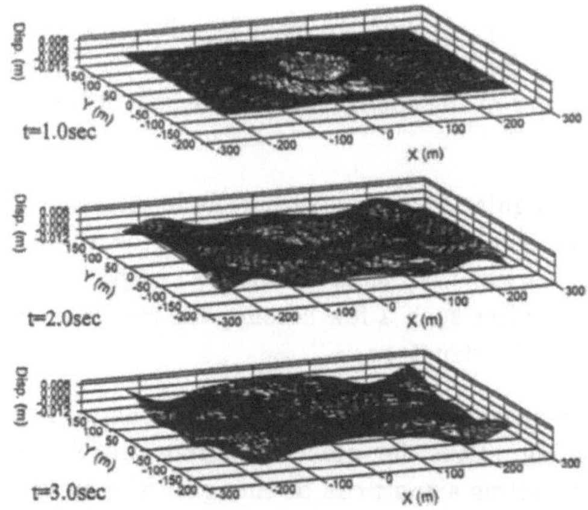

Fig. 12. Deflectional Configurations (Model 1)

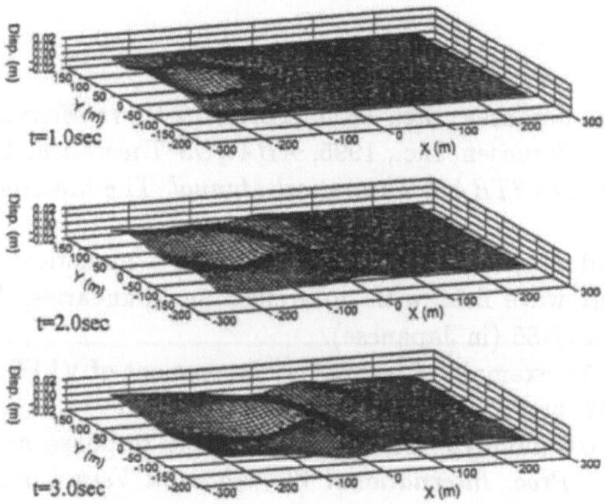

Fig. 13. Deflections due to Moving Aircraft (Model 1)

maximum response.

Next, the load is slided to see the effect of the landing of airplane in more practical manner for the Model 1. It is assumed that the plane maintains the speed of 50 m/sec after the landing, and the sequence of loading is applied at points at every 20 m after the landing along the runway.

The scenario of the deflection of the rectangular airport is shown in Fig. 13 up to 4 seconds after the landing. The maximum deflection of the deck is about 1.6 cm, which seems to be small enough.

4. CONCLUDING REMARKS

The following concluding remarks will be made:

1. The sponge layer is found to be very effective to dissipate the wave energy and is particularly suited for the conversion from a closed finite domain to the domain

with the open boundary. It must be kept in mind, however, that the volumetric drag coefficient must be kept not too large so that the reflection of waves will not occur at the boundary of the sponge, and the length of the sponge should be taken sufficiently long.

2. When a large rectangular floating structure is loaded with the impulsive landing load, a deflectional shape of concentric circular indentation occurs for the initial few seconds in such a way that the deflection can be predicted using an axisymmetric model; however, after such a few second, the response depends on the natural vibrational mode of the structure.

3. The maximum displacement response of the floating airport occurs within a few seconds after landing. Thus, for practical purposes of finding only the maximum displacement, the floating airport can be thought to be modeled as an axisymmetric disk model can be used. However, if the response analysis is required for a long period of time, then more exact modeling is required.

REFERENCES

Hibbitt, Karlsson and Sorensen, Inc., 1995, *ABAQUS Users Manual*, ver.5.5.

Hibbitt, Karlsson and Sorensen, Inc., 1995, *ABAQUS Theoretical Manual*, ver.5.5.

MacNeal, R.H., 1969, *NASTRAN Theoretical Manual*, The MacNeal-Schwendler Corporation.

Ohyama, T. and Nadaoka, K., 1991, Development of a numerical wave tank in nonlinear and irregular wave field with non-reflecting boundaries, *Proceedings of JSCE*, No.429/II-15, pp.77-86 (in Japanese).

Takarada, N., 1982, An example of technology assessment of VLFS (Part 5), *Proceedings of SNAJ*, No.642, pp.702-719 (in Japanese).

Watanabe, E. and Utsunomiya, T., 1996, Transient response analysis of a VLFS at airplane landing, *Proc. International Workshop on Very Large Floating Structures*, Hayama, Ship Research Institute, Ministry of Transport, pp.243-247.

Watanabe, E., Utsunomiya, T. and Taenaka, S., 1997, Transient hydroelastic analysis of a very large floating structure by finite element method, *J. of Structural Engineering*, JSCE, Vol.43A, pp.49-54 (in Japanese).

Wu, C., Utsunomiya, T. and Watanabe, E., 1997, Harmonic wave response analysis of elastic floating plates by modal superposition method, *Structural Eng./Earthquake Eng., JSCE*, Vol.14, No.1, pp.1s-10s.

Yamamoto, H., 1994, Concept for a super-large floating structure, *Proc. International Workshop on Floating Structures in Coastal Zone*, Hiroshima, Port and Harbour Research Institute, Ministry of Transport, pp.121-136.

THE SCALED BOUNDARY FINITE-ELEMENT METHOD: STATE OF THE ART

John P. Wolf and Chongmin Song

Institute of Hydraulics and Energy
Department of Civil Engineering
Swiss Federal Institute of Technology Lausanne
CH-1015 Lausanne
Switzerland

1 Introduction

This paper presents the state of the art of the scaled boundary finite-element method using elastodynamics as an example. In this *boundary-element method based on finite elements* only the boundary is discretized with surface finite elements yielding a reduction of the spatial dimension by one. No fundamental solution is necessary and thus no singular integrals must be evaluated and general anisotropic material can be analyzed. For an unbounded (semi-infinite or infinite) medium the radiation condition at infinity is satisfied exactly. No discretization of free and fixed boundaries and interfaces between different materials is required. Thus, the scaled boundary finite-element method not only combines the advantages of the finite-element and boundary-element methods but also presents appealing features of its own (Table 1).

After discretizing the boundary with finite elements the governing partial differential equations of linear elastodynamics are transformed to the scaled boundary finite-element equation in displacements, a system of linear second-order ordinary differential equations with the radial coordinate as independent variable. Introducing the definition of the dynamic stiffness, a system of nonlinear first-order ordinary differential equations in the dynamic-stiffness matrix with the frequency as the independent variable is obtained. For certain cases, such as the high-frequency asymptotic expansion for unbounded media and the low-frequency expansion for bounded media and statics, this system of first-order differential equations is solved *analytically*. The scaled boundary finite-element method is thus a semi-analytical procedure to solve partial differential equations with the polynomial approximation of finite elements in the two circumferential directions and a closed-form solution in the radial direction. The semi-analytical solution inside the domain leads to an efficient procedure to calculate the stress intensity factors accurately without any discretization in the vicinity of the crack tip.

In the accompanying paper "The scaled boundary finite-element method: future developments", published in the same proceedings, the scaled boundary finite-element equation in displacements is solved in the frequency domain analytically. The closed-form solution is expressed as a series in the radial coordinate which extends the efficiency and high accuracy to the analysis of unbounded and bounded media for any frequency.

This *semi-analytical fundamental-solution-less* boundary-element method applies to hyperbolic, parabolic and elliptic partial differential equations. For instance, the diffusion equation is addressed in Chapter 9 of the book [Wolf and Song, 1996a]. This paper concentrates on elastodynamics. In the following sections key concepts and equations are summarized and selected examples are presented. For a thorough derivation and in-depth discussion, the reader is referred to the cited references.

T.L. Geers (ed.), IUTAM Symposium on Computational Methods for Unbounded Domains, 305–314.

Table 1. Advantages of scaled boundary finite-element method compared with those of finite-element and boundary-element methods.

	boundary element method	finite element method	scaled boundary finite-element method
no fundamental solution required		X	X
reduction of spatial dimension by one	X		X
radiation condition satisfied exactly for unbounded medium	X		X
suitable for anisotropic material		X	X
symmetric static-stiffness and mass matrices or dynamic-stiffness matrix	(X)	X	X
no singular integrals to be evaluated		X	X
boundary conditions on free, fixed and inter-material surfaces satisfied without discretization			X
analytical solution in domain permitting efficient calculation of displacements, stresses and stress-intensity factors			X
no fictitious eigenfrequencies for unbounded medium		X	X

2 Galerkin's Weighted Residual Method

To derive the scaled boundary finite-element equation, the scaled boundary transformation is applied [Song and Wolf, 1997]. The scaling center O is chosen outside the unbounded medium, from which the total boundary must be visible (Fig. 1). By scaling the boundary in the radial direction with respect to O using a scaling factor larger than 1 the whole domain is covered. The boundary is discretized with curved surface finite elements. That part of the boundary whose extension passes through the scaling center is not discretized. Mathematically, the scaling corresponds to a transformation of the coordinates for each finite element, resulting in the two curvilinear local coordinates η, ζ in the circumferential directions on the surface and the dimensionless radial coordinates ξ representing the scaling factor (Fig. 2).

After applying the scaled boundary transformation to the governing equations of elastodynamics, Galerkin's weighted residual method is formulated in the two circumferential directions η, ζ. For vanishing body forces the *scaled boundary finite-element equation in displacements* in the frequency domain follows as

$$[E^0]\xi^2\{u(\xi)\}_{,\xi\xi}+((s-1)[E^0]-[E^1]+[E^1]^T)\xi\{u(\xi)\}_{,\xi}$$
$$+((s-2)[E^1]^T-[E^2])\{u(\xi)\}+\omega^2[M^0]\xi^2\{u(\xi)\}=0 \quad (1)$$

$\{u(\xi)\}$ are the displacement amplitudes at ξ corresponding to the degrees of freedom in the nodes on the boundary, s is the spatial dimension ($= 2$ or $= 3$). The coefficient matrices

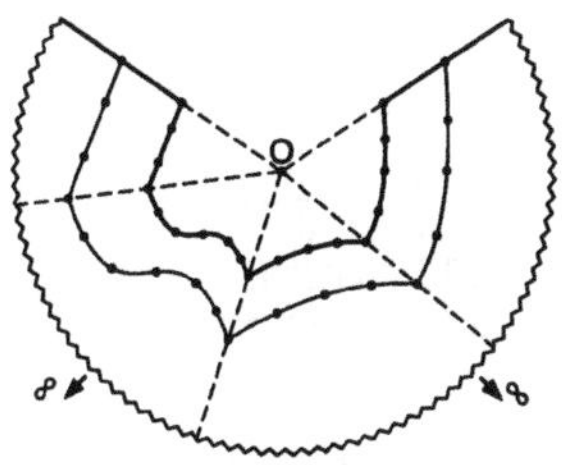

Fig. 1. Modeling of unbounded medium with surface finite elements and scaling of boundary (section).

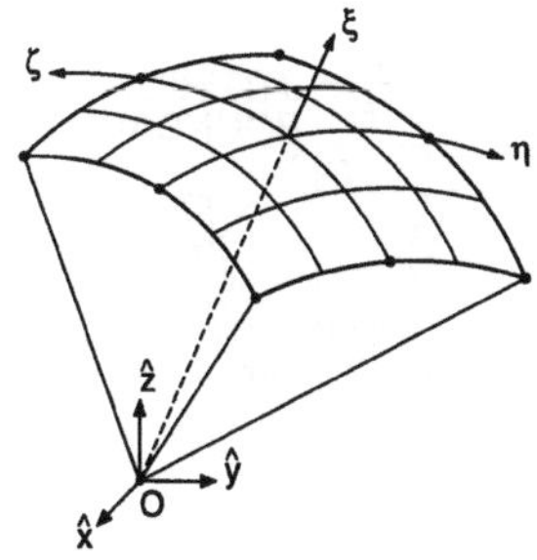

Fig. 2. Scaled boundary transformation of geometry of surface finite element.

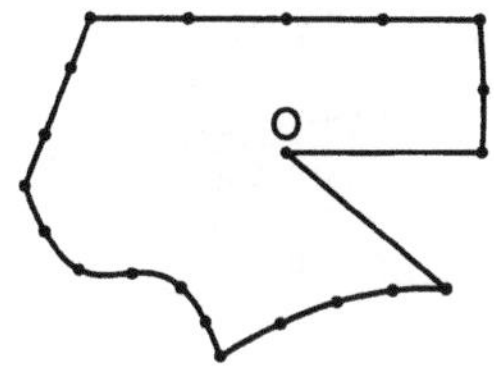

Fig. 3. Modeling of bounded medium with surface finite elements and scaling of boundary (section).

$[E^0]$, $[E^1]$, $[E^2]$ and $[M^0]$ are calculated and assembled similarly as the static-stiffness and mass matrices of finite elements on the boundary. Eq. (1) is a system of *linear* second-order ordinary differential equation.

Calculating the nodal forces and introducing the definition of the dynamic-stiffness matrix on the boundary $[S^\infty(\omega)]$ (superscript ∞ for unbounded), the *scaled boundary finite-element equation in dynamic stiffness* results

$$([S^\infty(\omega)]+[E^1])[E^0]^{-1}([S^\infty(\omega)]+[E^1]^T)-(s-2)[S^\infty(\omega)] \\ -\omega[S^\infty(\omega)]_{,\omega}-[E^2]+\omega^2[M^0]=0 \quad (2)$$

By scaling the boundary in the radial direction with respect to the scaling center O using a scaling factor smaller than 1, a bounded domain is covered (Fig. 3). Eq. (1) still applies, while the definition of the dynamic-stiffness matrix $[S^b(\omega)]$ (b for bounded) involves a sign change yielding

$$([S^b(\omega)]-[E^1])[E^0]^{-1}([S^b(\omega)]-[E^1]^T)+(s-2)[S^b(\omega)] \\ +\omega[S^b(\omega)]_{,\omega}-[E^2]+\omega^2[M^0]=0 \quad (3)$$

3 Mechanically Motivated Derivation

The scaled boundary finite-element equation in dynamic stiffness (Eq. (2)) can also derived using concepts of mechanics familiar to engineers. The procedure is based on the *assemblage* of an infinitesimal finite-element cell and on *similarity* (Fig. 4), followed by performing the *limit of the cell width analytically*. This approach is called the *consistent infinitesimal finite-element cell method* in [Song and Wolf, 1996]. The key element of the derivation consists of replacing the spatial derivative of the dynamic-stiffness matrix by the derivative with respect to frequency

$$\xi\frac{\partial[S^\infty(\xi,\omega)]}{\partial\xi}=(s-2)[S^\infty(\xi,\omega)]+\omega\frac{\partial[S^\infty(\xi,\omega)]}{\partial\omega} \quad (4)$$

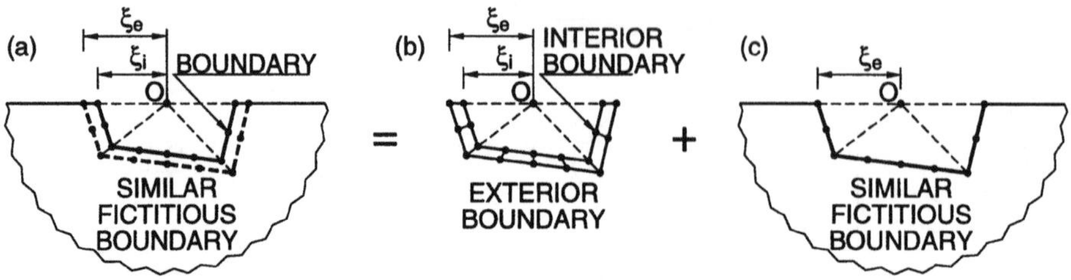

Fig. 4. Concept of consistent infinitesimal finite-element cell method for unbounded medium with infinitesimal cell width leading to finite-element discretization of boundary only.

The same mechanically motivated derivation is also applicable to a bounded medium [Wolf and Song, 1996a] (Fig. 5).

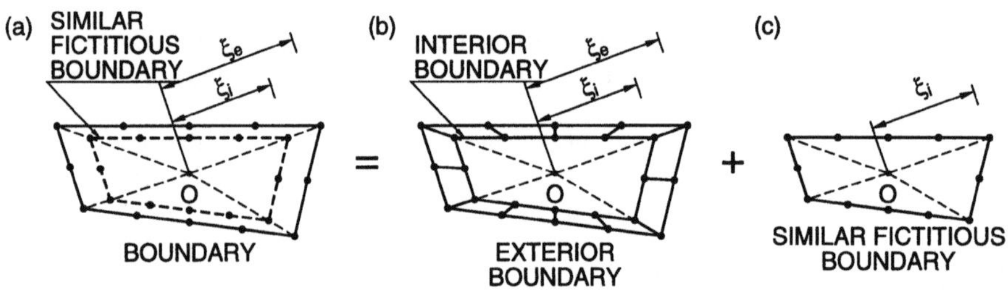

Fig. 5. Concept of consistent infinitesimal finite-element cell method for bounded medium with infinitesimal cell width leading to finite-element discretization of boundary only.

4 Solution in Frequency Domain

To start the numerical integration procedure of the scaled boundary finite-element equation in dynamic stiffness in frequency domain (Eq. (2)), a system of first-order non-linear differential equations, a boundary condition $[S^{\infty}(\omega_h)]$ is required, which consists of the dynamic-stiffness matrix at the selected frequency ω_h. A high-frequency asymptotic expansion which incorporates the radiation condition at infinity leads to $[S^{\infty}(\omega_h)]$ at a high but finite ω_h. A standard numerical integration procedure then yields $[S^{\infty}(\omega)]$ for decreasing ω.

The dynamic-stiffness matrix $[S^{\infty}(\omega)]$ at high frequency is expanded in a matrix polynomial of $\mathrm{i}\omega$ in descending order starting at one [Wolf and Song, 1996b]

$$[S^{\infty}(\omega)] \approx \mathrm{i}\omega[C_{\infty}] + [K_{\infty}] + \sum_{j=1}^{m} \frac{1}{(\mathrm{i}\omega)^j}[A_j] \tag{5}$$

The first two terms on the right-hand side represent the constant dashpot matrix $[C_{\infty}]$ and the constant spring matrix $[K_{\infty}]$ (subscript ∞ for $\omega \to \infty$). $[C_{\infty}]$, $[K_{\infty}]$ and the coefficient matrices $[A_j]$ are unknown. Substituting (5) in (2) and examining the terms of the matrix polynomial in descending order of the power of $\mathrm{i}\omega$ determine $[C_{\infty}]$, $[K_{\infty}]$ and $[A_j]$. The *radiation condition* is satisfied by selecting the positive definite solution of $[C_{\infty}]$.

5 Solution in Time Domain

For an analysis in the time domain, the unit-impulse response matrix $[S^{\infty}(t)]$ is required [Wolf and Song, 1998]. The dynamic-stiffness matrix is decomposed as

$$[S^{\infty}(\omega)] = \mathrm{i}\omega[C_{\infty}] + [K_{\infty}] + [S_r^{\infty}(\omega)] \tag{6}$$

Its inverse Fourier transform equals

$$[S^{\infty}(t)] = [C_{\infty}]\dot{\delta}(t) + [K_{\infty}]\delta(t) + [S_r^{\infty}(t)] \tag{7}$$

with the Dirac delta function $\delta(t)$. $[C_{\infty}]$ and $[K_{\infty}]$ are determined as described in Section 4. Substituting (6) in (2) and performing the inverse Fourier transformation leads to

$$\begin{aligned}\int_0^t [v_r^{\infty}(t-\tau)][v_r^{\infty}(\tau)]\mathrm{d}\tau + [c_{\infty}][v_r^{\infty}(t)] + [v_r^{\infty}(t)][c_{\infty}] + t[v_r^{\infty}(t)] \\ + ([k_{\infty}]+[e^1])\int_0^t [v_r^{\infty}(\tau)]\mathrm{d}\tau + \int_0^t [v_r^{\infty}(\tau)]\mathrm{d}\tau([k_{\infty}]+[e^1]^T) - (s-1)\int_0^t [v_r^{\infty}(\tau)]\mathrm{d}\tau \\ + t\left(([k_{\infty}]+[e^1])([k_{\infty}]+[e^1]^T) - [e^2] - (s-2)[k_{\infty}]\right) = 0\end{aligned} \tag{8}$$

with

$$[s_r^{\infty}(t)] = [\dot{v}_r^{\infty}(t)] \tag{9}$$

The lower-case coefficient matrices $[c_{\infty}]$, $[k_{\infty}]$, $[e^1]$ and $[e^2]$ follow from a linear transformation of the corresponding upper-case matrices $[C_{\infty}]$, $[K_{\infty}]$, $[E^1]$ and $[E^2]$. $[S_r^{\infty}(t)]$ is calculated from the corresponding inverse transformation of $[s_r^{\infty}(t)]$.

After discretization of (8) with respect to time $[v_r^{\infty}]_n$ is calculated for each time station n solving a (linear) Lyapunov equation.

6 Statics

Setting $\omega = 0$ in (2) yields the scaled boundary finite-element equation for statics for an unbounded medium [Wolf and Song, 1996a; Song and Wolf, 1997]

$$\begin{aligned}[K^{\infty}][E^0]^{-1}[K^{\infty}] + \left([E^1][E^0]^{-1} - \frac{s-2}{2}[I]\right)[K^{\infty}] \\ + [K^{\infty}]\left([E^0]^{-1}[E^1]^T - \frac{s-2}{2}[I]\right) - [E^2] + [E^1][E^0]^{-1}[E^1]^T = 0\end{aligned} \tag{10}$$

The static-stiffness matrix $[K^{\infty}]$ is calculated by solving this Riccati equation and selecting the positive definite solution.

The static displacements $\{u(\xi)\}$ follow from the Euler-Cauchy equation

$$\xi\{u(\xi)\}_{,\xi} = -[E^0]^{-1}([K^{\infty}] + [E^1]^T)\{u(\xi)\} \tag{11}$$

Its analytical solution consists of power functions of ξ.

As both the static-stiffness matrix and displacements are determined *analytically*, no approximation is introduced in solving the scaled boundary finite-element equation for statics.

The static-stiffness matrix $[K^b]$ of a bounded medium follows similarly from (3) for $\omega = 0$. As an analytical solution of the displacements within the domain results, the stress intensity factors are evaluated straightforwardly and accurately.

7 Mass Matrix for Bounded Medium

As in structural dynamics, the mass matrix $[M^b]$ of a bounded medium with respect to the degrees of freedom on the boundary only is determined based on the low-frequency expansion of $[S^b(\omega)]$

$$[S^b(\omega)] = [K^b] - \omega^2[M^b] \tag{12}$$

Substituting (12) in (3) yields the (linear) Lyapunov equation

$$\left((-[K^b]+[E^1])[E^0]^{-1} - \frac{s}{2}[I]\right)[M^b] + [M^b]\left([E^0]^{-1}(-[K^b]+[E^1]^T) - \frac{s}{2}[I]\right) + [M^0] = 0 \tag{13}$$

Note that in the scaled boundary finite-element method $[M^b]$ is calculated analytically without any additional assumption [Wolf and Song, 1996a].

8 Examples

To demonstrate the versatility and high accurate of the scaled boundary finite-element method, selected examples are addressed. For a complete description of the problems the references must be consulted.

8.1 Circular Cavity Embedded in Full-Plane [Wolf and Song, 1996a]

As a two-dimensional in-plane wave propagation problem with an analytical solution available, the circular cavity embedded in a full-plane is discussed (Fig. 6). On its boundary

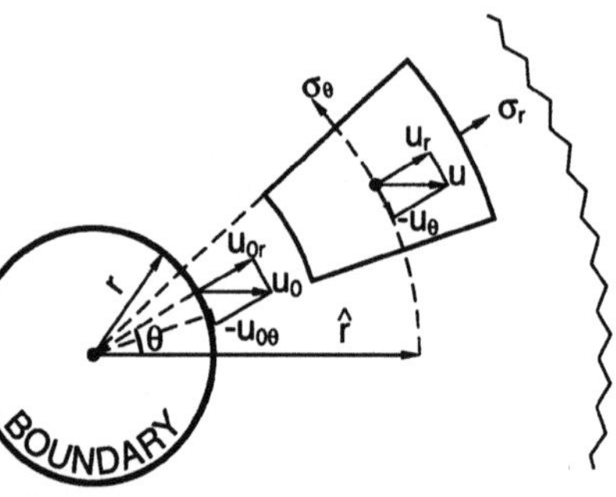

Fig. 6. Circular cavity embedded in full-plane with prescribed translational motion.

a constant horizontal displacement u_0 is enforced. Due to symmetry only one quarter of the boundary is discretized with 4 3-node line elements of equal length. The boundary condition of the dynamic-stiffness matrix $[S^\infty(\omega_h)]$ is determined from the asymptotic expansion with $m = 2$ (Eq. (5)) at $\omega_h = 6c_s/r$. $[S^\infty(\omega)]$ of order 16×16 is calculated for decreasing ω with the scaled boundary finite-element method. The dynamic-stiffness coefficient $S^\infty(\omega)$ follows from

$$S^\infty(\omega) = \{\phi\}^T[S^\infty(\omega)]\{\phi\} \tag{14}$$

with the vector $\{\phi\}$ determined from the translational motion of the rigid boundary. $S^{\infty}(\omega)$ non-dimensionalized with the shear modulus G is decomposed in $k(a_0)$ and $c(a_0)$

$$S^{\infty}(\omega) = G(k(a_0) + \mathrm{i}a_0 c(a_0)) \tag{15}$$

with the dimensionless frequency $a_0 = \omega r / c_s$. Excellent accuracy is observed for Poisson's ratio $\nu = 1/3$ (Fig. 7).

As an extension, the incompressible case can also be examined [Chapter 6 of Wolf and Song, 1996a]. From the unit-impulse response coefficient calculated in the time domain an inverse Fourier transformation yields the dynamic-stiffness coefficient (Fig. 8), which agrees well with the analytical solution.

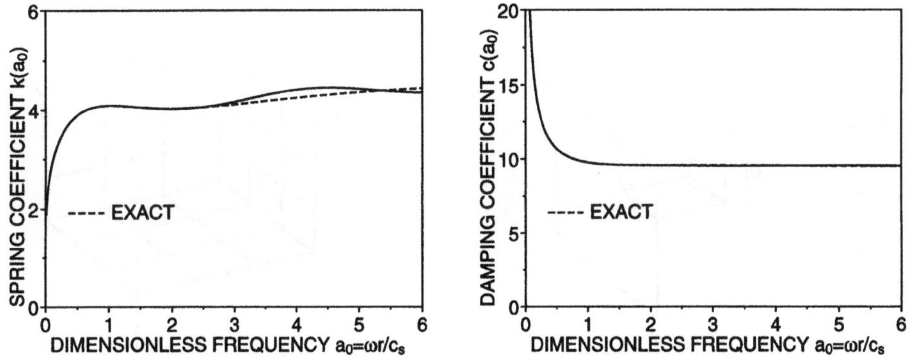

Fig. 7. Dynamic-stiffness coefficient of circular cavity embedded in compressible full-plane.

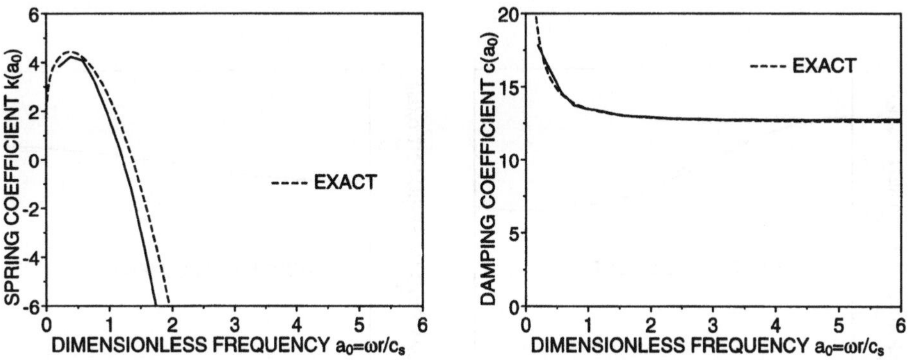

Fig. 8. Dynamic-stiffness coefficient of circular cavity embedded in incompressible full-plane.

8.2 Prism Embedded in Half-Space [Wolf and Song, 1998]

The three-dimensional motion of a prism of length $2b$ embedded with depth e in a half-space (Fig. 9) is calculated. The embedment ratio $e/b = 2/3$ and $\nu = 1/3$ are selected. The finite-element discretization of one quarter of the boundary is shown in Fig. 10. The free surface is not discretized. The dashpot matrix $[C_\infty]$, the spring matrix $[K_\infty]$ and $[S_r^\infty(t)]$ with $\Delta t = 0.02b/c_s$ up to $t = 3b/c_s$ are calculated. The uniform vertical displacement on the boundary is enforced. The equivalent coefficients C_∞, K_∞, and $S_r^\infty(t)$ then follow. The Fourier transform of $S_r^\infty(t)$ is equal to $S_r^\infty(\omega)$. The dynamic-stiffness coefficient $S^\infty(\omega)$ is determined as in (6). $S^\infty(\omega)$ is decomposed as in (15) but using the static-stiffness coefficient K^∞ instead of G for non-dimensionalization. Excellent agreement with a boundary-element solution is observed (Fig. 11).

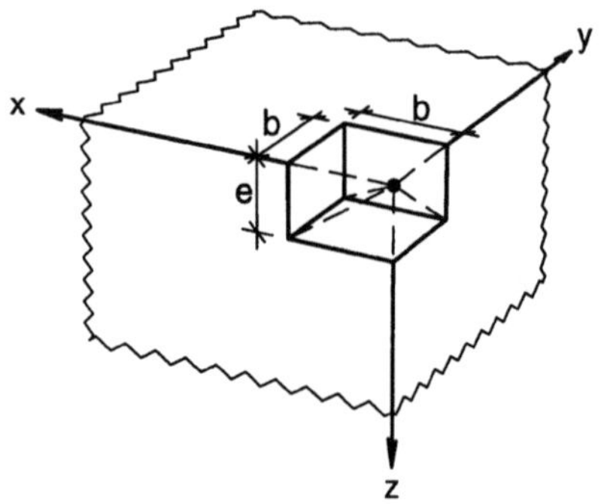

Fig. 9. One quarter of square prism embedded in half-space.

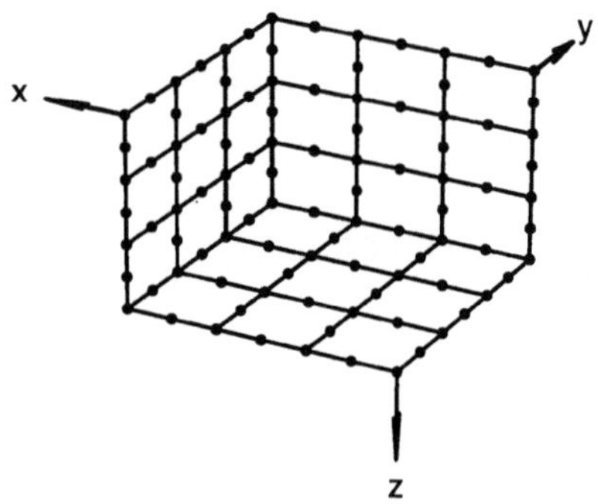

Fig. 10. Finite-element mesh of one quarter of boundary of square prism.

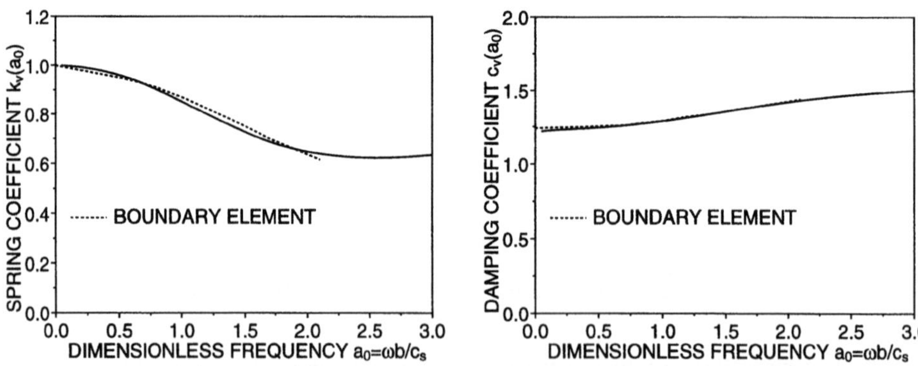

Fig. 11. Vertical dynamic-stiffness coefficient of rigid prism embedded in half-space.

8.3 Tunnel in Inhomogeneous Transversely Isotropic Unbounded Rock [Wolf and Song, 1996a]

As an example from actual practice, a railroad tunnel intersecting a fault is addressed (Fig. 12a). The fault (Zone C) exhibits isotropic behavior. The two zones of unbounded rock on both sides of the fault are modeled as transversely isotropic with a horizontal plane of isotropy.

The scaling center is selected on the middle line of the fault. Small deviations of the locations of the interfaces between the fault and the rock occur (dashed lines in Fig. 12a). The wall is discretized with 31 isoparametric 3-node line elements (Fig. 12b). The material interfaces are not discretized. The static displacements relative to the center of the track agree well with those of an extended mesh of finite elements (Fig. 12b).

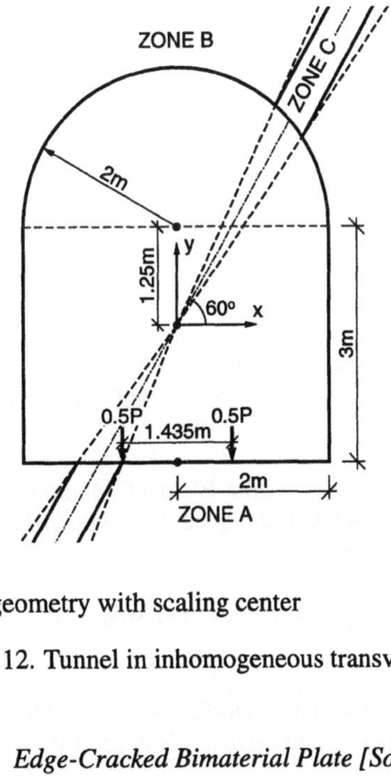

(a) geometry with scaling center

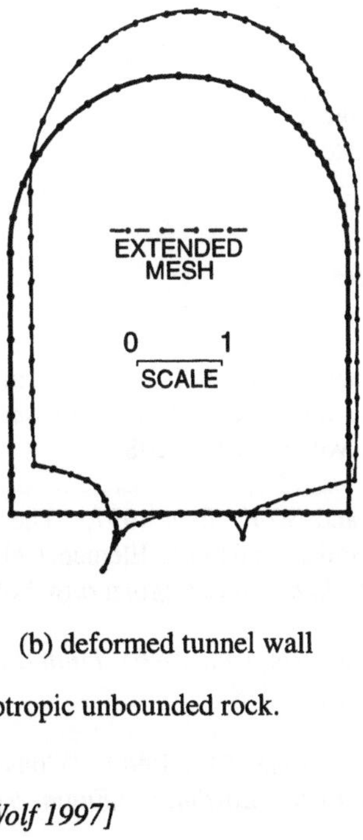

(b) deformed tunnel wall

Fig. 12. Tunnel in inhomogeneous transversely isotropic unbounded rock.

8.4 Edge-Cracked Bimaterial Plate [Song and Wolf 1997]

The scaled boundary finite-element method is well suited to analyze interface cracks between different materials. As an example, a rectangular plate composed of two materials characterized by Young's modulus E and Poisson's ratio ν with an interface crack of length a for plane stress is addressed (Fig. 13).

The scaling center O is chosen at the crack tip. The straight traction free crack faces are again not discretized. The same also applies to the material interface. The boundary is discretized with 14 4-node (cubic) line elements (Fig. 13).

The stress intensity factors $K_I + iK_{II}$ non-dimensionalized with $\sigma_0\sqrt{\pi a}$ are calculated for $\nu_1 = \nu_2 = 0.3$ varying the crack length a/b and the ratio of Young's moduli E_1/E_2. They are listed in Table 2. Good agreement is achieved with the results of the boundary-element method, which requires the discretization of the crack faces.

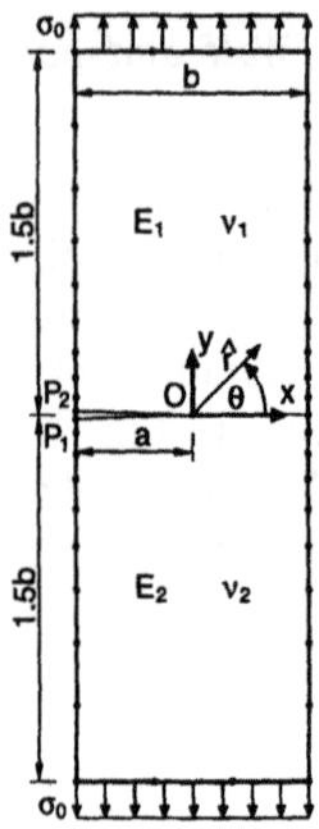

Fig. 13. Edge-cracked bimaterial plate.

Table 2. Stress intensity factors of edge-cracked bimaterial plate

a/b		RATIO OF YOUNG'S MODULI (E_1/E_2)			
		1	2	4	10
0.4		2.111	2.108 – i0.198	2.100 – i0.359	2.089 – i0.495
	BEM	2.113	2.109 – i0.197	2.100 – i0.358	2.089 – i0.493
0.5		2.824	2.818 – i0.267	2.805 – i0.484	2.787 – i0.664
	BEM	2.836	2.819 – i0.267	2.806 – i0.483	2.787 – i0.661
0.6		4.032	4.022 – i0.397	3.998 – i0.716	3.967 – i0.979
	BEM	4.037	4.027 – i0.396	4.001 – i0.714	3.971 – i0.974

9 References

Song, Ch. and Wolf, J. P. 1996, "Consistent Infinitesimal Finite-Element Cell Method: Three-Dimensional Vector Wave Equation", *International Journal for Numerical Methods in Engineering,* Vol. 39, 2189-2208.

Song, Ch. and Wolf, J. P. 1997, "The Scaled Boundary Finite-Element Method – alias Consistent Infinitesimal Finite-Element Cell Method – for Elastodynamics", *Computer Methods in Applied Mechanics and Engineering,* Vol. 147, 329–355.

Wolf, J. P. and Song, Ch., 1996a, *Finite-Element Modeling of Unbounded Media,* John Wiley & Sons, Chichester, UK.

Wolf, J. P. and Song, Ch., 1996b, "Consistent Infinitesimal Finite-Element Cell Method in Frequency Domain", *Earthquake Engineering and Structural Dynamics,* Vol. 25, 1307–1327.

Wolf, J. P. and Song, Ch., 1998, "Unit-Impulse Response of Unbounded Medium by Scaled Boundary Finite-Element Method", *Computer Methods in Applied Mechanics and Engineering,* (in press).

ABSORBING BOUNDARY CONDITIONS FOR DOMAIN DECOMPOSITION

BJORN ENGQUIST * AND HONG-KAI ZHAO †

1. Introduction. There are striking similarities in the requirements for two classes of artificial boundary conditions. One consisting of those conditions used in computer simulations over unbounded domains and the other of those used in domain decomposition. In both cases the original problem does not contain these artificial boundary conditions and the differential equation is valid across the location of the artificial boundaries. The boundary condition should thus be as "transparent" as possible and not introduce significant errors. As we shall see below there are some differences between these classes of boundary conditions, but in this paper we shall stress the similarities and describe how techniques which were developed in connection to simulation over unbounded domains can be beneficial as boundary conditions in domain decomposition.

We shall use the term absorbing boundary condition (ABC) for the artificial condition which is used to limit the computational domain in simulations over unbounded domains. Depending on application, these types of boundary conditions goes under many names, for example, radiation, far field or open.

The basic requirements for ABCs are, *stability, accuracy and computational efficiency.* The well posedness of the initial boundary value problem and the stability of the corresponding computational algorithm is clearly necessary.

With accuracy, we here mean that the true solution should well satisfy the ABC. The stability will then guarantee a good approximation over all.

Finally the ABC should be computationally efficient. This is, for example, why local ABCs are often preferred over global ones, at least for time dependent problems.

In domain decomposition the computational domain is subdivided into different parts which are connected by artificial boundary conditions. The purpose may be to reduce the computational complexity or to facilitate parallel processing. We shall here consider the iterative Schwarz alternating method (SAM). The problems will be elliptic but it is also possible to study time dependent problems and overlapping grids.

For SAM the subproblems with their boundary conditions must be individually stable. Computational efficiency is of course also useful.

* Department of Mathematics, University of California at Los Angeles, Los Angeles, CA 90095-1555, <engquist@math.ucla.edu>

† Department of Mathematics, Stanford University, Stanford, CA 94305-2125 <zhao@math.stanford.edu>

T.L. Geers (ed.), IUTAM Symposium on Computational Methods for Unbounded Domains, 315–324.

Accuracy, however, plays a slightly different role for SAM compared to ABC. The solution is not expected to satisfy the artificial boundary condition directly but only in the limit of an iterative process. The source term in the artificial boundary condition will be iteratively adjusted to achieve accuracy. The standard methodology is to use Dirichlet or Neumann conditions, which may give large errors in the first iteration step. We will show that "better" boundary conditions will give faster convergence, in particular, if the overlaps between the domains are small. Better boundary conditions will here include derivatives, which are tangential to the boundaries, as is common for ABCs.

2. Schwarz Alternating Method. In domain decomposition some compatibility conditions have to be imposed for the Dirichlet data and Neumann data at the artificial boundary between each subdomain. SAM is an important theoretical basis for domain decomposition method in both parallel computing and computations in complicated domains. The classical SAM method is an iterative method which couples the subproblems together by overlapping each subdomain and exchanging function value at the artificial boundaries. The classical SAM method is convergent for elliptic PDEs and its convergence rate is related to the size of overlap. This can be easily seen by either maximum principle or variational interpretation. (see [2],[1],[3],[9]) Our paper is motivated by the new modified SAM method, proposed by P.L.Lions ([8]), in which there are exchanges of a convex combination of Neumann and Dirichlet data at the artificial boundaries. Convergence was proved in a very general setting even with no overlap. The proof is based on delicate energy estimates. However no estimates on the convergence rate has been done. Here we will study convergence properties for even more general boundary conditions at the artificial boundaries motivated by the absorbing boundary conditions proposed by Engquist and Majda in [4]. Due to the 'positive' property of the Dirichlet to Neumann operator and the use of absorbing boundary condition both stability and improved convergence is achieved.

In [13] some analysis was done in linear algebra context and in [12] some analysis was done on the discretized equation. Similar analysis is carried out in [10]. In our paper [5] we are mainly interested in the better understanding of the effect of boundary conditions at the artificial boundary which is also useful in other situations such as overlapping grids and interface problems. In many practical applications, domain decomposition method has also been used as a preconditioner combined with a Krylov subspace method (see [11]), which could be a potential use for the techniques discussed here.

An interesting application for the generalized SAM is to interface problems where solutions to elliptic PDEs and their derivatives may be discontinuous across some interfaces due to the discontinuities in the coefficients or singular sources along the interfaces. This kind of interface problems occur very often, e.g. multiphase or free surface fluid problems. A domain decomposition along the interfaces is very natural since we can solve well behaved problems in each subdomain using any good numerical scheme and only couple the subproblems together by the prescribed interface conditions. In this situation the usual elliptic solver can not work well due to the discontinuities across the interface where the PDE is not defined. Even the traditional non-overlapping method in domain decomposition will not work since the Steklov-Poincaré operator or the Schur complement at the interface is not clear.

For simplicity let's first define the generalized SAM in the case where the whole domain Ω is decomposed into two subdomains Ω_1 and Ω_2 (i.e. $\Omega = \Omega_1 \cup \Omega_2$) with the two artificial boundaries Γ_1, Γ_2 intersecting $\partial\Omega$. Denote $\Omega_{11} = \Omega_1 \setminus \Omega_2, \Omega_{22} = \Omega_2 \setminus \Omega_1$ if there is overlap. (see figure 2)

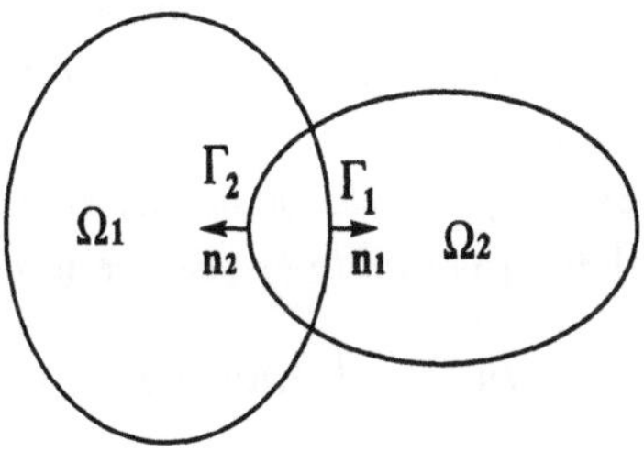

FIG. 1. *Domain decomposition with two subdomains.*

The elliptic PDE with Dirichlet boundary data is:

$$
\begin{aligned}
\mathcal{L}u &= f \quad in\ \Omega \\
u &= g \quad on\ \partial\Omega
\end{aligned}
\tag{1}
$$

where $\mathcal{L}$ is some elliptic differential operator. We generalize the classical additive SAM to the following formulation.

Set u_i^0 initially in Ω_i, $i = 1, 2$ and construct u_i^n in parallel

$$
\begin{aligned}
\mathcal{L}u_1^n &= f && in\ \Omega_1 \\
u_1^n &= g && on\ \partial\Omega_1 \backslash \Gamma_1 \\
\Lambda_1 u_1^n + \lambda_1 \frac{\partial u_1^n}{\partial n_1} &= \Lambda_1 u_2^{n-1} + \lambda_1 \frac{\partial u_2^{n-1}}{\partial n_1} && on\ \Gamma_1
\end{aligned}
\tag{2}
$$

and

$$
\begin{aligned}
\mathcal{L}u_2^n &= f && in\ \Omega_2 \\
u_2^n &= g && on\ \partial\Omega_2 \backslash \Gamma_2 \\
\Lambda_2 u_2^n + \lambda_2 \frac{\partial u_2^n}{\partial n_2} &= \Lambda_2 u_1^{n-1} + \lambda_2 \frac{\partial u_1^{n-1}}{\partial n_2} && on\ \Gamma_2 \\
& && n = 1, 2, \ldots,
\end{aligned}
\tag{3}
$$

The multiplicative version can be defined similarly. The solution u^n in Ω can be composed in many ways from u_1^n and u_2^n such that $u^n \in H^1(\Omega) \cap C^1(\Omega)$ and, $u^n = u_i^n$, in $\Omega_i \setminus (\Omega_1 \cap \Omega_2)$. The operator Λ_i in our case are pseudo differential operators and λ_i are constants. It is easily seen that if $\Lambda_1 = \Lambda_2 = I$(identity operator) and $\lambda_1 = \lambda_2 = 0$, then the above multiplicative version becomes the classical SAM. If $\Lambda_1 = \Lambda_2 = constant > 0$ and $\lambda_1 = \lambda_2 = 1$ then it is the modified SAM proposed by P.L. Lions in [8]. We can also change Λ_i, λ_i's with every iteration. This will give us an accelerated version of the generalized SAM which will improve the convergence rate. In particular if the Dirichlet data and Neumann data are exchanged alternately then we get the Dirichlet to Neumann type or the relaxation procedure in [9]. If the generalized SAM does not degenerate to classical SAM we will take $\lambda_1 = \lambda_2 = 1$ for the rest of the discussions. Details of the proof for most of the results below can be found in [5]

In order to have well-posedness for each subdomain problem, we need some requirement on the operators Λ_i, $i = 1, 2$

DEFINITION 1. *Let γ_0 be an open (relative to $\partial\Omega$) subset of $\partial\Omega$. A bounded linear operator $\Lambda : H^{1/2}(\gamma_0) \to H^{-1/2}(\gamma_0)$ is said to be positive if for*

$$\forall u \qquad \int_{\gamma_0} u \Lambda u \geq 0$$

It is said to be strictly positive if $\int_{\gamma_0} u\Lambda u > 0$, if $u \neq 0$.

Let $\Lambda(x, \eta)$ denote the symbol of the pseudo-differential operator Λ. If Λ is (strictly) positive then

$$\Lambda(x, \eta) \geq (>) 0, \qquad \forall x \in \gamma_0, \ \forall \eta$$

Dirichlet to Neumann operator for the Laplace equation

$$-\Delta u = 0, \ in \Omega \qquad u = 0 \ on \ \partial\Omega \setminus \gamma_0$$

$\Lambda u = \frac{\partial u}{\partial n}|_{\gamma_0}$ is a positive operator. This is also true for any second order coercive PDEs.

LEMMA 2.1. *If $\Lambda_i : H^{1/2}(\partial\Omega_i \setminus \partial\Omega) \longrightarrow H^{-1/2}(\partial\Omega_i \setminus \partial\Omega)$, $i = 1, 2$ are positive operators, then the two boundary value problems (2),(3) are well-posed in H^1.*

Theoretically the perfect choice of Λ_i's are the corresponding Dirichlet to Neumann operators which gives the convergence in two steps.

LEMMA 2.2. *A second order coercive elliptic PDE $\mathcal{L}u = f$ in Ω can be solved by the generalized SAM with or without overlap in finitely many steps if Λ_1(or Λ_2) is the Dirichlet to Neumann operator at the artificial boundary Γ_1(or Γ_2) for the corresponding homogeneous PDE in Ω_{22}(or Ω_{11}) with homogeneous boundary condition on $\partial\Omega_{22} \setminus \Gamma_1$(or $\partial\Omega_{11} \setminus \Gamma_2$). (see fig1)*

In general the Dirichlet to Neumann operator is a global operator and to find its action on u is equivalent to solving a Dirichlet problem. For ordinary differential equations with constant coefficients it can be easily found, (see [14]). For more general elliptic PDEs in simple geometries it is possible to incorporate this into appropriate finite element formulation, (see [7]).

Since we are concerned with the convergence, we only need to look at error function sequence $e_i^n = u - u_i^n$, $i = 1,2$ which satisfies the corresponding homogeneous equation and homogeneous boundary conditions at $\partial\Omega \cap \partial\Omega_i$. From now on we shall assume that $f = 0$, $g = 0$ in (1) and analyze how u_i^n converges to 0 in Ω_i, $i = 1,2$.

3. Convergence Analysis of the Generalized SAM for a model problem. This model problem let us see very intuitively why the generalized SAM works and how the convergence rate depends on Λ_i, the size of overlap and the size of the subdomains, since for the model problem the explicit formula of the convergence rate can be derived. Using the spectral equivalence of the elliptic differential operators we will see that the analysis for the model problem can be generalized to more general elliptic PDEs and geometries in the two subdomain case.

Model Problem:

$$\begin{cases} -\Delta u = f & in\ \Omega \\ u = g & on\ \partial\Omega \end{cases}$$

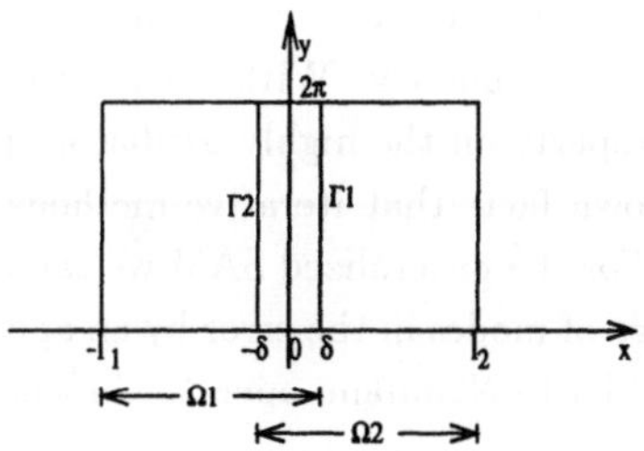

FIG. 2. *Rectangular domain with both subdomains touching the boundary.*

THEOREM 3.1. *For model problem (see figure 3) the generalized additive Schwarz alternating method, (2) - (3). ($\lambda_1 = \lambda_2 = 1$) is convergent if $\Lambda_1 = \Lambda_2 = \Lambda$ is strictly positive. If there is overlap, then there is a constant $0 < \rho < 1$ such that*

$$\| u - u_i^n \|_{H^1(\Omega_i)} \leq \rho \| u - u_i^{n-2} \|_{H^1(\Omega_i)} \qquad i = 1,2$$

where ρ depends on Λ, the size of $\Omega_i(l_i)$ and the size of overlap δ. If there is no overlap ($\delta = 0$) then

$$\lim_{n\to\infty} \| u - u_i^n \|_{H^1(\Omega_i)} \to 0 \quad i = 1,2$$
$$\lim_{n\to\infty} \| u - u_i^n \|_{H^{1/2}(\Gamma_i)} \to 0 \quad i = 1,2$$

After Fourier transform in y direction and using the absorbing boundary conditions at the interface we have the following contraction rate in each Fourier mode η is

$$r_G(\eta,\delta,l_1,l_2,\Lambda)$$

$$= \sqrt{\left|\frac{(\lambda(\eta)-\Lambda(\eta))e^{-\lambda(\eta)\delta}+(\lambda(\eta)+\Lambda(\eta))e^{-2\lambda(\eta)l_2}e^{\lambda(\eta)\delta}}{(\lambda(\eta)+\Lambda(\eta))e^{\lambda(\eta)\delta}+(\lambda(\eta)-\Lambda(\eta))e^{-2\lambda(\eta)l_2}e^{-\lambda(\eta)\delta}}\right|} \tag{4}$$

$$\times\sqrt{\left|\frac{(\lambda(\eta)-\Lambda(\eta))e^{-\lambda(\eta)\delta}+(\lambda(\eta)+\Lambda(\eta))e^{-2\lambda(\eta)l_1}e^{\lambda(\eta)\delta}}{(\lambda(\eta)+\Lambda(\eta))e^{\lambda(\eta)\delta}+(\lambda(\eta)-\Lambda(\eta))e^{-2\lambda(\eta)l_1}e^{-\lambda(\eta)\delta}}\right|}$$

From simple algebraic fact we have $r_G(\eta,\delta,l_1,l_2,\Lambda)<1 \quad \forall\eta$ *and* $\delta\geq 0$. Using Poincaré inequality and trace theorem we can finish the proof.

If we do some asymptotic analysis for the most interesting case where $l_1\approx l_2\gg\delta$, then

$$\begin{aligned} r_G(\eta) &\approx \left|\frac{\lambda(\eta)-\Lambda(\eta)}{\lambda(\eta)+\Lambda(\eta)}\right| e^{-2\lambda(\eta)\delta} \\ r_D(\eta) &\approx e^{-2\lambda(\eta)\delta} \\ r_N(\eta) &\approx e^{-2\lambda(\eta)\delta} \end{aligned}$$

Where r_D and r_N are the corresponding contration rates for passing Dirichlet and Neumann data at the artificial boundary. With small overlap the classical SAM only have a good convergence property on the highly oscillatory part in the solution, which corresponds to the well-known facts that iterative methods for elliptic PDEs smooth out the error very quickly. For the generalized SAM we can also have a good control of the smooth part or any range of modes in the error by an appropriate choice of operator Λ to approximate the Dirichlet to Neumann operator in that range.

4. Extension to General Equations in General Two Subdomain case. For the more general equation, we replace the simple Laplace operator by the following elliptic operator,

$$\mathcal{L}u(x) = -\nabla\cdot(a(x)\nabla u(x)) + \vec{b}(x)\cdot\nabla u(x) + c(x)u(x) \tag{5}$$

where $x\in R^N$, $a(x)$, $\vec{b}(x)$, $c(x)$ are smooth and $a(x)>0$, $c(x)\geq 0$. Though we can not do Fourier analysis for (5). But as long as we have the coercivity assumption as in [8], i.e. $\exists\ c>0$, such that $\forall u\in H_0^1(\Omega)$

$$\int_\Omega a(x)|\nabla u(x)|^2 + \vec{b}(x)\cdot\nabla u(x)u(x) + c(x)u^2(x)dx \geq c\int_\Omega|\nabla u(x)|^2dx \tag{6}$$

if $u(x)$ satisfies $\mathcal{L}u(x)=0$, then by divergence theorem we have

$$\int_\Omega a(x)|\nabla u(x)|^2 + \vec{b}(x)\cdot\nabla u(x)u(x) + c(x)u^2(x)dx = \int_{\partial\Omega}\frac{\partial u(x)}{\partial n}a(x)u(x) \tag{7}$$

By Poincaré inequality we can bound the H^1 norm of the solution by the boundary data. Now the only difference is that we have two bounded positive weight functions $a(x)$ and $c(x)$. Since elliptic differential operators are spectral equivalent to the Laplace operator in finite domain (see [6]), we also have the spectral equivalence for the corresponding Dirichlet to Neumann operator.

THEOREM 4.1. *The generalized additive or multiplicative SAM ($\lambda_1 = \lambda_2 = 1$) for elliptic differential operator (5) in two subdomain case is convergent if*

$$\Lambda_i = \Lambda : \; H^{\frac{1}{2}}(\partial\Omega_i) \longrightarrow H^{-\frac{1}{2}}(\partial\Omega_i) \qquad i = 1,2$$

is a strictly positive operator. If there is a uniform overlap, i.e. $dis(\Gamma_1, \Gamma_2) = \delta > 0$ then there is a constant $0 < \rho < 1$ such that

$$\| u - u_i^n \|_{H^1} \leq \rho^n \| u - u_i^0 \|_{H^1} \;\; in\ \Omega_i, \; i = 1,2$$

where ρ depends on Λ, the size of Ω_i, the size of the overlap δ and the partial differential equations.

If we look at the analysis in the model problems, we can actually see that the convergence property of the generalized SAM is better than the classical SAM due to two reasons,
(i) The generalized SAM inherits the same mechanism as the classical SAM, i.e. if there is overlap, maximum principle is true.
(ii) The "positivity" of the Dirichlet to Neumann operator, which gives convergence even without overlap.

Remark: If we change Λ_i's with the iteration, then we can damp out the error in different range of modes more efficiently (similar to the multigrid idea).

5. Local Operator Implementation of the Generalized SAM. Motivated by the absorbing boundary condition proposed by Engquist and Majda in [4], we will use a local operator to approximate the Dirichlet to Neumann operator by truncating an asymptotic expansion of a global pseudo differential operator. This operator will contain derivatives tangential to the artificial boundary. We again start with a model problem

$$\text{(8)} \qquad \begin{cases} -\Delta u + u = f & in\ \Omega \\ u = g & on\ \partial\Omega \end{cases}$$

where $\Omega, \Omega_1, \Omega_2$ and the generalized SAM are exactly the same as in the model problem in section3. From analysis for the model problem the convergence rate of the generalized SAM in each frequency mode is

$$r_G \approx \left| \frac{\lambda(\eta) - \Lambda(\eta)}{\lambda(\eta) + \Lambda(\eta)} \right| e^{-2\lambda(\eta)\delta}$$

in the case of $l_1 \approx l_2 \gg \delta$. Now $\lambda(\eta) = \sqrt{1+\eta^2}$. When solving elliptic equations the smooth part of the solution converges very slowly. We expand $\lambda(\eta)$ at 0.

$$\lambda(\eta) = \sqrt{1+\eta^2} = 1 + \frac{1}{2}\eta^2 + O(\eta^4)$$

Using the first order Taylor expansion we get the first order local approximation.

$$\Lambda(\eta) = 1 \;\Rightarrow\; \Lambda u = u \tag{9}$$

Using the second order Taylor expansion we get the second order local approximation assuming the boundary and the solution make $\frac{\partial^2 u}{\partial \tau^2}$ meaningful.

$$\Lambda(\eta) = 1 + \frac{1}{2}\eta^2 \;\Rightarrow\; \Lambda u = u - \frac{1}{2}\frac{\partial^2 u}{\partial \tau^2} \tag{10}$$

where τ is the tangential direction.

For the discretized problem suppose the grid size is $h \ll 1$ and has one grid size overlap, i.e. $\delta = h$. The highest possible frequency mode is $\eta_{max} = \frac{1}{h}$. After optimize the constants in the approximation we can have convergence rate $1 - ch^{\frac{1}{3}}$ for the first order scheme and $1 - ch^{\frac{1}{4}}$ for the second order scheme, where c is some positive constant that is independent of h.

6. Numerical Example.

$$\begin{aligned} -\Delta u(x,y) + u(x,y) &= xy \quad && in\ \Omega = [-1,1] \times [0,1] \\ u &= e^x y + xy && on\ \partial\Omega \end{aligned}$$

The exact solution is $u(x,y) = e^x y + xy$.
$\Omega_1 = [-1,\delta] \times [0,1]$, $\Omega_2 = [-\delta,1] \times [0,1]$ $\Gamma_1 = \{\delta\} \times [0,1]$, $\Gamma_2 = \{-\delta\} \times [0,1]$

In the computation we will change the size of overlap with the mesh size h, i.e. $\delta = h$. We use SOR with tolerance 10^{-6} for the system in each subdomain discretized by central difference. We tested the classical SAM, first order and second order generalized SAM schemes in both additive and multiplicative version of domain decomposition method. Figure3 shows the convergence of the multiplicative domain decomposition method in the L_2 norm as $h \to 0$. We see that the convergence of the generalized SAM is relative insensitive to the size of the overlap. Figure4 shows the convergence of the additive domain decomposition method.

Figure5 shows that both the multiplicative version and the additive version of the generalized SAM converges without overlap but not in any geometric fashion. However they can be dramatically improved by using accelerated version (see section4 Remark), i.e. using different Λ_i^n for each iteration. In this calculation, we use $\Lambda_i^n = 1 + n, n = 1, 2, \ldots, 20$ at nth iteration.

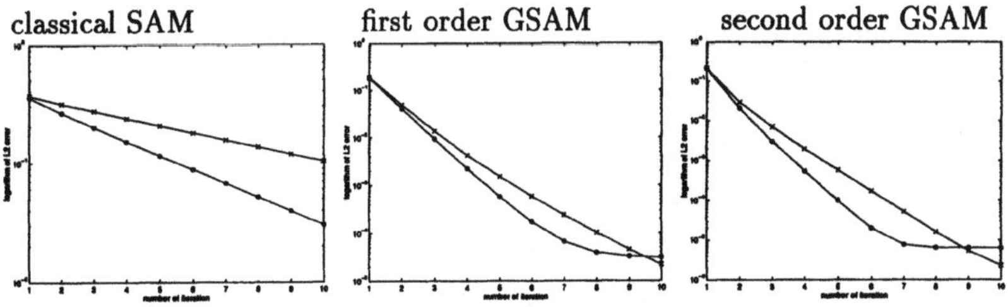

'o' for $\delta = h = 0.02$, 'x' for $\delta = h = 0.01$

FIG. 3. *Multiplicative SAM*

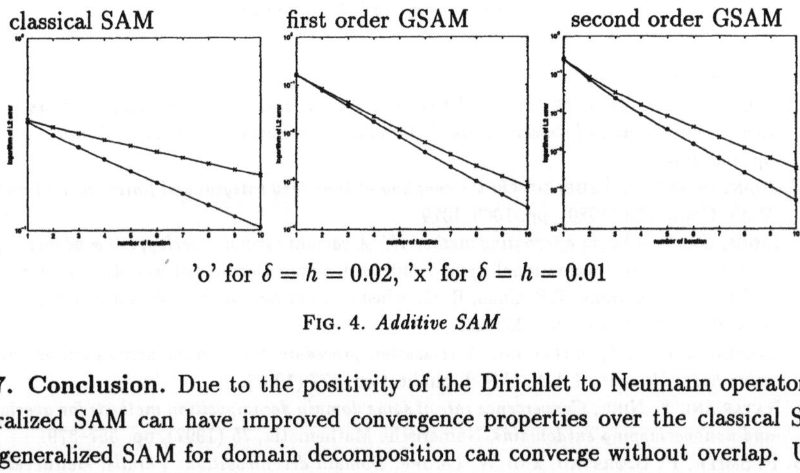

'o' for $\delta = h = 0.02$, 'x' for $\delta = h = 0.01$

FIG. 4. *Additive SAM*

7. Conclusion. Due to the positivity of the Dirichlet to Neumann operator the generalized SAM can have improved convergence properties over the classical SAM. The generalized SAM for domain decomposition can converge without overlap. Using the idea of absorbing boundary conditions, we can construct efficient local operator to approximate the Dirichlet to Neumann operator at the artificial boundary. Since a local operator can not approximate a global operator uniformly in general, we can not expect a convergence rate which is independent of the overlapping size for the generalized SAM.

We will discuss the accelerated version of generalized SAM, its extension to multidomain cases and the application to interface problems in some other papers.

REFERENCES

[1] P. BJORSTAD AND O. WIDLUND, *Iterative methods for the solution of elliptic problems on regions partitioned into substructures*, SIAM Num. Anal., 23 (1986), pp. 1097–1120.

[2] Q. DINH, R. GLOWINSKI, AND J. PÉRIAUX, *Solving elliptic problems by domain decomposition methods with applications.*, Elliptic Problem Solvers II, Academic Press, (1982).

[3] M. DRYJA AND O. B. WIDLUND, *Domain decomposition algorithms with small overlap*, SIAM J. Sci.Comput., 15 (1994), pp. 604–620.

[4] B. ENGQUIST AND A.MAJDA, *Absorbing boundary condition for the numerical simulation of waves*, Math. Comp., 31 (1977), pp. 629–651.

[5] B. ENGQUIST AND H.-K. ZHAO, *Analysis of generalized schwarz alternating procedure for domain*

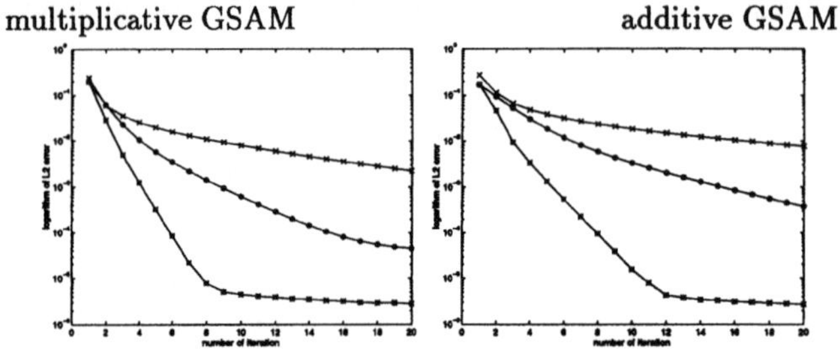

'x' for 1st order, 'o' for 2nd order, '*' for the accelerated version, $h = 0.02$

FIG. 5. *Convergence of GSAM without overlap*

decomposition, to appear.

[6] V. FABER, T. MANTEUFFEL, AND S. PARTER, *On the equivalence of operators and the implications to preconditioned iterative methods for elliptic problems*, Adv. in Appl. Math., 11 (1989), pp. 109–163.

[7] C. JOHNSON AND C. NEDELEC, *On the coupling of boundary integral and finite element methods*, Math. Comp., 35 (1980), pp. 1063–1079.

[8] P. LIONS, *On the schwarz alternating method III: A variant for non-overlapping subdomains*, Proceedings of Third International Symposium on Domain Decomposition Methods for Partial Differential Equations. T.F. Chan, R. Glowinski, J. Périaux, and O. Widlund, eds., Philadelphia, PA, 1990, SIAM., pp. 202–223.

[9] L. MARINI AND A. QUARTERONI, *A relaxation procedure for domain decomposition methods using finite elements*, I.A.N.-C.N.R. publication, 577 (1987).

[10] F. NATAF AND F. NIER, *Convergence rate of some domain decomposition methods for overlapping and nonoverlapping subdomains*, Numerische Mathematik, 75 (1997), pp. 357–379.

[11] B. F. SMITH, P. BJØRSTAD, AND W. GROPP, *Domain decomposition: Parallel multilevel methods for elliptic partial differential equations*, Cambridge University Press, (1996).

[12] K. TAN, *Ph.d thesis*, Universiteit Utrecht, (1995).

[13] W. TANG, *Generalized schwarz splittings*, SIAM J.Sci.Stat.Comput., 13 (1992), pp. 573–595.

[14] H. ZHAO, *Ph.d thesis*, UCLA, (1996).

Mechanics

SOLID MECHANICS AND ITS APPLICATIONS

Series Editor: G.M.L. Gladwell

Aims and Scope of the Series

The fundamental questions arising in mechanics are: *Why?*, *How?*, and *How much?* The aim of this series is to provide lucid accounts written by authoritative researchers giving vision and insight in answering these questions on the subject of mechanics as it relates to solids. The scope of the series covers the entire spectrum of solid mechanics. Thus it includes the foundation of mechanics; variational formulations; computational mechanics; statics, kinematics and dynamics of rigid and elastic bodies; vibrations of solids and structures; dynamical systems and chaos; the theories of elasticity, plasticity and viscoelasticity; composite materials; rods, beams, shells and membranes; structural control and stability; soils, rocks and geomechanics; fracture; tribology; experimental mechanics; biomechanics and machine design.

1. R.T. Haftka, Z. Gürdal and M.P. Kamat: *Elements of Structural Optimization*. 2nd rev.ed., 1990 ISBN 0-7923-0608-2
2. J.J. Kalker: *Three-Dimensional Elastic Bodies in Rolling Contact*. 1990 ISBN 0-7923-0712-7
3. P. Karasudhi: *Foundations of Solid Mechanics*. 1991 ISBN 0-7923-0772-0
4. *Not published*
5. *Not published.*
6. J.F. Doyle: *Static and Dynamic Analysis of Structures*. With an Emphasis on Mechanics and Computer Matrix Methods. 1991 ISBN 0-7923-1124-8; Pb 0-7923-1208-2
7. O.O. Ochoa and J.N. Reddy: *Finite Element Analysis of Composite Laminates*. ISBN 0-7923-1125-6
8. M.H. Aliabadi and D.P. Rooke: *Numerical Fracture Mechanics*. ISBN 0-7923-1175-2
9. J. Angeles and C.S. López-Cajún: *Optimization of Cam Mechanisms*. 1991 ISBN 0-7923-1355-0
10. D.E. Grierson, A. Franchi and P. Riva (eds.): *Progress in Structural Engineering*. 1991 ISBN 0-7923-1396-8
11. R.T. Haftka and Z. Gürdal: *Elements of Structural Optimization*. 3rd rev. and exp. ed. 1992 ISBN 0-7923-1504-9; Pb 0-7923-1505-7
12. J.R. Barber: *Elasticity*. 1992 ISBN 0-7923-1609-6; Pb 0-7923-1610-X
13. H.S. Tzou and G.L. Anderson (eds.): *Intelligent Structural Systems*. 1992 ISBN 0-7923-1920-6
14. E.E. Gdoutos: *Fracture Mechanics*. An Introduction. 1993 ISBN 0-7923-1932-X
15. J.P. Ward: *Solid Mechanics*. An Introduction. 1992 ISBN 0-7923-1949-4
16. M. Farshad: *Design and Analysis of Shell Structures*. 1992 ISBN 0-7923-1950-8
17. H.S. Tzou and T. Fukuda (eds.): *Precision Sensors, Actuators and Systems*. 1992 ISBN 0-7923-2015-8
18. J.R. Vinson: *The Behavior of Shells Composed of Isotropic and Composite Materials*. 1993 ISBN 0-7923-2113-8
19. H.S. Tzou: *Piezoelectric Shells*. Distributed Sensing and Control of Continua. 1993 ISBN 0-7923-2186-3

Kluwer Academic Publishers – Dordrecht / Boston / London

Mechanics

SOLID MECHANICS AND ITS APPLICATIONS

Series Editor: G.M.L. Gladwell

20. W. Schiehlen (ed.): *Advanced Multibody System Dynamics.* Simulation and Software Tools. 1993 ISBN 0-7923-2192-8
21. C.-W. Lee: *Vibration Analysis of Rotors.* 1993 ISBN 0-7923-2300-9
22. D.R. Smith: *An Introduction to Continuum Mechanics.* 1993 ISBN 0-7923-2454-4
23. G.M.L. Gladwell: *Inverse Problems in Scattering.* An Introduction. 1993 ISBN 0-7923-2478-1
24. G. Prathap: *The Finite Element Method in Structural Mechanics.* 1993 ISBN 0-7923-2492-7
25. J. Herskovits (ed.): *Advances in Structural Optimization.* 1995 ISBN 0-7923-2510-9
26. M.A. González-Palacios and J. Angeles: *Cam Synthesis.* 1993 ISBN 0-7923-2536-2
27. W.S. Hall: *The Boundary Element Method.* 1993 ISBN 0-7923-2580-X
28. J. Angeles, G. Hommel and P. Kovács (eds.): *Computational Kinematics.* 1993 ISBN 0-7923-2585-0
29. A. Curnier: *Computational Methods in Solid Mechanics.* 1994 ISBN 0-7923-2761-6
30. D.A. Hills and D. Nowell: *Mechanics of Fretting Fatigue.* 1994 ISBN 0-7923-2866-3
31. B. Tabarrok and F.P.J. Rimrott: *Variational Methods and Complementary Formulations in Dynamics.* 1994 ISBN 0-7923-2923-6
32. E.H. Dowell (ed.), E.F. Crawley, H.C. Curtiss Jr., D.A. Peters, R. H. Scanlan and F. Sisto: *A Modern Course in Aeroelasticity.* Third Revised and Enlarged Edition. 1995 ISBN 0-7923-2788-8; Pb: 0-7923-2789-6
33. A. Preumont: *Random Vibration and Spectral Analysis.* 1994 ISBN 0-7923-3036-6
34. J.N. Reddy (ed.): *Mechanics of Composite Materials.* Selected works of Nicholas J. Pagano. 1994 ISBN 0-7923-3041-2
35. A.P.S. Selvadurai (ed.): *Mechanics of Poroelastic Media.* 1996 ISBN 0-7923-3329-2
36. Z. Mróz, D. Weichert, S. Dorosz (eds.): *Inelastic Behaviour of Structures under Variable Loads.* 1995 ISBN 0-7923-3397-7
37. R. Pyrz (ed.): *IUTAM Symposium on Microstructure-Property Interactions in Composite Materials.* Proceedings of the IUTAM Symposium held in Aalborg, Denmark. 1995 ISBN 0-7923-3427-2
38. M.I. Friswell and J.E. Mottershead: *Finite Element Model Updating in Structural Dynamics.* 1995 ISBN 0-7923-3431-0
39. D.F. Parker and A.H. England (eds.): *IUTAM Symposium on Anisotropy, Inhomogeneity and Nonlinearity in Solid Mechanics.* Proceedings of the IUTAM Symposium held in Nottingham, U.K. 1995 ISBN 0-7923-3594-5
40. J.-P. Merlet and B. Ravani (eds.): *Computational Kinematics '95.* 1995 ISBN 0-7923-3673-9
41. L.P. Lebedev, I.I. Vorovich and G.M.L. Gladwell: *Functional Analysis.* Applications in Mechanics and Inverse Problems. 1996 ISBN 0-7923-3849-9
42. J. Menčik: *Mechanics of Components with Treated or Coated Surfaces.* 1996 ISBN 0-7923-3700-X
43. D. Bestle and W. Schiehlen (eds.): *IUTAM Symposium on Optimization of Mechanical Systems.* Proceedings of the IUTAM Symposium held in Stuttgart, Germany. 1996 ISBN 0-7923-3830-8

Kluwer Academic Publishers – Dordrecht / Boston / London

Mechanics

SOLID MECHANICS AND ITS APPLICATIONS

Series Editor: G.M.L. Gladwell

44. D.A. Hills, P.A. Kelly, D.N. Dai and A.M. Korsunsky: *Solution of Crack Problems*. The Distributed Dislocation Technique. 1996 ISBN 0-7923-3848-0
45. V.A. Squire, R.J. Hosking, A.D. Kerr and P.J. Langhorne: *Moving Loads on Ice Plates*. 1996 ISBN 0-7923-3953-3
46. A. Pineau and A. Zaoui (eds.): *IUTAM Symposium on Micromechanics of Plasticity and Damage of Multiphase Materials*. Proceedings of the IUTAM Symposium held in Sèvres, Paris, France. 1996 ISBN 0-7923-4188-0
47. A. Naess and S. Krenk (eds.): *IUTAM Symposium on Advances in Nonlinear Stochastic Mechanics*. Proceedings of the IUTAM Symposium held in Trondheim, Norway. 1996 ISBN 0-7923-4193-7
48. D. Ieşan and A. Scalia: *Thermoelastic Deformations*. 1996 ISBN 0-7923-4230-5
49. J. R. Willis (ed.): *IUTAM Symposium on Nonlinear Analysis of Fracture*. Proceedings of the IUTAM Symposium held in Cambridge, U.K. 1997 ISBN 0-7923-4378-6
50. A. Preumont: *Vibration Control of Active Structures*. An Introduction. 1997 ISBN 0-7923-4392-1
51. G.P. Cherepanov: *Methods of Fracture Mechanics: Solid Matter Physics*. 1997 ISBN 0-7923-4408-1
52. D.H. van Campen (ed.): *IUTAM Symposium on Interaction between Dynamics and Control in Advanced Mechanical Systems*. Proceedings of the IUTAM Symposium held in Eindhoven, The Netherlands. 1997 ISBN 0-7923-4429-4
53. N.A. Fleck and A.C.F. Cocks (eds.): *IUTAM Symposium on Mechanics of Granular and Porous Materials*. Proceedings of the IUTAM Symposium held in Cambridge, U.K. 1997 ISBN 0-7923-4553-3
54. J. Roorda and N.K. Srivastava (eds.): *Trends in Structural Mechanics*. Theory, Practice, Education. 1997 ISBN 0-7923-4603-3
55. Yu. A. Mitropolskii and N. Van Dao: *Applied Asymptotic Methods in Nonlinear Oscillations*. 1997 ISBN 0-7923-4605-X
56. C. Guedes Soares (ed.): *Probabilistic Methods for Structural Design*. 1997 ISBN 0-7923-4670-X
57. D. François, A. Pineau and A. Zaoui: *Mechanical Behaviour of Materials*. Volume I: Elasticity and Plasticity. 1998 ISBN 0-7923-4894-X
58. D. François, A. Pineau and A. Zaoui: *Mechanical Behaviour of Materials*. Volume II: Viscoplasticity, Damage, Fracture and Contact Mechanics. 1998 ISBN 0-7923-4895-8
59. L. T. Tenek and J. Argyris: *Finite Element Analysis for Composite Structures*. 1998 ISBN 0-7923-4899-0
60. Y.A. Bahei-El-Din and G.J. Dvorak (eds.): *IUTAM Symposium on Transformation Problems in Composite and Active Materials*. Proceedings of the IUTAM Symposium held in Cairo, Egypt. 1998 ISBN 0-7923-5122-3
61. I.G. Goryacheva: *Contact Mechanics in Tribology*. 1998 ISBN 0-7923-5257-2

Kluwer Academic Publishers – Dordrecht / Boston / London

Mechanics

***SOLID* MECHANICS AND ITS APPLICATIONS**

Series Editor: G.M.L. Gladwell

63. F.C. Moon: *IUTAM Symposium on New Applications of Nonlinear and Chaotic Dynamics in Mechanics*. Proceedings of the IUTAM Symposium held in Ithaca, NY, USA. 1998 ISBN 0-7923-5276-9
64. R. Wang: *IUTAM Symposium on Rheology of Bodies with Defects*. Proceedings of the IUTAM Symposium held in Beijing. China. 1999 ISBN 0-7923-5297-1
65. Yu. I. Dimitrienko: *Thermomechanics of Composites under High Temperatures.* 1999 ISBN 0-7923-4899-0

Kluwer Academic Publishers – Dordrecht / Boston / London